CLIMATE JUSTICE

Climate Justice

Vulnerability and Protection

HENRY SHUE

for Simon,
good neighbour,
Hey

OXFORD
UNIVERSITY PRESS

OXFORD
UNIVERSITY PRESS

Great Clarendon Street, Oxford, OX2 6DP,
United Kingdom

Oxford University Press is a department of the University of Oxford.
It furthers the University's objective of excellence in research, scholarship,
and education by publishing worldwide. Oxford is a registered trade mark of
Oxford University Press in the UK and in certain other countries

Published in the United States of America by Oxford University Press
198 Madison Avenue, New York, NY 10016, United States of America

British Library Cataloguing in Publication Data
Data available

Library of Congress Control Number: 2013956550

ISBN 978–0–19–871370–8

As printed and bound by
CPI Group (UK) Ltd, Croydon, CR0 4YY

For the women who made my life:

Catherine Harper Shue
who could not stay
1911–40

Violet East Harper
who did all of it all over again
1891–1971

Sallie Morris Shue
who cheerfully guided eight 'boys'
1876–1967

Vivienne Bland Shue
who, once found, has been there always

Preface

At the conclusion of the events portrayed in "Grand Hotel", one elegant denizen of the hotel lobby offers the marvellously unperceptive and obtuse summation, "People come, people go, nothing ever happens"! One hopes that something occasionally happens, especially regarding our headlong rush into humanly unprecedented climate change. But it is definitely true that people come, people go. Two of the people inadvertently most causally responsible for what follows here, though considerably younger than I, have sadly gone. First, Cornell agricultural economist Duane Chapman kept insisting, at around 1990, that since I claimed to know something about ethics, I must be able to provide some clues about what we ought to do about climate change. After vainly trying repeatedly to insist that whatever I might know about ethics, I did not actually know anything about climate change, Duane's lovable cantankerousness, combined with patient explanations by his Ph.D. students and the more gentle urgings of ag economist Tim Mount, finally persuaded me to try, with their help, to figure out what was happening. Secondly, meanwhile in England the far younger still John Vincent, who tragically was torn away even before the day of my lecture, urged Andy Hurrell to invite me to Oxford to give what turned out to be my first lecture on normative issues about climate and became the first paper here, "The Unavoidability of Justice". For me climate had become the unavoidable issue, as it ought to be, but still is not, for the whole world today. Decades later Andy was instrumental in my move to Oxford and continues to be supportive and appreciative.

As I scattered papers on questions about climate change around the world, including some somewhat out of the way places, in my efforts to encourage many different kinds of people to focus on the problem, both Chuck Beitz and Bob Goodin urged me to bring the papers together in some place in which they would all be accessible to all sorts. But in the early years of this century, as Washington continued to refuse to pick up the ball on climate, I turned my attention back to torture and war, topics on which I had worked prior to taking on climate change and which Washington seemed distinctly interested in.[1] After I arrived at Merton College, Oxford, in 2002 I found myself explaining to people that I sometimes worked on torture and war and sometimes worked on climate change. Whenever the Warden of Merton, Dame Jessica Rawson, heard me say this, she would respond, "how can you work on anything other than

---

[1] My papers on those subjects will follow in a second volume from Oxford University Press.

climate?!" Gradually, as the global situation became increasingly urgent and Washington continued to obsess over terrorism and engage in denial over climate, I concluded that there really was no alternative to trying harder on climate. Meanwhile, I had been joined at Oxford by Simon Caney whose own outstanding work on climate change became a focus of fresh activity and whose strong encouragement further inspired my efforts. Simon added his own question: "since you have written on human rights and written on climate, why haven't you written on human rights and climate together?", a suggestion to which I have recently been trying to respond. David Frame, while at Oxford, first introduced me to the trillionth ton and cumulative carbon budgets.

Some of my earliest work on climate change at Cornell was supported by the Rockefeller Foundation's Arts and Sciences Division, directed by Alberta Arthurs. And my research was then assisted by the passionate activist Gay Nicholson, who went on to lead the Finger Lakes Land Trust and found Sustainable Tompkins. Steve Gardiner has repeatedly invited me to speak on these issues at the University of Washington in Seattle and long been a valuable interlocutor. Dale Jamieson and Darrel Moellendorf too have long been comrades in what used to be the rather lonely endeavour of convincing people that climate change raises deep ethical issues. Dominic Byatt at the Press has been supportive, patient, and tolerant.

Most recently, Mary Robinson has done me the honour of including me in the High Level Advisory Committee for the Climate Justice Dialogue, an initiative of the Mary Robinson Foundation—Climate Justice and the World Resources Institute. That wonderfully diverse international committee's "Declaration on Climate Justice" is included here as an appendix.

Vivienne Shue has listened to more harangues about climate change than any one person could be expected to endure, but never wavered in her support.

Both the first person who wanted me to develop a view of climate change and the first person who thought any view I had might be worth hearing have passed on. So too, as I have indicated in the dedication, have most of the most important people who shaped me, as distinguished from my later philosophical positions. Meanwhile the government of my country has allowed opportunity after opportunity to take decisive action on climate change to pass by unseized. As one thinks about the climate, one cannot help but develop a longer-range sense of time and causation. The effects of some of our choices last far beyond our own lives. All the positive actions regarding the climate not taken by Washington twenty years ago have made action both more difficult and more urgent now for everyone. Some of the carbon dioxide we emit tomorrow will still be affecting sea-levels long after 2100. The energy regime we leave in power will dominate the lives and economies of generations to come because such entrenched institutions cannot be changed quickly.

What we choose to do and not to do now regarding the climate that makes it possible for us to live and the energy regime that is undermining that climate is well worth thinking about, for its effects will play out across the globe and over the centuries. I hope these essays promote hard thought and then decisive action, taken with a long view.

Henry Shue, Merton College

Table of Contents

Citations

1. 'The Unavoidability of Justice', in *The International Politics of the Environment: Actors, Interests, and Institutions*, edited by Andrew Hurrell and Benedict Kingsbury (Oxford: Oxford University Press, 1992), pp. 373–97.

2. 'Subsistence Emissions and Luxury Emissions', Henry Shue, in *Law & Policy*, 15:1, Wiley (January 1993), 39–59.

3. Copyright 1994, Henry Shue, as first published in *Indiana Journal of Global Legal Studies*, 1:2, pp. 343–66.

4. 'Avoidable Necessity: Global Warming, International Fairness, and Alternative Energy', in *Theory and Practice*, NOMOS XXXVII, edited by Ian Shapiro and Judith Wagner DeCew (New York: NYU Press, 1995), pp. 239–64.

5. From icipe Science Press, *Equity and Social Considerations Related to Climate Change*: Papers presented at the IPCC Working Group III Workshop on Equity and Social Considerations related to Climate Change, Nairobi, Kenya, 18–22 July 1994. World Meteorological Organisation/United Nations Environment Programme. 1st Edition. Copyright © 1995 by Secretariat of Nairobi Workshop/Kenya Meteorological Department. Reprinted with permission from icipe Science Press, Nairobi.

6. Reprinted, Henry Shue, 'Environmental Change and the Varieties of Justice', in *Earthly Goods: Environmental Change and Social Justice*, ed. Fen Osler Hampson and Judith Reppy, pp. 9–29. Copyright 1996, Ithaca, N.Y.: Cornell University Press, used by permission of the publisher Cornell University Press.

7. 'Eroding Sovereignty: The Advance of Principle', in *The Morality of Nationalism*, edited by Robert McKim and Jeff McMahan (New York: Oxford University Press, 1997), 340–59.

8. 'Bequeathing Hazards: Security Rights and Property Rights of Future Humans', Henry Shue, in *Limits to Markets: Equity and the Global Environment*, ed. Mohammed Dore and Timothy Mount (Malden, Mass.: Wiley Blackwell Pubs, 1998), pp. 38–53.

9. 'Global Environment and International Inequality', Henry Shue, *International Affairs*, vol. 75, no. 3 (1999), Wiley, pp. 531–45.

10. 'Climate', in *A Companion to Environmental Philosophy*, ed. Dale Jamieson (Malden, MA.: Blackwell Pubs., 2001), pp. 449–59.

11. 'A Legacy of Danger: The Kyoto Protocol and Future Generations', in *Globalisation and Equality*, ed. by Keith Horton and Haig Patapan (London and New York: Routledge, 2004), pp. 164–78.

12. 'Responsibility to Future Generations and the Technological Transition', in *Perspectives on Climate Change: Science, Economics, Politics, Ethics*, ed. Walter Sinnott-Armstrong and Richard B. Howarth (Amsterdam and San Diego: Elsevier, 2005), pp. 265–83.

13. 'Making Exceptions', Henry Shue, *Journal of Applied Philosophy* 26:4, Wiley, (2009), 307–22.

14. 'Deadly Delays, Saving Opportunities: Creating a More Dangerous World?', in *Climate Ethics*, ed. by Stephen M. Gardiner, Simon Caney, Dale Jamieson, and Henry Shue (New York: Oxford University Press, 2010), 146–62.

15. 'Face Reality? After You! A Call for Leadership on Climate Change', in *Ethics & International Affairs*, 25:1 (2011), Carnegie Council, 17–26.

16. 'Human Rights, Climate Change, and the Trillionth Ton', by Henry Shue, in Denis G. Arnold (ed.), *The Ethics of Global Climate Change* (Copyright Cambridge: Cambridge Univ. Press, 2011), 292–314. Reprinted with Permission.

17. 'Climate Hope: Implementing the Exit Strategy', copyright: *Chicago Journal of International Law*, 13:2 (Winter 2013), 381–402.

Shorthand Names of Chapters

In order to be able to refer to various chapters economically by means of parenthetical references in the text, I have given one-word shorthand names, which are here listed alphabetically, to each chapter:

Climate	Climate
Contingent	After you: may action by the rich be contingent upon action by the poor?
Delays	Deadly delays, saving opportunities: creating a more dangerous world?
Exceptions	Making exceptions
Hazards	Bequeathing hazards: security rights and property rights of future humans
Hope	Climate hope: implementing the exit strategy
Inequality	Global environment and international inequality
Justice	The unavoidability of justice
Kyoto	A legacy of danger: the Kyoto Protocol and future generations
Nairobi	Equity in an international agreement on climate change
Necessity	Avoidable necessity: global warming, international fairness, and alternative energy
Reality	Face reality? After you! A call for leadership on climate change
Rights	Human rights, climate change, and the trillionth ton
Sovereignty	Eroding sovereignty: the advance of principle
Subsistence	Subsistence emissions and luxury emissions
Transition	Responsibility to future generations and the technological transition
Varieties	Environmental change and the varieties of justice

Introduction

In the early 1990s I observed one of the sessions in New York City of the Intergovernmental Negotiating Committee for the drafting of an international treaty to confront the reality that human beings are by our own actions rapidly and radically modifying the climate to which we—and all other living species—have over millennia become adapted. Consequently I began to wrestle with what I took to be some of the central issues inherent to any response to the undeniable need to arrange somehow for wide-ranging international action. Here is my record of those attempts to get a grip on the moral issues at the heart of an unprecedented threat and to add them to the philosophical agenda. I continue to believe that the questions I struggled with over the next two decades were and, unfortunately, remain today, critical, although naturally I did not take on all the important moral and political questions. I am more uneasy with some of the early answers I proposed to those questions, although it is difficult for me to judge when I could not see the then best exit route from our self-inflicted dangers and when the problems have simply fed on themselves so voraciously for the two additional decades until some of what might have worked then would be too little, too late now. The appalling failure to this day of our politicians and political institutions, especially at the national level in the United States, where I initially hoped to see 'superpower leadership', to respond in even minimally adequate ways has allowed the dangers to mushroom for more than twenty additional years. The problem now is far worse and much more urgent than it was then.

Because it was perfectly clear from the beginning that the primary source of climate change was the injection into our planet's atmosphere of carbon from the burning of fossil fuels—that is, coal, oil, and gas, all of which are carbon-based—it was obvious that vigorous steps needed to be taken to reduce carbon emissions enough to slow, if not stop, the expansion of the atmospheric concentration of carbon dioxide blanketing the globe and holding additional heat here on the surface.[1] It would be possible to continue to burn fossil fuels if

[1] Tyler Volk, CO_2 *Rising: The World's Greatest Environmental Challenge* (London and Cambridge, MA: MIT Press, 2008).

technology had been developed for the capture and secure, indefinite sequestration of the carbon emitted. The carbon is safest in the ground as part of the
coal, oil, and gas, where it rested for millennia without harming a soul. But
these substances could be extracted and burned as fuels, if the carbon were
recaptured immediately and held indefinitely so that the carbon would not
reach the atmosphere in the form of the carbon dioxide that is now the
primary 'forcing factor' in climate change. But the giant corporations who
sell the fossil fuels have never acknowledged the reality of the environmental
havoc their products are wreaking and have chosen not to invest very much in
research and development of the sequestration technology that would be
necessary to make their products safe to continue to use on such a large
scale. Because of this corporate neglect, the available sequestration technology
is primitive and unusable.

   Given the indifference of the fossil fuel corporations to the safety of the
products they market, the dominant approach to climate change explored so
far has been to rely on political leaders to design institutions that will prohibit
or discourage the use of fossil fuels by politically raising their prices—by
'putting a price on carbon'. The fuels themselves have prices, of course, but
those prices do not reflect in the least the dangers for all living beings that the
burning of the fuels is creating by transforming the harmless embedded
carbon into carbon dioxide, an extremely long-lasting greenhouse gas
(GHG). The US government still even provides fossil fuel firms massive tax
breaks and other subsidies that artificially hold the fuel prices even lower than
the failure to internalize their treacherous externalities already makes them.[2]
That fossil fuel prices are driven down by government subsidization, as well as
by the failure of the prices to reflect the damage done to the planet, has made it
next to impossible so far for non-carbon fuels to compete successfully without
offsetting subsidies of their own.

---

[2] For oil subsidies, see Hanlon, Seth. 'Big Oil's Misbegotten Tax Gusher: Why They Don't
Need $70 Billion from Taxpayers Amid Record Profits', (Washington: Center for American
Progress, 2011) <http://www.americanprogress.org/issues/tax-reform/news/2011/05/05/9663/
big-oils-misbegotten-tax-gusher/> and US Congress, Joint Committee on Taxation, 'Estimated
Budget Effects of S. 2204, The "Repeal Big Oil Tax Subsidies Act"', 23 March 2012, JCX-29-12
<https://www.jct.gov/publications.html?func=startdown&id=4415>.
   On coal subsidies, see Goad, Jessica, and Stephen Lacey, 'Top Three Ways That American
Taxpayers Subsidize Dirty Coal Development', *Climate Progress* (Washington: Center for
American Progress, 2012) <http://thinkprogress.org/climate/2012/04/13/463874/top-three-
ways-that-american-taxpayers-subsidize-dirty-coal-development/>.
   For the global picture, see International Energy Agency, *IEA Analysis of Fossil-Fuel Subsidies*,
World Energy Outlook 2011 (Paris: International Energy Agency, 2011) <http://www.iea.org/
media/weowebsite/energysubsidies/ff_subsidies_slides.pdf>.
   For a general argument concerning developing countries, see Maria Vagliasindi, *Implementing
Energy Subsidy Reforms: An Overview of the Key Issues*, Policy Research Working Paper 6122
(Washington: World Bank, 2012).

So the most thoroughly discussed part of such plans as there are to get control of climate change rely on the following means/end sequence: stop climate change (1) by reducing or, far better, stopping the injection of carbon dioxide into the earth's atmosphere (2) by sharply reducing emissions of carbon dioxide from the burning of fossil fuels (3) by sharply reducing or eliminating the burning of fossil fuels (4) by 'putting a price on carbon' by (5) either (a) carbon taxes or (b) a market in required emissions permits—'cap-and-trade', 'cap-and-dividend', etc. The choice between taxes and permit trading is important, but is one of the issues on which I have not contributed. Either way the first step would have to be imaginative and courageous political action. If there are to be taxes, governments must pass them; if there is to be permit trading, governments must put a ceiling on—'cap'—emissions and arrange the trading, including the important initial distribution of permits and the decision about the use of the potentially vast funds that would be collected from the marketing of emissions permits.

A third alternative mechanism, besides carbon taxes and emissions permit trading, would depend on even more direct political action: emissions can be limited or prohibited by law like other pollutants—regulation to levels, if any, that are safe. The fossil fuel industry is uniquely indulged by the world's governments in being allowed to pour out its most dangerous pollutant— carbon emissions—without restriction while enjoying large subsidies from public funds.[3] My work on the moral issues has assumed that, rightly or wrongly, regulation is least likely and that therefore either taxes or cap-and-trade will be used to establish and drive up the price for emitting carbon, if effective political action is ever in fact taken.

Institutional arrangements for pricing carbon, however, deal with only one piece of the problem of climate change. People are burning fossil fuels to obtain energy, and energy is a necessity for any remotely modern form of life. It is just barely conceivable that we could cease harnessing any energy other than, say, the energy produced by the metabolism of human and animal bodies—go back to manual labour and draft animals—but for very good reasons, such as the vast reduction in human numbers that would require, no one is contemplating it. Contemporary societies depend on huge amounts of energy. Suppose that either taxes or trading start to make the use of fossil fuel as a source of energy more expensive. At present fossil fuels are the cheapest sources of energy—this is why they are the dominant sources. It would not take very long at all before rising prices for fossil fuel would drive the poorest members of humanity out of the energy market and into deeper poverty.

---

[3] McKibben, Bill. 'Global Warming's Terrifying New Math', *Rolling Stone*, 19 July 2012. <http://www.rollingstone.com/politics/news/global-warmings-terrifying-new-math-20120719>.

The central question I have explored is: How can we limit the dangers resulting from climate change without driving additional hundreds of millions of people into poverty? We must limit climate change for many reasons, not least for the sake of the most vulnerable members of future generations. But if one link in the chain of means to that end is to be step (4) above, instituting and then steadily raising a price for the carbon content of the fossil fuels on which many of the most vulnerable members of the current generation depend for subsistence, we could seriously harm the current vulnerable while attempting to protect the future vulnerable (and ourselves). But this is getting ahead of the story.

During the two decades of failed international attempts to arrive at an effective and widely accepted multilateral treaty to bring the dangers of climate change under control, the US government has devoted itself to the impossible task of making justice go away. The negotiations occur within the framework provided by the United Nations Framework Convention on Climate Change (UNFCCC), adopted by the nations of the world and immediately ratified by the US Senate in 1992. In the last two rounds of negotiations, culminating respectively in the annual conferences in Doha (2012) and Durban (2011), the USA insisted that none of the references to principles of justice that are even in the UNFCCC itself, such as the conception of 'common but differentiated responsibilities', be mentioned in the official statements adopted at the conclusions of the sessions.

My first article, 'The unavoidability of justice' (*Justice*), is devoted to explaining why the still unvarying US negotiating strategy is both wrong morally and almost certain to continue to block a widely accepted agreement, as it so far has. The fundamental reason is that when one proposes to a large number of other sovereign agents treaties that treat them unfairly, they have no good reason to accept the treaties unless they can be coerced; and in this case the sovereign agents cannot be coerced. The fuller explanation involves what I called 'compound injustice' (*Justice*, 36, 38, 41), which comes in various species. 'Compound injustice' occurs when an initial injustice paves the way for a second, as when colonial exploitation weakens the colonized nation to such an extent that the colonizer can impose unequal treaties upon it even after it gains independence.

In the case of climate change, practically everyone acknowledges that the largest cumulative emissions have come from the nations which were the early industrializers and which thereby gained great wealth from the energy consumption that produced their damaging emissions, while the nations which will under business as usual suffer most from the climate change driven by those emissions will be the poorer countries that have not fully industrialized (and have emitted very little). This is primarily because these less industrialized nations control less wealth that can readily be used for coping with the effects of climate change as it occurs, and much of the wealth consumed by their adaptation to these effects may have to be diverted from investment in their own development. Many of us view these extensive international inequalities as embodying one injustice already; others view the situation as basically having involved bad luck for the now disadvantaged nations.

Either way, to insist, against the background of these great international inequalities in nations' respective capacity to protect their own people against the effects of climate change, that every nation take on binding commitments to contribute to a joint plan to mitigate climate change, thus sharing the costs of mitigation, when there is no prospect of wealthy countries taking on binding commitments to contribute to a joint plan for everyone to adapt to the climate change that will in fact occur, is clearly unfair in demanding similar contributions from nations in radically dissimilar circumstances—treating unequals equally. The wealthier countries, which have been the main contributors to the occurrence of climate change, are not offering to assist the poorer countries, which are much less able already than the wealthy countries to adapt to that climate change, in adapting to it. Nevertheless, some of the wealthier, led by the unbending USA, insist adamantly that the poorer contribute to mitigation by pursuing more costly, lower-emission forms of development, which may slow the development that is their only prospective source of resources for adaptation to the disruptions that will be, and are already being, produced by climate change. The burdens of mitigation are to be shared by all, including those who lack the resources to adapt, but the burdens of adaptation are not to be shared by all. This is the US government's position.

The contention of the USA has consistently been, in effect, 'climate first, justice maybe later'. Negotiations will have two tracks: a fast track for climate, meaning mitigation, and a slow track for justice, including adaptation. General binding commitments to share in the costs of national adaptation would, according to the US government, constitute the dreaded phenomenon of international redistribution, and that might conceivably come later (although purely voluntarily, not as any duty of justice). Every nation ought first to make binding commitments to share in the costs of mitigation—including those nations who do not have adequate resources to deal with the problems of adaptation that climate change will certainly confront them with. This is patently unfair, the poorer nations have no good reason to sacrifice their own interest in this way, and they have quite reasonably for two decades refused to agree to cut their own throats by diverting scarce current resources from the development that is their only realistic source of increased future resources (since they do not believe the empty promises of justice deferred).

The US notion of two-track negotiations—climate (meaning mitigation) on the fast track, justice on the (very) slow track—is incoherent, because while issues about binding commitments to share in the costs of adaptation are indeed issues of justice, those issues are nevertheless inherent in the issues about the fair sharing of the costs of mitigation among parties radically unequal in their capacity to adapt. However long the US government refuses to face this reality, the question of justice remains unavoidable.

In 'Subsistence emissions and luxury emissions' (*Subsistence*), which became my most influential and most cited article, I make an analytic suggestion and a substantive suggestion. The first suggestion was that it would greatly reduce confusion if we analytically separated four interrelated questions of justice instead of trying to swallow 'justice' all in one gulp: (1) the allocation of the costs of mitigation, (2) the allocation of the costs of adaptation, (3) the background allocation of resources as it affects fair bargaining, and (4) the allocation of emissions, both in the transitional period and in the end. Obviously the point about compound injustice in *Justice* had been that injustices in (3), the background resources, would badly distort negotiations about everything else because of some nations' far weaker bargaining positions. And the thesis about the necessity of considering mitigation and adaptation together had been that what is a fair allocation of the burdens of mitigation is substantively dependent on relative capacities for adaptation. But one can spell out the material connections lucidly only by being explicit about the conceptual differences. The Intergovernmental Panel on Climate Change (IPCC) drew on these distinctions among the four questions of justice in its 1995 report.[4]

The long-term value of the second, substantive suggestion, I now think, is more doubtful. It was obvious that if GHG emissions in general, and carbon emissions in particular, had to be limited in order to cap the atmospheric concentrations of them that drive climate change, it would no longer be possible for everyone to emit whatever they pleased. Emissions would be limited, and thus by political choice made zero-sum across at least contemporaries. So they would have to be allocated one way or another, leading to the fourth question of justice. In the classic working paper that was the wellspring of contemporary discussions of ethical questions concerning climate change, Anil Agarwal and Sunita Narain had in passing distinguished 'survival emissions' and 'luxury emissions'.[5] Slightly modifying their terminology to be consistent with my account of rights, I adopted their distinction and attempted to mobilize it to help answer the fourth question of justice.[6] The thought was that if we are going to be creating institutions to restrict emissions, the last emissions that ought to be restricted are emissions that are vital to the

---

[4] T. Banuri, K. Göran-Mäler, M. Grubb, et al., 'Equity and Social Considerations', in James P. Bruce, Hoesung Lee, and Erik F. Haites (eds), *Climate Change 1995: Economic and Social Dimensions of Climate Change*, Contribution of Working Group III (Cambridge: Cambridge University Press for the Intergovernmental Panel on Climate Change, 1996), 79–124, at 100.

[5] Anil Agarwal and Sunita Narain, *Global Warming in an Unequal World: A Case of Environmental Colonialism* (New Delhi: Centre for Science and Environment, 1991), 5.

[6] Henry Shue, *Basic Rights: Subsistence, Affluence, and U.S. Foreign Policy*, 2nd edn (Princeton, NJ: Princeton University Press, 1996), 22–9.

production of the basic necessities for subsistence, and we ought instead to start with the purely wasteful, frivolous, and superfluous emissions of the affluent engaging in activities they do not need to engage in. Eliminate luxury emissions, while leaving subsistence emissions alone: 'The poor in the developing world would be guaranteed a certain quantity of protected emissions, which they could produce as they choose' (*Subsistence*, 66).

My specific suggestion about how to protect the basic subsistence rights of the poor was: 'if there is to be an international market in emissions allowances, the populations of poor regions could be allotted inalienable—unmarketable—allowances for whatever use they themselves consider best' (*Subsistence*, 66). This was an attempt to protect the autonomy as well as the subsistence of the poor by preventing the affluent who would have lost any right to their accustomed unlimited emissions from simply coercing the poor into selling them their emission rights rather than using them to fulfil their own needs, and I continued to struggle with the notion of inalienable emissions rights (or 'allowances' or 'permits') for years (for example, *Necessity*, 98–9 and 100; *Varieties*, 136; *Climate*, 202; *Nairobi*, 115; and *Rights*, 312 and 314, n. 41).

I now think that, apart from difficulties of institutionalization, the suggestion of unmarketable emission allowances was less helpful than it might have been, for at least three reasons. First, what the poor need to be guaranteed in order to satisfy their subsistence rights is energy, not emissions, as Tim Hayward pointed out.[7] As long as the world is dominated by a fossil fuel regime, being able to use energy of course involves being able to emit carbon. But, second, for reasons I have explored in my most recent articles (*Rights* and *Hope*, discussed below), it has become urgent, thanks in part to the political failures of the previous two decades, to escape from the fossil fuel regime entirely and advance to energy sources that do not emit carbon. No one, rich or poor, can for much longer rely on carbon-based energy. Third, while the allocation of emissions certainly ought to take ethical factors very seriously, it is simplistic to think that this allocation can be determined entirely by them while ignoring, in particular, the alternative incentive effects of various different allocations, as persuasively argued by Michael Grubb.[8] We need institutional arrangements other than guaranteed carbon emissions to protect the poorest and most vulnerable members of humanity during the necessarily rapid transition away from fossil fuels and the familiar energy regime that has caused our problems (see *Hope*).

---

[7] Tim Hayward, 'Human Rights Versus Emissions Rights: Climate Justice and the Equitable Distribution of Ecological Space', *Ethics & International Affairs*, 21 (2007), 431–50, at 440–3.

[8] Michael Grubb, 'Seeking Fair Weather: Ethics and the International Debate on Climate Change', *International Affairs*, 71 (1995), 463–96.

'After you: may action by the rich be contingent upon action by the poor?' (*Contingent*) was a direct response to the attitude being taken by the USA in the international negotiations over an implementing agreement within the UNFCCC, which was 'simple bullying' (85) in the form of threatening to continue to do what was wrong and damaging to all concerned until all other parties agreed to terms for stopping that satisfied the US negotiators (an instance of injustice of the third procedural kind identified in *Subsistence*). The US government acted then, and still acts now, as if there is no right or wrong outside legal agreements to which the USA has acceded—a kind of extreme conventionalism, or relativism, of standards: nothing is wrong unless we have agreed that it is. *Contingent* made the case for believing that excessive emissions are wrong prior to legal agreements about how to reduce them, shifting my focus decisively from concern in *Justice* with allocating costs of mitigation (first question of justice, in the terms of *Subsistence*) to allocating emissions (fourth question of justice) and coining the slogan 'excess en-croaches' (86). The main point of the article is to explain how excess encroaches.

It was abundantly obvious that if climate change is ever to be brought under control, carbon emissions will have to be limited. If they are limited, they become zero-sum: any of the emissions under the limit that I use, you cannot use (unless we together are to exceed the limit). With emissions, there is no extra space—no cushion and no slack. It was also evident that whatever exactly the limit on emissions would need to be, it was utterly impossible for every member of humanity to emit at the current extremely high per capita levels of the wealthiest people—such a global total would obviously be wildly beyond anything compatible with slowing the rate of climate change. If the poorest people on the planet were to be allowed enough emissions to use the energy necessary to a decent life *and* we were all together to stay within some reasonable total, the wealthiest people would need to reduce their actual share to some reasonable share. Of course, no one had established the precise quantity of a 'reasonable share', although a number of moral philosophers were defending equal shares of emissions per capita. I took no position on that.

My thesis was that individuals clearly ought no longer simply to emit whatever they pleased and could afford (as we still now continue to do years later). People ought to restrict their emissions to some (still unspecified) reasonable share that was compatible with at least two constraints: the poorest people being allowed at least the minimum necessary emissions for a decent life and dangerous climate change being avoided. And while the reasonable share was yet to be specified with any precision, we already knew the two features about it that we needed to know, namely that a reasonable individual share is much smaller than the amount of emissions for which wealthy individuals are now responsible and that using more than your own reason-able share means grabbing someone else's share (no extras, no slack, no cushion). Excess encroaches. When one exceeds a reasonable share, one not only behaves unfairly, which is of course wrong enough, but also inflicts damage by either taking someone else's share or contributing to violating the total limit set to prevent dangerous climate change. One does damage to either the planet, if others meanwhile use their reasonable shares and the total is exceeded, or one takes away part of, or all, of the shares of one or more others. Either way one violates a fundamental negative duty to do no harm in addition to being unfair (and, one could add, greedy).

In the complementary article, 'Avoidable necessity: global warming, international fairness, and alternative energy' (*Necessity*), I try to confront what I reluctantly take to be the brute fact that if reducing individual emissions to a reasonable share means radical reductions in energy use for the wealthiest, they are, however wrongly, going to continue to refuse to do it. In a volume on theory and practice, this is preceded by brief reflections on the importance to moral life of not restricting one's efforts to so-called 'ideal theory'. Consequently, and sadly, the only hope of in fact preventing dangerous climate change depends on a rapid transition to energy sources that do not generate emissions harmful to the planetary system. In the course of formulating the arguments in this chapter I was driven to the conclusion that the development of alternative, non-carbon-based energy sources is neither superfluous nor merely a helpful ancillary to some strategies for slowing climate change, but a central and vital component of every strategy.

At present carbon emissions are a necessity for anyone who wishes to enjoy a decent or better than decent standard of living. This is because fossil fuels are by far the cheapest and most abundant sources of energy, and there simply is not enough alternative energy available for everyone to have a decent life using only it, not to mention not enough for the wealthier members of humanity to have the standard of living to which they are accustomed. But this is an 'avoidable necessity' because the dominance of the fossil fuel regime can be broken by price-competitive non-carbon energy when the necessary research, development, and dissemination is done. The apparent necessity of carbon emissions is contingent on our having only primitive energy technology such as burning coal. What we need is energy—we can through technological innovation get it without causing carbon emissions.

I could—and can still—see only three options: '(1) go ahead and produce global warming, thereby avoiding any decision to make carbon emissions zero-sum; (2) reduce our economic activity and standard of living, thereby living within our fair share of a zero-sum emissions total; or (3) develop non-carbon-based energy sources, thereby making carbon emissions non-essential and making any zero-sum total set to resist global warming irrelevant to most people's lives, thereby defusing this issue of fairness. I believe the first option is dangerous and irresponsible toward future generations . . . [T]he second option is politically and psychologically impossible... That leaves the third option as the only practical hope' (104).

'Equity in an international agreement on climate change' (*Nairobi*) was a paper presented in a workshop held in Nairobi by Working Group III of the IPCC in preparation for the writing of *Climate Change 1995*, vol. III. Consequently, besides reformulating some theses for which I had already argued in the academic papers above, I offer a critique of one type of specific mechanism then being proposed for future treaties and now found in the Clean Development Mechanism (CDM) of the Kyoto Protocol (which sounds good but is largely a tragically wasted opportunity—see *Kyoto* below). My familiar theses included 'adoption of a global ceiling on GHG emissions will radically transform the international context for fairness' (109), and 'overconsumption—that is, use of more than one's share—always wrongs someone once a total has become zero-sum' (111).

The mechanism in question is in some respects a bilateral version of trading in emissions permits. Often it would be cheaper for a firm in a wealthier country to reduce emissions in a poorer country than to bring about the same reduction in emissions in its own country. If the firm can be rewarded for acting in the poorer country by being released from any obligation to bring about those reductions in its own country, this produces a kind of gain in global efficiency: the same emissions reduction at less cost (or a greater reduction for the same cost—the latter would obviously be preferable for the world). So far, so good, but as usual the devil is in the details. One kind of project a wealthy country firm can conduct is to build an additional new facility, say an electricity generating plant, that produces more electricity than an existing plant produces but with the same emissions. This too is a certain gain in the efficiency of electricity production, but it is crucial to see that in itself it not only does nothing to mitigate climate change but instead increases the emissions. This may well contribute to the economic development of the poorer nation, which could be a good thing, but if no compensating reductions in emissions are brought about elsewhere to balance the additional emissions from the new plant, more emissions are simply injected into the earth's atmosphere, and climate change is made worse. The point is not to maximize efficiency, but to reduce emissions. From this and other examples, I argued that such projects are acceptable only if they are conducted within an international institutional framework that (a) specifies which nations are sufficiently poor that they are entitled to increase their emissions in spite of the pressing global need to reduce total emissions, in order to abide by international justice and (b) requires emissions reductions of at least an equal amount somewhere—the global total must go down if there is to be mitigation of climate change.

After observing that economists' inveterate habit of reducing all desires to 'preferences' is incompatible with taking justice seriously because it impoverishes our vocabulary by discarding utterly basic distinctions such as the difference between a desire for what one objectively needs and a desire for something one simply wants, 'Environmental change and the varieties of justice' (*Varieties*) explores some variations on the trilemma in *Necessity*. It is theoretically conceivable, but I hope morally inconceivable, that one could open up a fourth option by, like fascists and other racists, not allowing all humans to count equally. Who ought to count? is one of the most fundamental questions in political morality. Energy problems are so difficult only if one assumes that every person is entitled to at least the minimum energy for a decent life. One could instead admit openly a belief 'that there are two kinds of people, people worthy of life and people not necessarily worthy of life, with the "preferences" of the latter to be excluded from our calculations if need be' (137). We would then not worry about how to satisfy the latter's desires for minimal energy for an adequate standard of living. If, by contrast, we are committed to everyone's counting, we can see that although during a transition period ultimate standards of justice need not be satisfied, minimum standards ought always to be satisfied. Otherwise lives lived entirely during the transition effectively do not count.

Assuming we are still restricted to roughly the three options discussed above, I conceded, on the one hand, that a more plausible version of the first would be, not 'Don't worry about climate change', but 'Don't worry about climate change yet' in the following sense: Don't invest heavily in mitigation, but instead continue to amass wealth, and when the climate change becomes severe, the resources to adapt to it may have been accumulated. Versions of this option have been advocated by a number of economists. On the other hand, without international institutional arrangements to guarantee it, there is, for one thing, no reason to think that enough of the accumulated wealth will be available to the most severely threatened, such as the Bangladeshis. So, if we are to mitigate, that is, cut emissions, we face in the end the same dilemma: reduce emissions by reducing energy use or reduce emissions by developing energy sources that do not produce climate-threatening emissions.

'Eroding sovereignty: the advance of principle' (*Sovereignty*) is not about climate change specifically, but its general thesis, if well founded, is the basis for a strong critique of the US national governmment's approach to climate change, which, as I have already mentioned, seems to be that it is free to do whatever it pleases in its own national interest until it willingly signs an international agreement giving it specific duties. *Sovereignty* argues that while the external sovereignty of a state may permit *exclusive* promotion of the interests of its own citizens, no reasonable kind of sovereignty allows any state the *unrestrained* promotion of the interests of its own citizens. This means that there are 'external limits on the means by which domestic economic ends may be pursued by states' (151). Those limits, as in the case of any kind of autonomy, include prohibitions on the infliction of damage on others.[9] Powerful state actors have a responsibility to take due care to avoid policies that cause certain kinds of harm, and one sufficient set of conditions for a prohibited type of harm is:

'1. The policies contribute substantially to harm to people living outside the territory of the state that controls the policies.

2. The states that govern the territories in which the people harmed live are powerless to block this harm.

3. The harm is to a vital human interest such as physical integrity (a physically sound body).

4. An alternative policy is available that would not harm any vital interest of anyone inside or outside the state that controls the choice among policies.' (156)

The stubborn resistance of politicians in the United States to the adoption of any overall national limit on carbon emissions, lest it interfere with further domestic enrichment for the USA, is precisely such a destructive policy against which other nations (and other generations, as we shall see below) are utterly defenceless. These flagrant emissions are causing weather phenomena that damage the health and take the lives of people around the world. I continue this focus on physical threats to individual persons in the next article.

An aspect of justice between generations becomes my explicit topic for the first time in 'Bequeathing hazards: security rights and property rights of future humans' (*Hazards*). I continue to assume a 'practically undeniable basic right to physical security' (164), 'a (non-marketable) fundamental right not to have their bodies damaged by the actions of others, when the damage is preventable' (167); and I explore what difference, if any, it ought to make if damage to a person's physical security will occur in the future rather than the present—if the distance is temporal, not spatial. Climate change will of course produce illness, injury, and death for large numbers of people through disruptions to agriculture and other sources of food, disruptions to drinking water supplies, more violent storms, and stressful weather conditions such as heatwaves. But most discussions of the possible significance of differences in the time at which events occur are conducted in economic terms and focus exclusively on money. If one had to choose between enjoying a certain benefit now and enjoying twice that benefit after a few years, a comparison simply between the amounts of benefit involved would lead one to conclude that one should

---

[9] This was the dawning of a realization, also mentioned in *Hazards* and made more explicit in *Climate* and subsequently, that right action concerning climate change involves not only fair sharing of burdens but also avoidance of the infliction of harm. For a valuable current development of this distinction, see Simon Caney, 'Two Kinds of Climate Justice: Avoiding Harm and Sharing Burdens', *Journal of Political Philosophy* (2013), forthcoming. doi: 10.1111/jopp.12030.

choose the double benefit later. But many people would think that it is unreasonable not to take the difference in time into account, even if the two benefits are equally likely. Similarly if one can either use a certain amount of money for consumption now or invest it and enjoy double the amount of consumption later, it might seem obviously better to choose the double benefit later from investment over the consumption now if one ignored the difference in time. But again many would think one ought to give some weight to the difference in time, even if the two benefits had the same probability. Economists tend to deal with this by recommending that the 'present value' of the future benefit be 'discounted' in a manner that reflects the difference in time.

My contention here is that whatever may be the case about discounting when one is discussing money and choices between current consumption and investment for the future, it is wrong to 'discount' physical damage to and deaths of persons. Thus, for example, other things equal, an event that kills one hundred people 150 years from now is exactly as serious as an event that kills one hundred people today. And if one could choose between the death of one person today and the deaths of ten people in fifty years, it is absolutely no good reason to be less concerned about the ten deaths than about the single death that they will occur in the future, other things equal. Of course it might be unlikely that we could be as certain that the ten will die later as that the one will die now, in which case we should adjust for the differences in probability. But the pure difference in time matters not at all.

In *Varieties* I mentioned the proposal made by some economists that, instead of spending as much money now to prevent future climate change and its resulting damage, we should invest much of the money productively and leave future generations with more financial resources with which to adapt to whatever damage occurs. But we do not believe that if I pay for medical insurance for someone, I am then free to do something that will break his leg. This is because physical security is not marketable, and so one cannot 'trade' financial benefit for physical safety. But choosing not to prevent avoidable future physical harms to people on the rationale that one has provided for financial resources that they can use to deal with the harms seems to follow the same logic. The injuries, illnesses, and death of persons in the future do not matter any less than the same numbers now, and one cannot 'compensate' in advance through investment for choosing to cause or allow preventable injuries, illnesses, and deaths.

'Global environment and international inequality' (*Inequality*) has been the other most anthologized and cited of my articles. The classic paper by Agarwal and Narain had begun from the undeniable fact that whatever international institutional arrangements we create to deal with climate change, we will be operating in a context of radical inequality among nations. Returning to the first question of justice, the allocation of the costs of mitigation, I suggest that three alternative common-sense principles can reasonably be applied to the existing international situation:

1. Unequal burdens—'When a party has in the past taken an unfair advantage of others by imposing costs upon them without their consent, those who have been unilaterally put at a disadvantage are entitled to demand that in the future the offending party shoulder burdens that are unequal at least to the extent of the unfair advantage previously taken, in order to restore equality' (183).

2. Greater ability to pay—'Among a number of parties, all of whom are bound to contribute to some common endeavour, the parties who have the most resources normally should contribute the most to the endeavour' (186).

3. Guaranteed minimum—'When some people have less than enough for a decent human life, other people have far more than enough, and the total resources available are so great that everyone could have at least enough without preventing some people from still retaining considerably more than others have, it is unfair not to guarantee everyone at least an adequate minimum' (190).

Two dialectical features about these three principles are dramatically striking.

On the one hand, although all reflect ordinary convictions and can plausibly be argued to be appropriate to guide the building of fair international institutions from an initial situation of extreme inequality, the three differ greatly among themselves. For example, the first assumes that the existing inequality is unjustified because of how it arose. While the third also assumes that existing inequality is unjustified, this is independent of how it arose and simply because it is so extreme. The second principle, meanwhile, makes no assumption either way about whether existing inequality is justified. The first assumes that there ought to be equality; neither the second nor the third does. The third in particular is much less demanding than the first. And so on.

On the other hand, all three 'converge upon the same practical conclusion: whatever needs to be done . . . about global environmental problems such as ozone destruction and global warming, the costs should initially be borne by the wealthy industrialized states' (194). The content of the criteria implied respectively by the three principles as well as the reasons for those criteria are quite different, but whichever criterion one applies, one selects largely the same states to be the major burden bearers, at least in the beginning. The practical answer, theoretical differences notwithstanding, is always the same.[10]

---

[10] This accords generally with spirit of the division of states in 1992 by the FCCC into Annex I and non-Annex I. Simon Caney has argued powerfully that the three-way convergence that I defend is not quite as neat as I portray it—see 'Human Rights, Responsibilities, and Climate

'Climate' (*Climate*) is a brief overview of the range of moral issues raised by climate change, initially emphasizing that besides the kinds of injustice that have concerned me from the beginning, our continuing to act in ways that exacerbate climate change also violates the fundamental prohibition against inflicting harm, in two different ways. First, as emphasized in *Sovereignty* and *Hazards*, causing climate change produces 'straightforward, if indirectly caused, physical harm of many kinds' (196) to spatially and temporally distant others. Second is 'the harm of preventing people from obtaining the necessary minimum of a resource vital for their survival' (196). The vital resource is the planet's capacity to absorb carbon dioxide without dangerous change in the climate. Access to safe absorption of carbon dioxide is vital for as long as the use of fossil fuels remains an 'avoidable necessity', and we deprive others by using more than our own fair share of the rapidly diminishing zero-sum supply of absorptive capacity compatible with avoiding dangerous climate changes. (The planet will of course absorb whatever we emit, one way or another, but not without changes in the climate that are dangerous for humans and other living beings.)

This deprivation of the poorest people of absorptive capacity is especially likely if we combine excessive emissions on our own part with the imposition of a global ceiling on total emissions. At this point I continued to be wedded to the notion of inalienable emission rights as the way to protect the poorest, proposing that 'the most obvious reason why every person born into a fossil fuel-based world economy would be entitled to a guaranteed minimum of the emissions essential to life is quite simply that to make the political choice to impose a ceiling on total emissions, while not guaranteeing a minimum to each person, would condemn to death the poorest people on the planet' (202). This makes the fair distribution of emission rights necessary to the avoidance of harm.

And this is true as far as it goes, but I had already recognized in *Necessity* that another option was to emphasize rapid development and dissemination of affordable sources of energy with no emissions of GHGs—that the current necessity for decent life of emissions is avoidable. There, however, I saw non-carbon energy as the only way to deal with the refusal of the wealthiest to accept lower living standards as the means to reducing their emissions. I was yet to see, as I later did in *Hope*, that a rapid exit from the fossil fuel regime into a non-carbon-based energy regime is the best way to try to guarantee the energy needs of the poorest.

The next pair of articles take up questions of responsibility to future generations that I had so far considered explicitly only in *Hazards*. 'A legacy of danger: the Kyoto Protocol and future generations' (*Kyoto*) does so as

Change', in Charles R. Beitz and Robert E. Goodin (eds), *Global Basic Rights* (Oxford: Oxford University Press, 2009), 227–47.

background to a critique of the Clean Development Mechanism in that 1997 treaty; that critique develops further the line of argument opened up in *Nairobi*. One assumption unquestioned by national governments is that it is simply rational always to pursue least-cost-first solutions to climate mitigation, but these solutions embody two dangers to future generations. First, unless the level of expenditure is firmly fixed independently, least cost first is liable to lead to doing 'the same mitigation for a lower expenditure rather than accomplishing more mitigation for the same expenditure' (209), thus saving this generation money but leaving more to be done by future generations. Second, this obviously means that the higher-cost tasks will be faced later 'unless technological advances that reduce the costs of the currently more expensive tasks at least keep pace with the passage of time' (210).

'Like the cavemen, we still generate a great deal of our energy by setting fire to lumps of coal' (211). We need to reach the date of technological transition, by which I mean 'the year in which the burning of fossil fuels ceases to add greenhouse gas to the earth's total atmospheric concentration of greenhouse gases' (212–13). This could be accomplished by technological changes of various kinds. But these technological developments will not just happen, and another unquestioned assumption of many governments seems to be that it is not reasonable to abandon fossil fuel technologies until they become too expensive for the present generation and that therefore there is no pressing need for robust research on and development of alternative forms of energy. However, the less our generation does to reach the date of technological transition, the greater the burden we push off onto future generations. And of course for every year that the total atmospheric concentration grows, the larger it will be when it stops growing. A later date for the technological transition means more severe climate change. So we do not simply pass on the problem but magnify it.

The Clean Development Mechanism (CDM) in the Kyoto Protocol continues to be a great lost opportunity for speeding the date of technological transition. Like the mechanism discussed in *Nairobi*, it allows firms to get credit for supposed emissions reductions that contribute to development in poorer nations. One major fault is that proposals that credit be granted under the CDM only if non-fossil fuel technologies were used was defeated at the convention of the parties to the FCCC that established it, so the CDM is in fact a mechanism for subsidizing the further installation in the poorer nations of the type of technology that is the source of climate change. The rationale for allowing this is that the (fossil fuel) technology that is transferred must be superior in emissions to the other technology for which it supposedly substitutes—hypothetically. For example, if a nation would, or at least might, have otherwise built a coal-burning electricity generating plant and a CDM project will build a gas-burning plant, the CDM project receives credit for emissions reductions equivalent to the difference in emissions between the hypothetical coal-burning plant and the actual gas-burning plant (because gas is a much cleaner fossil fuel than coal). Such a new gas-burning plant might indeed contribute to development, but such a scheme suffers from two large defects. First, it is far from clear that the hypothetical more polluting plant must in fact otherwise have been going to be built. Second, and much worse from the perspective of mitigation of climate change, the relative reduction in emissions (based on comparing coal with gas) is nevertheless an absolute increase in emissions. Once the plant is in operation, there will be more carbon emissions than there were before. Perhaps this can be justified by benefits to the people of the country in question, who may indeed need more electricity, but it is still a net addition to carbon emissions, for which the firm receives credits for emissions reductions that could be used against any requirement that it otherwise reduce emissions absolutely. So relative reductions in emissions, which are in fact absolute increases, are being counted as equivalent to absolute reductions! This is a farce.

'Responsibility to future generations and the technological transition' (*Transition*) follows on directly and underlines 'the irretrievability of lost historical opportunities' (238). Returning to the central conclusion of *Necessity* in 1995, I offer this summation: 'there is no allocation of GHG emissions specifically in the form of carbon dioxide that is both morally tolerable and, at present, politically feasible as long as most economies are dependent for energy upon carbon-based fuels' (225). And 'climate policy is energy policy' (225). While I had argued in *Kyoto* that our generation has a responsibility to advance the date of the technological transition nearer in time, the main suggestion here is that the usual picture of responsibilities to future generations is misleading in respects that minimize the responsibilities, for two reasons.

'First, the usual picture of responsibility is distortingly static' (231). We tend to assume that a fixed amount of effort is necessary so that if our generation does less of it, later generations will simply have to do more. But some problems, almost certainly including climate change, grow more intractable, or even become insoluble, over time. For example, if the climate changes become worse as the result of delay, not only is mitigation likely to be as difficult as ever, but adaptation will have become more challenging. Further, the climate system has over geological time not always changed in a smooth linear fashion but has undergone abrupt changes, including abrupt reversals (rapid warming producing rapid cooling around 10,000 years ago in the Younger Dryas period, for example). Delay in ceasing the pressure our emissions are now putting on the system may take the system past a threshold into an abrupt change that could release positive feedbacks to make the change worse still and would certainly make adaptation far harder for some and impossible for others. These are instances of irretrievability, historical possibilities that will not return. 'Suppose you know that you are walking through a fog toward a cliff, but you do not know how many steps lie between you and the cliff—can you think of a good policy? Yes: stop as soon as you can' (232).

The other misleading feature of many accounts of responsibilities toward future generations is that they leave the impression that the only moral issue at stake is fairness: is our generation doing its share? Fairness is of course fundamentally and supremely important. But as long as we keep expanding the atmospheric concentration of GHGs, which drives climate change, we are inflicting damage on future generations by making the environmental conditions they face more difficult and more threatening. The conditions for future generations will be worse than they are for us, thanks to our emissions, and they will be worse than they need to be—worse than they will be if we take robust measures promptly. So we violate not only the imperative 'Be fair' but the prohibition 'Do no harm'. And we inflict this grief on people who are utterly at our mercy because they live beyond us in time. What kind of people will they think their ancestors were?

'Making exceptions' (*Exceptions*) provides an explicit defence of the methodology employed—sometimes more fully, sometimes less—in all my other articles, which will likely strike many philosophers as bogged down in empirical details, such as the dynamics of the atmospheric concentration of carbon, and empirical disputes, such as whether the most effective means for protecting the lives of the poorest during the global transition out of the now dominant fossil fuel energy regime into a non-carbon-based energy regime is recognition of an inalienable right to emissions or temporary subsidies to speed a rapid exit away from fossil fuels (more on the latter below in *Hope*). I argue that if one is concerned, as anyone in practical ethics is, to justify judgements of practical possibility, not merely judgements of conceptual conceivability, one must focus on the messy details of real cases rather than the cleaned-up imaginary cases that philosophers find naturally congenial. Practical judgements need to be all things considered, and this often means that they must consider possible exceptions to general rules. Whether an exception can be justified in a particular case will normally depend on the features of that case, which one must therefore to some extent investigate. This article looks briefly at three kinds of cases, all of which I have discussed more fully elsewhere, that are argued to involve, in different ways, exceptional circumstances: climate change, preventive attack, and torture.

'Deadly delays, saving opportunities: creating a more dangerous world?' (*Delays*) examines the weight that ought to be given to scientific uncertainty in the case of climate change. Unless our entire understanding of how this planet works is wildly misguided, a number of aspects of climate change involve virtually no uncertainty, for example sea levels will continue to rise for several centuries even if we were to stop all GHG emissions immediately, creating the possibility of higher storm surges like those in Hurricane Sandy in 2012.[11] On the other hand, much else is uncertain: how high will sea level in the Atlantic Ocean rise in the twenty-first century? How high will it rise before it stops rising several centuries from now at the soonest? Of course, one primary reason these questions are currently unanswerable is that we do not know when humans will get a grip on their GHG emissions. How worried should we be about outcomes that would be severe but the probabilities of which are incalculable? Should we delay taking any action until the results of more research are known?

Often it is only reasonable to take some precautions against an outcome that may or may not occur but that would be awful, or irreversible, or both. Various versions of a precautionary principle have been formulated, often by people attempting to be responsible toward the environment. This piece is my attempt to formulate a version of a precautionary principle that is generally reasonable but directly applicable to uncertainties about climate change and its effects on humans.

The thesis defended is that we ought not to discount the seriousness of an outcome at all on the basis of its probability or uncertainty if the following three conditions are met: '(1) *massive loss*: the magnitude of the possible losses is massive; (2) *threshold likelihood*: the likelihood of the losses is significant, even if no precise probability can be specified, because (a) the mechanism by which the losses would occur is well understood, and (b) the conditions for the functioning of the mechanism are accumulating; and (3) *non-excessive costs*: the costs of prevention are not excessive (a) in light of the magnitude of the possible losses and (b) even considering the other important demands on our resources' (265). Many elements of climate change clearly satisfy these conditions, and consequently we ought to take the dangers they are likely to produce very seriously.

'Face reality? After you! A call for leadership on climate change' (*Reality*) notes that with its splendid cadre of scientists and its history of technological innovation, not to mention its being the 'sole surviving superpower', one might have expected the USA to be the leader of a global effort to tackle climate change. On the contrary, American politicians at the national level have been a dismal failure at even facing the facts, much less offering an inspiring vision for humanity's response. Perversely, the debates over justice and the appropriate sharing of responsibility that have always been at the centre of the international negotiations have been twisted by Washington into an excuse for foot dragging and belligerent obstructionism. Instead of the

---

[11] See Susan Solomon et al., 'Irreversible Climate Change Due to Carbon Dioxide Emissions', *Proceedings of the National Academy of Sciences of the USA*, 106 (2009), 1704–09; and Michiel Schaeffer et al., 'Long-Term Sea-Level Rise Implied by 1.5°C and 2°C Warming Levels', *Nature Climate Change*, 2 (2012), 867–70.

leader's approach of 'I went first, so now you,' the USA has followed the pitiful and failing defensive line of 'After you (because I am afraid you will otherwise treat me as a sucker)'. China says it is waiting for the USA to get serious; the USA says it is waiting for China to get serious. Few nations are actually getting serious; many economies continue haemorrhaging GHGs; and the atmospheric concentration of carbon dioxide swells like a malignancy on the skin of the planet. What pass for negotiations consist largely of what pass for leaders reasoning in circles, trying to dodge blame.

Even if the USA lacked the first special historical responsibility that it clearly bears as the world's largest cumulative emitter of GHGs, it would like everyone else bear a share of an additional kind of somewhat Good Samaritan-like responsibility: we have come upon an urgent problem (actually we, unlike the Good Samaritan, were main agents in its creation, but never mind!). Why is it our problem? Because we are the ones who are here now, and action needs to be taken here and now. Further, the USA also bears a second special responsibility to help to lead out of the obvious collective action problems that plague attempts to arrive at a global treaty: it is powerful, rich, and scientifically and technologically talented. To whom much has been given, of them will much be required. Aren't three kinds of responsibility enough reason for robust action?

The final two essays collected here are explorations of the moral implications of the recent suggestion by atmospheric scientists that we can most fruitfully conceive of the injection of carbon dioxide into the atmosphere as exhausting a specific budget of cumulative carbon emissions compatible with a specific surface temperature.[12] If we 'consume' a larger budget of emissions, life on the surface, including humans and human agriculture, must endure a correspondingly higher average global temperature and the other climate changes that will accompany it. 'Human rights, climate change, and the trillionth ton' (*Rights*) and 'Climate hope: implementing the exit strategy' (*Hope*) reflect on the calculation that 'total emissions of more than one trillion tonnes of CO_2 between 1750 and 2500 will most likely result in an emissions-induced rise in average global temperature more than $2°C$ above the pre-industrial average' (*Hope*, 325).

Rights notes that the fact that what matters about carbon emissions is the cumulative total means that 'carbon emissions are *zero-sum across all emitters throughout foreseeable time*' (308). That emissions are zero-sum means of course that each emission leaves one less emission compatible with the

---

[12] The original suggestion is in Myles Allen, David Frame, Katja Frieler, et al., 'The Exit Strategy', *Nature Reports Climate Change*, 3 (May 2009), 56–8, at 57. Other reflections on its significance are: Niel H.A. Bowerman, David J. Frame, Chris Huntingford, et al., 'Cumulative Carbon Emissions, Emissions Floors and Short-term Rates of Warming: Implications for Policy', *Philosophical Transactions of the Royal Society A*, 369 (2011), 45–66; and R.T. Pierrehumbert, 'Cumulative Carbon and Just Allocation of the Global Carbon Commons', *Chicago Journal of International Law*, 13 (2013), 527–48.

likelihood of any given temperature available for others to produce. If, for example, the rich help themselves to all the emissions available for a given temperature target, such as no more than 2°C above the pre-industrial average, then no emissions will be left inside the cumulative budget for that temperature for the poor to use. But the poor are able to afford only the cheapest energy, which for now is in fact precisely the fossil fuel that cannot be used without emitting carbon dioxide. I defend a general thesis and make a specific suggestion about the challenge of limiting the temperature while not preventing the poor from leading decent lives.

The general thesis is that the vulnerability of the poor to being deprived of material resources essential for a decent life, which for as long as the earth is dominated by a fossil fuel regime includes the freedom to emit carbon dioxide from burning fossil fuel, is precisely the kind of vulnerability that institutions of human rights are ordinarily constructed to protect people against. The threat of being deprived of the use of the only energy that one is able to afford seems to qualify as what Charles Beitz has recently called a 'predictable danger' and I earlier called a 'standard threat'. Any social guarantee, or institutional protection, against the threat of the deprivation of minimally essential energy would need to be international, intergenerational, and immediate. At the same time that we ought to protect people against deprivation of essentials, we also must restrain total cumulative emissions within a cumulative budget compatible with some amount of temperature change such as 2°C above pre-industrial levels that will limit climate change. How do we limit a zero-sum ('avoidable') necessity, carbon emissions, while protecting the vulnerable poor against being deprived of affordable energy?

My specific suggestion is that if the mechanism that is chosen to limit carbon emissions is some kind of emissions permit trading, the permits for the emissions that remain within the budget for some generally humanly 'tolerable' temperature rise[13] must have progressively rising prices for most consumers, but permits for the poorest ought to remain free: '*all* the free emissions should at least tentatively and temporarily be reserved entirely for the market-dependent poor' (314). This is, I realize in retrospect, a distant echo of the proposal in *Subsistence* that the poorest should have an inalienable right to produce carbon emissions as long as the only affordable energy is carbon-based energy. And I now think that what it proposes is too passive, for it is now strikingly clear that everyone's emissions, the poor's and the rich's, must be constrained within a cumulative carbon budget compatible with a non-awful temperature rise. At present everyone, poor and rich, emits freely, and emissions are soaring. In future at most the poor could emit freely, and

---

[13] Many species are already being driven into extinction by the smaller rises that have already occurred, and significant numbers of climate change-related human deaths have occurred, so the question is very much a matter of 'tolerable for whom?'.

that too must soon stop, or the cumulative carbon budget would be exceeded entirely by the emissions of the poor, and the temperature will be driven by the emissions to dangerously high levels.

In *Hope* I have come to grips with the fact that not only may the rest of us not continue to emit freely in as great quantities as we please, but even the poorest members of humanity must soon cease to engage in most carbon emissions. Humanity as a whole must not exceed the cumulative carbon budget for a non-awful temperature rise. But not only is carbon-based energy the only energy that the marginal poor can afford, but about two billion of the poorest members of humanity can afford little or no energy even now. These two billion need considerably more energy in order to escape dire poverty. But the supreme consideration is that this additional energy must not produce carbon emissions. How can we increase energy for the poorest and avoid depriving the marginally poor of affordable energy? How can we prevent dangerous climate change? Above all, is it possible to do both—increase energy and eliminate emissions—at once?

Yes, it is. We have robust grounds for hope. Manifestly, we must move as quickly as possible to non-carbon-based energy—alternative energy. And we have ways to make this happen. First, we must eliminate the huge public subsidies now paid to fossil fuel industries. Second, we should use those funds, some of the proceeds from whatever is done to raise the price of fossil fuels in accord with their carbon content—carbon tax or emissions permit trading, for example—and some of our development assistance for temporary subsidies for alternative energy until the increase in productive capacity for alternative energy brings its price down until it crosses over the rising price of carbon-based energy, driven up by the carbon tax, permit trading, or whatever is used to price its carbon. Third, the subsidies for alternative energy should be invested in the countries with the most poor people so that the increasing amounts of alternative energy are accessible and affordable to the poorest—for example, funds from carbon taxes in wealthy countries can be partly used to subsidize the installation of capacity for alternative energy in poorer countries such as India. This will both divert the Indians from burning their vast supply of dirty coal, the very worst source of energy as measured by carbon dioxide released, and make non-carbon energy available to those who could otherwise not afford it and would remain in poverty. Such an institutional innovation would protect the subsistence rights of the poor against deprivation of affordable energy and protect the climate against dangerous carbon-driven change. *Hope* sketches one feasible illustrative plan of this kind that might enable the transition to alternative energy in around two decades.

Included as an appendix is a *Declaration on Climate Justice* released at the United Nations on 23 September 2013 by Mrs Mary Robinson, former President of Ireland and President of the Mary Robinson Foundation—Climate Justice, and Dr Andrew Steer, President of the World Resources Institute, on behalf of the 'Climate Justice Dialogue' jointly sponsored by MRFCJ and WRI. I have the honour of being a member of the High Level Advisory Committee for the 'Climate Justice Dialogue', a committee including a marvellous variety of people ranging from several former heads of state, to activists from China, India, Africa, and small island states, to an international labour leader and an international banker, to a couple of scholars. I include it here, not to claim that everyone in the Dialogue would agree with every word of mine, as they definitely would not, but to show that the general spirit of my approach, however unpopular in the increasingly isolated Washington, DC, is widely shared around the globe by some of the mighty and some of the humble, who join in the cry that 'the "fierce urgency of now" compels us to act'.

OUR FOUR ALTERNATIVES

On 23 March 2013, the lights on Big Ben, the Empire State Building, the Kremlin, and the Eiffel Tower went off for the annual 'Earth Hour', a brief period for reflection on the urgency of climate change, sponsored by WWF (World Wide Fund for Nature).[14] We must choose one of four paths. First, we can fail to take climate change seriously. This has so far been largely the choice of my own national government, Washington. Various cities, states, and regions in the United States have kept climate hope alive,[15] but the federal government in Washington, especially the Senate and the House, has been a wretched failure at leadership. One US senator has even published a book mocking climate science.[16] Big Ben and the Empire State were dark for an hour yesterday. The US Senate chamber has been dark as far as climate change is concerned for the best part of a quarter century—ignorance and apathy, and very likely the pure corruption of being bought by lobbyists for fossil fuels, obscure the mounting scientific evidence. The Senate's last notable acknowledgement of climate change was its ratification in 1992 of the UNFCCC, which Senators have mostly subsequently treated as not worth the paper it is written on, contemptuously flouting all its principles. Unbelievably, more than twenty years later the USA has no national limit whatsoever on carbon

[14] Christopher F. Schuetze, 'Can Earth Hour Help Save the Planet?' *New York Times*, 25 March 2013. <http://rendezvous.blogs.nytimes.com/2013/03/25/can-earth-hour-help-save-the-planet/?src=recpb>.

[15] Joe Garofoli, 'California, China to Link Climate-Change Efforts', *San Francisco Chronicle*, 13 September 2013.

[16] US Senator James Inhofe, *The Greatest Hoax: How the Global Warming Conspiracy Threatens Your Future* (WND Books, 2012).

emissions.[17] One option is for such obstinate reality defiance to continue to reign in Washington and to obstruct multilateral agreement in the world at large.[18] Rarely have so few done so much to block obviously sensible measures to guard against danger for so many—the people of all nations and many generations. If incumbents who are mindless defenders of the climate-undermining fossil fuel regime are to be challenged and defeated, we need to be clear about the alternative we seek. We have three more options.

Second, we can rest our measures to deal with climate change on the backs of the world's poorest, many of whom already today suffer from 'energy poverty', the inability to afford enough energy to support decent lives. Those just above the victims of 'energy poverty' can afford energy for now only because fossil fuels are as cheap as they are. But they are so cheap because governments subsidize their price and because the costs of their threat to the climate and their pollution of the environment are not internalized into their price but socialized onto everyone. Governments such as Washington can act on climate change by withdrawing carbon subsidies and by imposing either carbon taxes or a system of cap-and-trade, cap-and-dividend, etc. to drive up the price of fossil fuel and wean us off it before its carbon emissions exceed the cumulative budget compatible with any affordably manageable rise in surface temperature. That would be so far, so good, as far as climate is concerned. But if we do nothing but price carbon in accord with its dangers, we will drive many more of the marginally poor into absolute poverty. The climate may be preserved, but on the backs of the worst off members of humanity upon whom those of us with the political power would have imposed what is a solution for us but by no means a solution for them.

Third, the richer ones of us could possibly, at least in theory, acknowledge that the budget of cumulative carbon emissions injected into the atmosphere is strictly limited for any given temperature on the surface and 'make space' within the emissions budget for the emissions of the marginally poor who can afford only fossil energy by purchasing more expensive non-carbon energy ourselves (as urged in *Rights*). If we had moved aggressively in this direction in 1992 by 'taking the lead' as solemnly promised in the UNFCCC, we could have

---

[17] The USA has recently had some 'serendipitous success in reducing greenhouse gas emissions'—see Eduardo Porter, 'A Model for Reducing Emissions', *New York Times*, 19 March 2013. But the conclusion suggested is: 'Putting a price on emissions of CO_2 that reflects the burden they impose on the environment and the threat excessive amounts pose to future generations would almost certainly be the most effective strategy.'

[18] The ignorance is not restricted to the US Senate. The US Department of State accepted a cost–benefit analysis of a proposed oil pipeline that only compares the carbon emissions that will result from the burning of the oil if the pipeline is constructed with the burning of other oil and—amazingly—does not even consider not burning this much oil at all—see US Department of State, *Draft Supplementary Environmental Impact Statement*, (2013), Appendix W, 'Life-Cycle Greenhouse Gas Emissions of Petroleum Products from WCSB Oil Sands Compared with Reference Crudes'. <http://keystonepipeline-xl.state.gov/documents/organization/205563.pdf>.

extended the life of any given carbon budget for many years longer than it can possibly last at this point after absorbing two additional decades of rising carbon emissions. But two points are crucial. First, 'making space' could never have been more than a temporary expedient because the cumulative carbon budget for any specific surface temperature is limited and would be exhausted relatively soon even if only the marginally poor produced carbon emissions. Second, we more well off failed in fact to take the lead and instead have consumed much of the cumulative budget ourselves since 1992 by continuing to burn carbon fuels profligately, leaving much less for the poor who can afford nothing else. The remaining budget compatible with, for example, the unambitious target of a 2°C rise in temperature will be gone well before mid century without immediate radical steps and will be gone in a few decades in any case.[19]

Fourth, we can couple together two complementary initiatives (as urged in *Hope*). On the one hand, we can do what is necessary to produce urgent reductions in carbon emissions, which appears to require pricing carbon progressively more highly—through either carbon taxes or emissions permit trading—until carbon emissions fall sharply. We must quickly exit the profoundly ambiguous era of fossil fuels that gave many of us our greatest wealth and all of us our greatest threats and move into a different energy regime. The burning of lumps of coal is not the way forward. Nor is the fracturing of the earth's bedrock in order to rip out more oil and gas. At the level of carrots and sticks, pricing carbon is the stick against fossil fuel.

On the other hand, we must do what is necessary to make alternative fuels widely affordable as soon as possible. This initiative has two grounds. First, it can be the carrot that positively moves societies toward the alternative energy regime that accompanies the stick of pricing carbon that moves them away from fossil fuels. It would be dumb to rely only on negative incentives if positive incentives are affordable and effective as well. Second, we have a negative duty not to drive the marginal poor into absolute poverty by pricing them out of the energy market. Unless governments choose simply to regulate the amount of fossil fuels burned, carbon will need to be priced and to become much more expensive. If we did that and nothing else, we would by means of the political actions we designed to slow climate change also foreseeably drive hundreds of millions of poor people out of the energy market. This would be a

---

[19] A recent analysis primarily based not on climate models but on paleoclimate data from an evidently comparable interglacial period, offers powerful and disturbing reasons to believe that aiming for a 2°C rise in temperature on the scientific assumptions that I have relied upon in *Rights* and *Hope* is much too unambitious and that movement away from the fossil fuel regime must be much more rapid even than I have said—see James Hansen, Pushker Kharecha, Makiko Sato, et al., 'Assessing "Dangerous Climate Change": Required Reduction of Carbon Emissions to Protect Young People, Future Generations and Nature', *PLoS ONE*, 8:12 (December 2013): e81648, doi:10.1371/journal.pone.0081648 (online open access).

morally reprehensible, unnecessary, harmful act in itself, and it would arouse the opposition of any political agents who are genuinely committed to representing the interests of the poor against the climate measures—certainly they would not cooperate with them. So it is politically necessary as well as morally necessary not to try to make the marginally poor pay in deprivation for the measures needed to slow climate change. Plans to subsidize temporarily the installation of additional capacity for alternative energy within poorer countries seem well designed to accomplish both important goals: drive down the price of alternative energy and make it accessible to the poorest.[20] We still have time to retrieve the situation, to avoid the worst, and to provide invaluable protection for living beings yet to be born.[21] What greater opportunity could one wish?

Henry Shue
Merton College, Oxford
Autumn 2013

---

[20] For a concrete global plan, see, in addition to the illustrative plan sketched in *Hope*, the splendid study by Tariq Banuri and Niclas Hällström, 'A Global Programme to Tackle Energy Access and Climate Change', *Development Dialogue*, 61 (September 2012), 264–79. <http://www.dhf.uu.se/wordpress/wp-content/uploads/2012/10/dd61_art18.pdf>.

[21] Two excellent complementary studies that will be appearing at roughly the same time as this collection are: Dale Jamieson, *Reason in A Dark Time: Ethics and Politics in a Greenhouse World* (New York: Oxford University Press, 2014) and Darrel Moellendorf, *The Moral Challenge of Dangerous Climate Change: Values, Poverty, and Policy* (Cambridge: Cambridge University Press, 2014). Another valuable recent work is Stephen M. Gardiner, *A Perfect Moral Storm: The Ethical Tragedy of Climate Change* (New York: Oxford University Press, 2011).

1

The unavoidability of justice*

In the end this chapter is about justice, specifically the justice of the international allocation of the costs of dealing with a global environmental problem such as global warming. In the beginning it looks briefly and informally at what kinds of allocations of costs between rich nations and poor nations it would be reasonable for a poor nation to accept. The later comparison of what would be acceptable to a poor nation and what would be just to a poor nation turns out, I think, to be instructive.

TWO TRACKS

Perhaps questions about international justice would not have to be considered by negotiations aimed at producing international accords concerning global climate change. One thing at a time, experts on negotiation will suggest. Lawrence Susskind and Connie Ozawa, for example, observe: 'This underlying North–South conflict is a serious impediment to concerted action, and it is aggravated by the traditional approach to environmental diplomacy that emphasizes the gap between the haves and the have-nots.'[1] They go on to caution against the making of unnecessary 'linkages'. Similarly, James K. Sebenius has recently warned that 'issues should be linked with caution' and recalled that as the regime for the seabed contained in the Convention on the Law of the Sea 'took on more of an NIEO-like character, industry opposition grew'.[2]

---

* 'The Unavoidability of Justice', in *The International Politics of the Environment: Actors, Interests, and Institutions*, edited by Andrew Hurrell and Benedict Kingsbury (Oxford: Oxford University Press, 1992), pp. 373–97.

[1] Lawrence Susskind and Connie Ozawa, 'Environmental Diplomacy: Strategies for Negotiating More Effective International Agreements' (MIT-Harvard Public Disputes Program, 1990), 1.

[2] James K. Sebenius, 'Designing Negotiations Toward a New Regime: The Case of Global Warming', *International Security*, 15 (1991), pp. 126 and 128. The New International Economic Order (NIEO), he has explained, 'involved a series of proposals advocated by LDCs during the

In the end, I shall argue, some issues of justice need not come up now, but others are unavoidable. In order to see why, we will later need to distinguish different issues about justice from each other. For now let us pursue the suggestion that we leave aside 'the gap between the haves and the have-nots', that is, leave aside all issues about international justice.

Negotiators could, then, say to each other: 'We acknowledge that some parties to these climate negotiations believe that the current international situation is unjust and that this pre-existing injustice might infect and corrupt our negotiations here. Nevertheless, we all agree to set aside within this forum all doubts about the circumstances from which we begin and agree simply to negotiate with scrupulous fairness from this point forward.' Let us, the suggestion could be, have two negotiating tracks unlinked to each other: a climate track and a justice track.

From the point of view of poor nations, who might have been expected to favour always keeping injustice on the agenda, the two-track approach has at least two strong considerations in its favour. First, assuming that what the elimination of injustice would turn out to require, in the crudest terms, is transfers of wealth from rich nations to poor nations, one can see that even negotiations that were merely rational bargaining about how to divide the costs of slowing the rate of global climate change, even if they left consider-ations of justice entirely to one side, might quite independently require that rich nations bear a far higher proportion of the costs than poor nations. Fundamentally, this could be because, with much more to lose materially from increases in the rate of climate change, rich nations could spend much more on cooperative action to slow the rate of climate change and still be better off than they would have been if cooperative action had not been taken to slow the rate. Those with more to lose from inaction can rationally choose to pay more to bring about action, even when they are basing their choices exclusively upon their own national interest.[3] Therefore, while it may be that proper consideration of international injustices would lead to the conclusion that rich nations ought to transfer large sums to poor nations, simple rational consideration of national interest, based upon de facto holdings, may in any case lead to the conclusion that rich nations are well advised to contribute the vast preponderance of whatever is needed to prevent severe damage to, if not destruction of, those very holdings by rapid climate change. If what poor nations want from negotiations about global climate is for rich nations to pay most of the bill for action, they may be able to get it, even if justice does not

---

1970s which included significant wealth redistribution, greater LDC participation in the world economy, and greater Third World control over global institutions and resources' (p. 128).

[3] This is convincingly demonstrated in Jeremy Waldron, 'Bargaining and Justice: A Simple Model', in *Ethical Guidelines for Global Bargains* (Ithaca, NY: Cornell Program on Ethics and Public Life, 1990), photocopy.

come into the discussions. The rich nations have the most to lose—and, given current holdings of the world's wealth and resources, are the only ones able to pay the amounts probably required—so either they pay most of the freight or no cooperative effort will be undertaken. So it may not in fact be in the interest of the poor nations to complicate and embitter the negotiations by insisting upon keeping the issue of injustice on the agenda, however serious the injustices may actually be.

The second step of this defence of the two-track approach—climate now, justice later—from the poor nations' point of view builds upon the first. It may be in the material interest of the rich nations to reach accords to deal with global climate change even if those accords will be quite expensive for those rich nations (because rising temperatures and thinning ozone will be far more expensive still). Even from negotiations that are nothing but narrow rational bargaining, the poor nations are likely to receive substantial transfers to avoid future problems concerning climate.

It is, on the other hand, presumably not in the material interest of the rich nations over the short term, nor perhaps even the medium term, to correct international injustices if that simply requires substantial additional transfers of wealth to the poorest nations, with no additional quid pro quo. If one is profiting from injustice, it is hardly going to be in one's interest to pursue justice. Consequently, while rich nations could be expected to accept an accord that dealt strictly with climate change, they might bridle at an accord that, in effect, added the bill for the elimination of international injustices to the bill for the prevention of climatic disaster and demanded payment of both with, so to speak, one cheque. The poor nations are not likely to receive substantial transfers to correct *past* problems of injustice. Better for the poor nations to put climate on a fast track (and secure the transfers called for on that track), one might reckon, even at the price of leaving justice on a slower one (since those transfers are unlikely to occur in any case).

DOUBLE OR NOTHING

One consideration against the two-track approach, from the point of view of the poor nations, is the following. The growing political crises about the atmosphere and the climate may provide unprecedented leverage for the most populous of the poor nations on non-environmental issues, including justice, on which the rich nations have already proved beyond all doubt to be recalcitrant, and the poorest of the poor nations may be desperate enough to want to see this leverage used. I assume that while it may be in the material interest of the rich nations to foot most of the climate bill (because they may have so much more to lose materially from rapid climate change than do the poor nations), the rich nations must have the thoroughgoing cooperation of the most populous of the poor nations—because so much of what needs to be done must be done inside the borders of the most populous nations that are yet to develop. The mountains of coal that must not be burned in future with the same technology with which the USA and the UK burned most of their own coal in the past are in China; and while only nations as wealthy as the United States and Japan can afford to underwrite alternative technologies, the large poor nations such as China, India, and Indonesia must agree to use the improved but more expensive technologies within their own respective territories if their emissions are not to add enormously to whatever magnitude the problem would otherwise have. Thus, the problem of global climate is one of the very few on which the rich nations actually need the cooperation of major groups of the poor in the implementation of a solution.[4]

If the rich do in fact need the poor, the populous poor nations may have leverage on the issue of global climate to an unusual, if not unique, degree. They would be sorely tempted to make political linkages of the kind that Susskind and Ozawa seem to advise against, even if there were no ethical linkages of the kind that I hope eventually to show. And James K. Sebenius, who himself advises in the end against such moves, nevertheless reluctantly judges that 'especially given current levels of distrust, as well as the steep energy requirements of vital development, a threat by key developing nations not to cooperate with an emerging climate regime—although it might ultimately be mutually destructive, and its effects more severe in the developing world—could have a clear rationale and a measure of credibility'.[5]

For poor nations have no particular reason to believe that any progress would in fact ever be made on a politically unlinked justice track. A separate international justice 'track' has, after all, been available for a long time without any movement having occurred along it, in spite of loud and urgent cries from the South to the North in support of a New International Economic Order and various other proposals, none of which has ever been taken seriously by any of the richest nations (in spite of the endorsements by a few of the Willy Brandts and Olaf Palmes of the North). How do we imagine the de-linked justice track is to be activated? Can those who pride themselves upon being 'realists' about climate negotiations consistently assume that the wealthy nations will one day be seized by a spontaneous desire to do justice? Since the rich nations have so far been unwilling to redistribute voluntarily when there was not known to be an urgent and expensive global climatic crisis, what reason is there to believe that they will—without exceptionally strong pressure—step forward to do

[4] Terrorism may be one of the few other problems on which the rich need the poor—and it is not going particularly well so far, in spite of doing almost infinitesimal harm compared to the potential catastrophes that the alteration of the global climate may hold in store.

[5] Sebenius, 'Designing Negotiations', 129.

justice in the midst of an ongoing crisis? If the rich would not agree to pay both the climate bill and the justice bill in order to get a climate accord they badly need, when the two subjects were politically linked, why should anyone expect that they will spontaneously volunteer to deal with injustice later, after they have already secured cooperation in protecting their own holdings against a threatening disaster that may continue to require expensive countermeasures throughout the next millennium?

Negotiations that were no more than rational bargaining intended to advance narrowly construed national interests might indeed yield some transfers of wealth or technology from rich to poor. Those transfers, however, would merely be calculated to cover some of the additional costs resulting from an agreement to take action to deal with some threat such as global warming. The transfers from rich to poor nations would be calculated, perhaps, to guarantee that the poor nations would not have to bear much of the additional expense of the additional efforts to slow climate change. This might leave at least some poor nations little or no worse off from having cooperated with the climate initiative than they would have been from refusing to cooperate. No reason at all has been given, however, to think that the transfers produced by the single-track agreement about climate would exceed the additional costs produced by the agreement. These 'transfers of wealth' would be wholly distinct from any 'transfers of wealth' that might be required for the elimination of injustice. If the transfers required to undo existing injustices are ever to occur, leverage will need to be employed to bring them about. Poor nations should not fail, according to this counterargument to the two-track approach, to use their leverage in this rare situation in which they actually have some. The strategy for poor nations being considered is double or nothing: demand from the richest nations redress of the grievances of injustice as well as performance of a fair share of the task of slowing destructive environmental changes, and, by that very linkage, risk securing neither. This is a gambler's strategy, but the poorest of the poor are in precisely the kind of circumstances that can encourage the adoption of such strategies.[6]

POOR NATIONS WITH THE LEAST LEVERAGE

Unfortunately, the alternative of double or nothing is, upon a second look, a less promising alternative for the poor nations than it initially seems. First, the poorest of the poor nations and the most populous of the poor nations are by

---

[6] Indeed, one of the genuinely self-interested reasons for the wealthiest nations to correct injustices is to remove desperate situations specifically in order not to have to deal with such daring strategies.

no means all the same.[7] The prospect of the Indians and Indonesians all installing refrigeration using CFCs, or of the Chinese burning all their coal in an industrialization process as polluting as the one that occurred in Europe and North America (*and* installing refrigeration using CFCs), is a monumental threat to the value of the material holdings, not to mention the health and safety, of the people of the wealthy nations. That threat might provide considerable bargaining leverage to these large nations, whose development could produce far more emissions of dangerous gases than the wealthy nations can possibly eliminate from their own economies without virtually eliminating the economies themselves. But how would all this help Ethiopia, Sudan, Chad, Mali, Haiti, or any of the rest of the very poorest nations whose populations (and consequent potential for emissions) are relatively insignificant? If there were some sort of general Third World solidarity stretching across the continents, the leverage of the most populous nations might be mobilized politically on behalf of justice for the poorest nations. Yet I see no such solidarity between the poor with leverage and the poor without it.

Further, even the poorest nations with populations too small to have significant leverage of their own might be split among themselves. Desperation produces radically divergent reactions: the desperately poor may either anxiously grasp the least they are offered or, reckoning that they have next to nothing to lose in any case, hold out for the most they believe they might get.[8] Some might do each, further undermining any potential Third World solidarity. Consequently, the prospects of a general double-or-nothing strategy appear dubious at best for all but the most populous nations whose potential emissions provide them with the greatest leverage.

One must be careful about what is taken to constitute bargaining strength or leverage in this case. Wealth provides bargaining strength because it constitutes resources that can be drawn upon in the absence of any agreement, making the wealthy bargainer willing to pay relatively less in order to reach agreement. The wealthy party can draw upon her wealth to survive non-agreement. In this case, however, non-agreement means 'dirty development' rather than 'clean development' by some very large countries that have a lot of developing still to do. That is, if there is no agreement between wealthy nations and poor nations upon a plan under which the wealthy subsidize, to some extent, the choice by the most populous poor nations of a relatively less polluting, but therefore more costly, strategy of development, the largest of the poor nations will presumably proceed with what they would have done anyway: the most cost-effective available development strategy in which the costs of, for example, emissions of greenhouse gases are ignored

[7] I am grateful to Andrew Hurrell for impressing the significance of this fact upon me.

[8] See Waldron, 'Bargaining and Justice'.

('externalized'), as they were during the development of the now wealthy countries. This means that without necessarily intending anyone else any harm and without engaging in any non-standard economic calculations, the most populous poor nations would on the option of non-agreement add gigantic increments to the gases that cause global warming by simply proceeding with their own economic development. Accordingly, while the wealthy nations might have sufficient wealth to protect themselves against the effects of any climate changes caused by the choice of 'dirty development' over 'clean development', the additional expense would presumably be huge—and probably incalculable in advance. The costs that the most populous poor nations might unintentionally impose on the wealthy nations with a relatively 'dirty' development strategy (even one less dirty than those the wealthy nations followed in their own times) would provide these poor nations with considerable bargaining strength, even while questions of justice were being ignored. Because of the leverage provided by the threat of 'dirty' development, the most populous of the poor nations might be able to negotiate fairly favourable terms for themselves in any international environmental agreement.

For the even poorer but smaller nations, as we have already noted, no such threat is available. Haiti can emit all the carbon dioxide it can afford to emit for a very long time before it creates any increase in the rate of global warming; I would think the same would be true even if all the poorest nations without large populations, which would include much of the continent of Africa, acted in (improbable) concert. They cannot pollute enough to gain any bargaining power.

That the double-or-nothing strategy faces difficulties does not, of course, establish that the pure two-track approach is a good idea. A third alternative is that rich nations should, contrary to the two-track approach, take some aspects of justice into account from the beginning but, unlike the use of the double-or-nothing strategy, without having to be forced to consider justice by the poor nations with leverage. This would involve approaching climate negotiations not as rational bargaining in the narrow sense but as a process constrained all along by some consideration of justice. It is this third alternative for which I shall in the end argue. Meanwhile, it is useful to return to the two-track approach, on which negotiations are rational bargaining unconstrained by justice, to see what it would be reasonable for a poor nation populous enough to have a great deal of leverage to agree to.

POOR NATIONS WITH THE MOST LEVERAGE

To make the issue more concrete, suppose that if the Chinese cooperate on an unlinked climate track, on which the United States (simply because it is rationally in its interest to do so) foots many times as much of the bill as

China does by developing and transferring pollution-abating technology that China then implements, global warming can at best be held to a temperature that produces only a 'moderate' rise in sea level. Suppose further that a nation with the wealth then still remaining to the United States can, in one way or another, provide for its coastal inhabitants so that their lives are not ruined (or taken) by the expanding seawater, but that a nation with only the wealth that China can by then expect to have, in spite of having been able to continue developing as rapidly as possible thanks to the subsidization of its environmental efforts by the US technology transfers, will not be able to save Shanghai and other seaports. Is it clear that China could live with a cooperative global environmental initiative that was *successful*? For the two-track option is implicitly assuming a future in which each nation handles *its own* problems with *its own* resources.

For, consider the question that China confronts after any international environmental agreement has been carried out by all sides. It seems evident that some destructive climate changes have already today been set in motion and can no longer be prevented; for example, it seems likely that average global temperatures are going to rise during the next century because of (among other things) carbon dioxide that we have already released into the atmosphere by burning fossil fuels, so that all that the best possible accord upon international action can possibly accomplish now is to slow the rate of an already inevitable temperature rise of unknown magnitude. A poor nation's agreeing to proceed on the climate track without any preconditions about what happens on the justice track means, then, its agreeing to try to ameliorate the effects upon its own citizens of the climate changes that can no longer be prevented even by implementation of the concerted action agreed to in the climate negotiations, with only the wealth and resources that would be left to that nation after the performance of its share as specified in the agreement emerging from the climate negotiations. The nation must live with the results of even the full implementation of the best attainable climate accord entirely within whatever then happen to be its own assets.

Agreeing to a climate accord is certainly going to entail, at least for rich nations, making economic sacrifices in the sense of putting wealth and resources that might have been invested otherwise into less polluting but not necessarily more productive (perhaps even less productive) technology. What about poor nations? Suppose the rich nations will agree to no climate accord that does not also require (no doubt smaller but) genuine sacrifices by the poor nations as well. Continue to think of China as the case in point in order to make matters slightly more concrete. A proposed final agreement might specify that every year from 2000 to 2050 China was to divert $n billion out of its own economic development and into the production and installation of pollution-abatement technology. Fifty years of diverting significant sums away from increased production and into reduced carbon dioxide emissions would leave China considerably less wealthy in 2050 than it otherwise could have been entirely with its own resources, although a much smaller contributor of carbon dioxide to the greenhouse effect than it otherwise would have been. Let us call China's choices 'Cooperation' (with such a proposed climate accord) and 'Isolation' (which means proceeding with its own development at the maximum sustainable rate, ignoring the effects on global climate that do not undermine the sustainability of its own strategy). Now obviously no determinate answer to the question whether Cooperation or Isolation is more in China's interest is available, because a strategic interaction between China's choice and the choices made by other nations would determine the outcomes. And clearly there are various quite different possibilities. All that we need to notice, however, is uncertainty—on one specific issue. The issue is whether it might be in China's national interest to choose Isolation, even if its own defection from the proposed accord would doom the accord and prevent concerted action to slow climate change, rather than to choose Cooperation and enable the accord to be implemented at the price of reducing the level of its own wealth in 2050 by the amount that its compliance with the accord would require. And the critical factor is that China has no guarantee that it would be better off by cooperating irrespective of the amount represented by n billion per year. On the contrary, there is surely some number n such that if China were required to divert n billion per year from development to pollution abatement, it would be worse off by choosing Cooperation than by choosing Isolation. And there is no a priori guarantee that there is any number m such that if China were required to divert m billion per year from 2000 to 2050, it would be better off in 2050.

The fundamental difficulty is that what matters is a ratio. What counts is the ratio of China's wealth in 2050 to China's cost for dealing with the climatic effects the Chinese people face in 2050. We may call the two possibly quite different ranges of levels of wealth 'Cooperation Wealth' and 'Isolation Wealth', and we may similarly label the two ranges of levels of costs of dealing with greenhouse effects 'Cooperation Costs' and 'Isolation Costs'. We may readily grant that the Isolation Costs for China (or any other nation except an extremely small one) would be more severe than the Cooperation Costs, on the reasonable assumption that the international effort would be more successful with China's participation. Consider, however, that Isolation Wealth would be considerably greater than Cooperation Wealth on the equally reasonable assumption that China can develop further if it does not divert resources away from development.[9] The critical information that we do not

---

[9] I realize that one strategy for cooperators to take against defectors in desperate circumstances would be economic warfare designed to guarantee that no nation does better by refusing to cooperate. Indeed, given the awful magnitude of the social consequences of even some of the intermediate climatic disasters, I think we must seriously contemplate nuclear blackmail, nuclear

have is whether the ratio of Cooperation Wealth to Cooperation Costs is better or worse than the ratio of Isolation Wealth to Isolation Costs. This uncertainty means, unfortunately, that a populous poor nation has no guarantee that Cooperation with an agreement on the climate track is more in its interest than Isolation from the agreement. Obviously it matters how costly the terms of Cooperation are, and, other things being equal, the more costly Cooperation is, the more reason to suspect that Isolation would be in its national interest as normally construed.

TWO KINDS OF COMPOUND INJUSTICE

This fundamental uncertainty about whether it would be in the interest of a populous poor nation to cooperate in international attempts to prevent harmful climate change in the absence of any complementary agreement for international cooperation in coping with the harms that turn out not in fact to have been prevented means, it is clear, that it would not be reasonable for such a poor nation to agree to a two-track approach according to which the climate track covered only prevention and did not cover coping as well. Any such nation should insist that any agreement include cooperative coping with the harms that in fact occur later, as well as earlier cooperative attempts to prevent harm. I do not see how anyone could show that it would be reasonable for a populous poor nation to agree to the combination of cooperative prevention and non-cooperative coping. Since a populous poor nation whose 'dirty' development will involve extensive emissions of greenhouse gases has considerable leverage, it might well be able to insist in fact upon a fairly favourable agreement covering coping as well as prevention, at least for itself. On the other hand, as we have already noted, a poor but not very populous nation— say, Haiti or Mali—would lack the leverage to insist upon such a complementary agreement for dealing with the harms that will occur and so could easily be left to fend for itself, as far as its bargaining strength goes.

This, of course, is where justice comes in. Justice is about not squeezing people for everything one can get out of them, especially when they are already much worse off than oneself. A commitment to justice includes a willingness to choose to accept less good terms than one could have achieved—to accept only agreements that are fair to others as well as to oneself. Justice prevents negotiations from being the kind of rational bargaining that maximizes self-interest no matter what the consequences are for others. There are some bargains too favourable for a just person to accept. Justice means sometimes

---

attack, and massacre of refugees fleeing disaster. I am simply trying to deal with less horrible cases at this point.

granting what the other party is in no position to insist upon. In this case, it means sharing the costs of coping with the Haitis and Malis of the world, which cannot insist, as well as with the Chinas and Indias, which probably can insist. Or so I shall now try briefly to show.

Three general reasons would be worth considering, although I shall rely upon the third. The first reason involves the first of two kinds of what I shall be calling compound injustice. I have referred above to the situation in the absence of a complementary agreement about how to share the costs of coping with the unprevented harmful effects of climate change as 'a future in which each nation handles *its own* problems with *its own* resources'. If 'its own' merely means 'those it happens to have', this description is unobjectionable: each nation would handle the problems it happened to have with the resources it happened to have. This description, however, has no moral force; and, in particular, it provides no basis for concluding: 'and so things are arranged as they should be'. That conclusion would be supported only if 'its own resources' meant 'the resources to which it is entitled' or 'what really belongs to it'. Since *all* questions of justice would be left unexamined by the pure two-track approach, it would be totally groundless to assume that each nation miraculously just happens to have exactly what it would turn out to be entitled to if the justice of holdings were ever discussed and settled. So we should not comfort ourselves by misreading 'Haiti has whatever Haiti has' as if it meant 'Haiti has received its due (and so is entitled to nothing else'. Until the justice of the national holdings that happen to exist today is discussed, there would be no basis for assuming that, without even trying, we had inadvertently arrived at the most just of all possible worlds, in which each nation has exactly the wealth and resources it ought to have.

So far I have put the point negatively: that it is groundless simply to assume that whatever international distribution turns up over the course of history is fully just. I would think, although I shall not argue for it here, that, quite to the contrary, we have lots of good reasons to think that the existing international distributions of wealth and resources are morally arbitrary at best and the result of systematic exploitation at worst. Insofar as the existing distribution is unjust toward, say, Haiti, it may turn out that some of Haiti's 'own' resources are elsewhere (for example, in France). If we are a little more precise about some of the varieties of justice, we can formulate and assess more positive assertions.

Issues of justice arise, of course, in more than one way, and it is an open question whether the standards of justice applicable to the various different contexts in which questions of justice arise are the same; and if they are not the same, how the standards for one kind of case relate to the standards for other kinds of cases and whether, in particular, some standards refer back to the fulfilment or violation of other standards.

When two or more parties agree to work together toward the solution of a common problem, we often want to ask whether the terms of their agreement are fair to all the parties concerned, even in cases in which all parties have in fact accepted, or would in fact accept, the agreement. We believe that a party may in fact acquiesce in an unfair agreement for any of several different reasons: for example, through a failure to understand how unfair the arrangements really are, or through a lack of any good alternative. The lack of any good alternative might itself reflect a different case of injustice: an injustice in the background conditions within which the agreement is being made. I shall refer to the justice of the terms of the agreement itself—the justice among the parties inside the agreement—as *internal justice*; and I shall refer to the justice of the circumstances within which the agreement is being made as *background justice*.

I believe that it is uncontroversial to note that, in general, it is perfectly possible for an instance of internal injustice to be the result of a background injustice: someone may in fact accept unconscionable terms in an agreement because an independent background injustice has, for example, left her with no good alternative to the agreement.[10] Without the agreement she would be even worse off than she will be with the agreement—in this specific sense, she may even be rational, in the circumstances, to accept the unfair terms of the agreement—but the reason she would be worse off without the agreement is a prior injustice, independent of the agreement in question.

It is an open question, of course, whether any of the parties to a particular agreement was a perpetrator of a relevant background injustice; *ex hypothesi* the victim of the background injustice is a party to the agreement. If Wrong is a beneficiary of the internal injustice against Right *and* a perpetrator of the background injustice against Right as well, the plot thickens. One then confronts the first kind of compounded injustice. Any attempt to avoid taking up any questions of justice in negotiations over future action creates, generally speaking, the ideal situation for inflicting this kind of compound injustice. Some parties are liable to press for the acceptance of internally unjust agreements using bargaining strength that results from background injustice against the very other parties now being pressed.

One would of course have to look at cases to establish when and where injustice is being compounded: for example, is the reason why Haiti has so little bargaining strength on environmental issues its level of poverty (and if so, is that in turn the result of background injustice in the form of, say,

---

[10] One can, of course, argue about whether in that case the acceptance of the agreement is voluntary or about whether the case would embody genuine consent. I suspect that those matters are less completely independent from the question of fairness than one might hope. Be that as it may, I think we sometimes consider even a consensually or voluntarily accepted agreement to be unfair to one or more of the parties who in fact accepted it.

landholdings shaped by French colonialism?)? I shall not pursue any case here. Whatever the truth about, say, French colonialism in Haiti, I have already made clear that I think the main reason for Haiti's having so much less leverage on environmental matters than, say, China is not in fact its (undeniable and appalling) poverty, but its relatively small population and consequent relatively small potential for pollution (even if it somehow finds the resources with which to develop). In a perverse way, the more of a nation's development that lies ahead of it—that is, the poorer it is now—the greater the threat of future pollution it can bring to bargaining about the terms for preventing environmental damage. Accordingly, the fact that Haiti cannot offer much of a positive threat if the terms of an agreement do not accommodate Haitian requirements must be blamed less on any past colonial exploitation than upon the small size of Haiti's population compared to China, India, or Indonesia.

Nevertheless, Haiti's position is also weak on the other side of the bargaining coin, namely its poverty gives it so little capability itself for coping with the harmful results of climate change that it might have no better alternative than to grasp at any terms that would leave it even only slightly better off than it would be with no agreement. This is not a lack of an ability to make non-agreement worse for others (such as the lack of a capability to pollute during the 'dirty' development of a country with a large population), but a lack of an ability to survive non-agreement oneself. But it still makes one's bargaining position weak relative to those in a better position to survive. In sum, Haiti can neither threaten to make non-agreement worse for others nor deal well with non-agreement itself. The former is more the result of population size than of poverty, but the latter is the result of poverty. Insofar as that poverty can be shown to be, in turn, the result of a background injustice (such as colonial exploitation), the potential for compound injustice during climate negotiations exists. For any rich nation which has benefited from exploitation to insist that poor nations which have suffered from exploitation must cope with unsolved problems using only 'their own resources' would constitute this first kind of compounded injustice.

The second reason for thinking that the Haitis of the world should receive more assistance in coping with unprevented harms than they are in a position to insist on involves the second of the two kinds of compound injustice. This kind will become clear through a progression of examples. Suppose, first, that you agree to help me wash my car even though I make a point of announcing in advance that I have no intention of helping you to wash your car (or in any other way repaying the favour). This would be a totally one-sided arrangement, and it is not at all obvious why you would choose to assist me instead of someone else who at least might possibly reciprocate. In general, I think we would have some doubts about the character of anyone who consistently offered only such one-sided arrangements. In a specific case, however, we might accept that if you actually had agreed to help me with my car knowing that I was intending not to help you with yours, you had no grounds on which later to insist upon my help (although at least some grounds to wonder about the character of someone who seemed unmoved by any notion of reciprocity). The conclusion that you had no grounds to insist upon reciprocal assistance is, of course, supported by the triviality of the whole business: cars can go for years without being washed and do not seem to rust much faster than shiny cars (so you do not lose much by not getting your car washed), and the example implicitly suggests that you were not making any great sacrifice to help me wash mine—presumably you had nothing better to do anyway.

Now consider a second example. An arsonist has set fire to my house, and you agree to help put out the fire. While you are away from your house helping to fight the fire in my house, the arsonist sets fire to your house. You then ask me to help you to fight your fire. I say: 'You know me—do I wash your car just because you wash my car? No way.' This would of course be outrageous. For a start, the case is no longer trivial: whether cars get washed does not matter, but whether fires get put out does; and fighting fires is dangerous, while washing cars is not. In addition, this time you need assistance only because you had chosen to provide assistance to me. If you had simply stayed at home to watch me fight my fire, the arsonist would not (let us assume) have been able to set fire to your house. It may be that my lack of reciprocity in car washing is tolerable—partly because trivial—if unadmirable. The lack of reciprocity in firefighting is inexcusable, however, because the favour from you that I fail to return is the cause of your needing it returned: it is only because you did what you could to help me to fight my fire that you now have a fire of your own to fight.

In the third example there is no evil arsonist to cause the problems. I myself am innocently burning brush in a field that seems to be a safe distance from all houses. Unexpectedly the wind shifts and stiffens, and the fire gets away from me and begins moving through the tall dry grass toward the houses. You and I agree to do what we can to stop the fire from reaching the houses, and we each do our best. My house is saved, but yours burns. I say: 'Thanks for helping to save my house.' You say: 'What about my house?' I reply: 'Our agreement was about preventing houses from burning, not about coping with burnt ones.' The absurdity of my response partly results from the implicit assumption built into the example that during the prevention phase you did not restrict your efforts exclusively to protecting your own house but did your part in the general enterprise of trying to protect all the houses. Perhaps you could have saved your own house if you had restricted your efforts to keeping the flames away from it and had ignored the edge of the fire near my house. The primary reason my response is outrageous, however, is that I started the fire that burned down your house. I did not intend to burn it, or any other house—this was an accident—but I did cause it. At a minimum, then, I cannot simply wash my hands of the actual consequences of my own action.

Since the analogy with prevention of, and coping with, climate change such as global warming is built explicitly into the third example, there is no need to belabour it. I chiefly want to note that a second kind of compounding of injustice occurs here. If, in the second example, it would already have been unfair of me to ignore difficulties of yours that arose from your assisting me with my difficulties caused by a third party, this is compounded in the third example by the fact that I myself am the cause of everyone's difficulties. No one doubts that the overwhelmingly preponderant cause so far of ozone depletion, which is certainly occurring, is the activity of the rich nations. One can doubt whether global warming is already occurring, but if it is, the industrial activity of the now rich will again have been the principal cause.[11] Any harms caused were unintentional, but it has surely been the same industrial activity that has made the rich nations rich that will have been the source of the most serious climate problems, even if it is not sure how seriously the gases added to the atmosphere will affect the climate at ground level. We built the fires of industrialization, doubtless ignorant in the beginning of the damage they might do, but they are our fires and we continue to benefit from them.

About specific alleged instances of the first kind of compound injustice controversy continues: to what extent is the poverty that contributes to the weak position of Haiti the result of French colonialism? to what extent is the poverty that contributes to the weak position of Indonesia the result of Dutch colonialism? and so forth. These are important but intractable debates about causal mechanisms. Concerning the second kind of compound injustice, in contrast, there are far fewer debates about the causal mechanism. The rich nations have indisputably so far caused most of whatever problems there are in the cases of ozone depletion and global warming. Consequently, any attempt by rich nations to wash their hands of any resultant harms that are not prevented would be doubly unfair.

---

[11] For extraordinarily specific calculations, see Yasumasa Fujii, 'An Assessment of the Responsibility for the Increase in the CO_2 Concentration and Inter-generational Carbon Accounts', Working Paper WP-90-55 (Laxenburg, Austria: IIASA, 1990), processed. Fujii has calculated cumulative carbon emissions per capita by region back to 1800. Fujii's calculations suggest that North America and Western Europe are responsible for, respectively, 35 per cent and 26 per cent of the total cumulative man-made increases in atmospheric CO_2 since 1800 (Fig. 9, p. 20). For this reference I am grateful to Dale Rothman. However, causal responsibility does not translate smoothly into moral responsibility. Peyton Young has observed: 'There is also a theoretical difficulty: Not only did producers benefit from unregulated past emissions, so did consumers and investors—including consumers and investors in the nonindustrialized world.... Indeed, the entire economic order has been based on a regime of free CO_2 disposal. The accumulated benefits (and blame) are widely diffused, so past emissions are *not* a measure of liability.' See H. P. Young, *Sharing the Burden of Global Warming* (College Park, MD.: School of Public Affairs of the University of Maryland, 1990), p. 8.

VITAL INTERESTS

Before we turn to the third, and strongest, reason for rejecting the separation of prevention and coping, let us look back over what has been argued so far. The first half of the chapter attempted to show that it cannot be established to be in the interest of a poor nation—even so populous a poor nation that its cooperation with a climate agreement might be essential to the success of the agreement—to cooperate with an agreement that covers only the attempted prevention of climate changes that will be harmful and fails to cover coping with the harms that do occur. This does not show that, from the point of view of poor nations, the two-track approach per se is objectionable but only that it is objectionable if it is arbitrarily restricted exclusively to the terms of cooperative prevention and ignores the terms of cooperative coping. This is a point not directly about justice but about interest, and it is a point about national interests, not about the interests of specific persons taken individually. Justice, on the other hand, is ultimately, I think, about the interests of specific human individuals (even if some principles concerning justice are stated in terms of nations or other groups).

When in the second half of the chapter we turned directly to questions of justice, we asked, in effect, whether, given that it would be in the interest of a poor nation not to agree to be deserted by its erstwhile partners when the time came to cope with unprevented problems, it would be fair to demand the sacrifice of this interest by those nations with too little bargaining strength to resist the demand. Negotiators often ask each other to sacrifice some of their interests—there is nothing in principle wrong with that, and there would be little to negotiate about without it. The question of justice is: is there anything special about the interest in question, or about the circumstances in which its sacrifice is demanded, which makes a specific demand unfair? The first two reasons for thinking that the demand by the rich nations for the sacrifice of this interest on the part of poor nations is unfair have been based upon the circumstances in which the demand is made, rather than upon the nature of the interest that would have to be sacrificed in order to satisfy the demand. These have been the two points about compound injustice: first, that if background injustices have produced the weak bargaining position of the poor nations, it is doubly unfair to exploit that bargaining weakness in order to insist that the poor nations sacrifice the interest in question; and second, that if the rich nations have caused, albeit unintentionally, the impending harms that cooperation would attempt to prevent, it is doubly unfair to leave poor nations that have pitched in on the prevention effort to cope on their own with what the effort fails to prevent. I have not tried to show that background injustices have in fact produced the weak bargaining position of the poor, because I consider that much too controversial to deal with in a single chapter; and I have not tried to show that the rich nations have caused any impending harms, because that is conceded on all sides. So far I have also said nothing about the inherent nature of the interest of a poor nation that it would be asked to sacrifice if asked to cope with '*its own* problems with *its own* resources' once the prevention effort had achieved whatever it can from this point forward still achieve. I would now like to correct this last omission.

The third reason for thinking that the exclusion of problems of coping with unprevented harms from the agreement to try to prevent harms is unjust turns upon the nature of the interests ignored if coping is ignored. 'Coping' has so far been given no concrete content. What would 'coping' with the unresolved problems actually amount to? Clearly, a genuinely concrete account would depend upon exactly which problems remained, and we do not know exactly which problems will remain after our best prevention efforts. However, since we are talking about very poor nations, we do know some things about the general character of the 'unresolved problems', most notably that they are highly likely to be life-threatening. Why? Because in very poor nations almost all big problems are life-threatening. This is what it means to be 'very poor': it means having no cushion to fall back upon, no rainy-day fund, no safety net, no margin for error. Being very poor means living on the edge, and having a big problem—sometimes, even, having a small problem—means going over the edge: losing one or two of the children, for example. In spring 1991 the Kurds in Northern Iraq and the Bangladeshis near the Bay of Bengal each provided another demonstration of the meaning of being very poor: one big disruption to the normal routine and people start starving left and right. For such a group, dealing with '*its own* problems with *its own* resources' means: sitting and watching loved ones while they die.

Climate changes mean problems for agriculture, and problems for agriculture mean that, even if there are no aggregate shortages, food is in the wrong place at the wrong time. Everyone understands by now that climate changes need not produce cyclones or droughts to disrupt agriculture severely; a shortening of the growing season by a week or two, or a modification of the average temperature by a degree or two, can produce a local crop failure. The sort of thing that 'unprevented' climate changes are most likely to produce are local crop failures of a kind that rich nations can easily adjust to and that poor nations cannot. In rich nations such as the USA and the countries of the EC, food routinely travels hundreds and thousands of miles from field to table—there are virtually no local crops; in large areas of poor nations, where transportation is inadequate, the local crops are all there is.

So, what is the nature of the individual interests that the poorest nations would be asked to sacrifice if they were asked to ignore provisions for coping? They are, in a word, vital interests—survival interests. This means, I think, that it is unfair to demand that they be sacrificed in order to avoid our sacrificing interests that are not only not vital but trivial. If this is correct, we are now in a position to see a little about what justice may actually require in our case of international climate agreements.

A MODEST PRACTICAL IMPLICATION FOR THE
NEGOTIATING AGENDA ON GLOBAL WARMING

I have been arguing that justice requires that coping, not merely prevention, be part of any package negotiated. This does not mean that no variant of a two-track approach to negotiations could be used. Issues about justice that are truly extrinsic should, perhaps, not be given purely political linkages to unrelated climate issues. The point, however, has been that a number of considerations about justice are intrinsic, not extrinsic, to any negotiations that are not simply going to create new injustices, in some cases compounding old ones. These aspects of justice are unavoidable, except at the price of committing fresh injustices.[12]

Far short of taking the positive action needed to allow the poorest nations to cope fully with the climate changes that will occur despite everyone's best efforts—positive action that I assume would involve considerable international transfers of wealth and resources from rich to poor nations—rich nations could begin by adopting a far weaker, but still significant, guideline that is, I would suggest, an absolutely minimal requirement of justice. Poor nations ought not to be asked to sacrifice in any way the pace or extent of their own economic development in order to help to prevent the climate changes set in motion by the process of industrialization that has enriched others. 'Their own' resources for coping later should be the most that they can sustainably develop between now and then.

It is not unusual, even on the part of writers relatively sensitive to the concerns of developing nations, for it to be taken for granted that development must be to some extent slowed or diverted. Durwood Zaelke and James Cameron assume interference with development and urge subsequent compensation:

> A central issue for many developing states will be how to obtain compensation for the opportunity costs they will incur from forgoing or altering their development. This compensation must include the transfer of more efficient industrial technologies and sufficient financial aid and debt relief to allow developing states to achieve sustainable development.[13]

---

[12] The measures to enable the poorest nations to cope with climate changes might rectify some, but certainly not all, background injustices.

[13] Durwood Zaelke and James Cameron, 'Global Warming and Climate Change: An Overview of the International Legal Process', *American University Journal of International Law and Policy*, 5 (1990), 283.

I suggest, on the contrary, that any technology transfers and financial relief ought to be sufficiently timely and substantial that no opportunity costs are incurred by the poor nations. All the economic cost of cooperative action by poorer nations should be borne by wealthier nations. This requires what is called in UN circles 'additionality'.[14] The external funds for 'clean' development should be in addition to whatever funds already ought for other reasons (including other aspects of justice genuinely external to climate agreements) to be provided for development, so that sustainable development is neither forgone nor altered. If, for example, Chinese coal or Brazilian rainforests must not be burned, then cooperation means not burning them. But Chinese and Brazilian development—and, I must say, internal redistribution in the case of Brazil—should not be set back as a result.[15] International transfers of wealth and/or technology should subsidize cooperation by poorer nations with environmental actions, as was already to some degree recognized explicitly in the London amendments to the Montreal Protocol on Substances that Deplete the Ozone Layer in June 1990; and to some extent was acknowledged, for the case of the Brazilian rainforests, by the summit meeting of the Group of Seven in Houston in July 1990.

I want to underline the extreme modesty of the weak minimum guideline proposed here for climate negotiations. There are, I believe, other completely independent reasons why affluent nations such as the United States, Japan, Germany, and Saudi Arabia ought to transfer quite separate resources to poor nations such as China and Haiti so that they can not only develop as rapidly as possible but also act as decisively as possible against pollution of the earth's atmosphere.[16] It is being maintained here, by contrast, only that it surely ought not to be demanded—as part of any international agreement for cooperation in dealing with global climate—that such poor nations *divert resources already in their possession* from their own development as their contribution to the common effort to save the atmosphere. If, as a matter of rational self-interested choice, rich nations can spend large sums (smaller than the amount they stand to lose if the common effort fails), and if, as a matter of

---

[14] See Peter S. Thacher's chapter in *The International Politics of the Environment*.

[15] The precise relationship between domestic injustice, such as the extremes of wild affluence and crushing poverty in Brazil, and international justice obviously requires further discussion.

[16] I have argued elsewhere—and continue to believe—that the fulfilment of at least some of people's most basic needs entitles them, if necessary, to the use of some resources currently held by *other* people. The judgement put forward in the text is far, far weaker (although, of course, not in any way inconsistent), yet has quite powerful implications for the case at hand. Since the weaker thesis is presumably acceptable to more people, it is worthwhile to endorse it, and explore its implications, even though it is not the most that one might reasonably defend. For the stronger thesis, see Henry Shue, *Basic Rights: Subsistence, Affluence, and U.S. Foreign Policy* (Princeton, NJ: Princeton UP, 1980), ch. 5, although the argument there failed to take up the important issue of the significance of the difference between situations of full, or strict, compliance, and situations of only partial compliance.

efficiency, large portions of those sums are most effectively spent in developing nations, the funds for pollution abatement within the poor nations can reasonably come from outside their borders. These transfers would, however, merely be what is necessary to avoid committing fresh injustice in future if the effort to protect the atmosphere is to go forward in the places where, as a matter of efficiency, some of the biggest differences can perhaps be made: in the poor but developing countries.

If unprevented climate changes eventually become rapid enough to destroy world agriculture as we know it, everyone, rich and poor, could in the end starve to death. The question is where to begin in order to see to it that the human threat never becomes that severe, and the answer being suggested is that justice requires that one not begin by slowing the economic development of the countries in which considerable numbers of people are already close to the edge of starvation just so that the affluent can retain more of their affluence than they could if they contributed more and the poor contributed less. Poor nations, therefore, ought not to be required to make sacrifices in their sustainable development. Even in an emergency one pawns the jewellery before selling the blankets. The weak guideline being proposed as a start merely reflects that, whatever justice may positively require, it does not permit that poor nations be told to sell *their* blankets in order that rich nations may keep *their* jewellery.[17]

---

[17] The most explicit recognition of a version of this modest guideline in the London amendments to the *Montreal Protocol* came in the new Article 10, which was adopted in June 1990 and begins: 'The Parties shall establish a mechanism for the purposes of providing financial and technical cooperation, including the transfer of technologies, to Parties operating under paragraph 1 of Article 5 of this Protocol [poor states] to enable their compliance with the control measures set out in Articles 2A to 2E of the Protocol. The mechanism, contributions to which *shall be additional* to other financial transfers to Parties operating under that paragraph, *shall meet all agreed incremental costs* of such Parties in order to enable their compliance with the control measures of the Protocol' [emphasis added]. See *Montreal Protocol on Substances That Deplete the Ozone Layer*, as amended, Article 10, para. 1.

2

Subsistence emissions and luxury emissions*

INTRODUCTION

The United Nations Framework Convention on Climate Change adopted in Rio de Janeiro at the United Nations Conference on Environment and Development (UNCED) in June 1992 establishes no dates and no dollars: no dates are specified by which emissions are to be reduced by the wealthy states and no dollars are specified with which the wealthy states will assist the poor states to avoid an environmentally dirty development such as our own. The convention is toothless because throughout the negotiations in the Intergovernmental Negotiating Committee during 1991 to 1992 the USA played the role of dentist: whenever virtually all the other states in the world (with the notable exceptions of Saudi Arabia and Kuwait) agreed to convention language with teeth, the USA insisted that the teeth be pulled out.

The Clinton administration now faces a strategic question: should the next step aim at a comprehensive treaty covering all greenhouse gases (GHGs) or at a narrower protocol covering only one, or a few, gases, for example only fossil-fuel carbon dioxide (CO_2)? Richard Stewart and Jonathan Wiener (1992) have argued for moving directly to a comprehensive treaty, while Thomas Drennen (1993) has argued for a more focused beginning. I will suggest that Drennen is essentially correct that we should not try to go straight to a comprehensive treaty, at least not of the kind advocated by Stewart and Wiener. First I would like to develop a framework into which to set issues of equity or justice of the kind introduced by Drennen.

* 'Subsistence Emissions and Luxury Emissions', Henry Shue, in *Law & Policy*, 15:1 (January 1993), Wiley, pp. 39–59. This chapter has endnotes rather than footnotes.

A FRAMEWORK FOR INTERNATIONAL JUSTICE

Four kinds of questions

It would be easier if we faced only one question about justice, but several questions are not only unavoidable individually but are entangled with each other. In addition, each question can be given not simply alternative answers but answers of different kinds. In spite of this multiplicity of possible answers to the multiplicity of inevitable and interconnected questions, I think we can lay out the issues fairly clearly and establish that commonsense principles converge to a remarkable extent upon what ought to be done, at least for the next decade or so.

Leaving aside the many important questions about justice that do not have to be raised in order to decide how to tackle threats to the global environment, we will find four questions that are deeply involved in every choice of a plan for action. (1) What is a fair allocation of the costs of preventing the global warming that is still avoidable? (2) What is a fair allocation of the costs of coping with the social consequences of the global warming that will not in fact be avoided? (3) What background allocation of wealth would allow inter-national bargaining (about issues such as (1) and (2)) to be a fair process? (4) What is a fair allocation of emissions of greenhouse gases (over the long term and during the transition to the long-term allocation)? Our leaders can confront these four questions explicitly and thoughtfully, and thereby hope to deal with them more wisely, or they can leave them implicit and unexam-ined and simply blunder into positions on them while thinking only about the other economic and political considerations that always come up. What leaders cannot do is evade taking actions that will in fact be just or unjust. The subject of justice will not go away. Issues of justice are inherent in the kinds of choices that must immediately be made. Fortunately, these four issues that are intertwined in practice can be separated for analysis.

Allocating the costs of prevention

Whatever sums are spent in the attempt to prevent additional warming of the climate must somehow be divided up among those who are trying to deal with the problem. The one question of justice that most people readily see is this one: who should pay for whatever is done to keep global warming from becoming any worse than necessary?

One is tempted to say: 'to keep it from becoming any worse than it is already going to be as a result of gases that are already in the air'. Tragically, we will in fact continue to make it worse for some time, no matter how urgently we act. Because of the Industrial Revolution, the earth's atmosphere now contains far more accumulated CO_2 than it was normal for it to contain during previous centuries of human history. This is not speculation: bubbles of air from earlier centuries have been extracted from deep in the polar ice, and the CO_2 in these bubbles has been directly measured. Every day we continue to make large net additions to the total concentration of CO_2.

Several industrial nations have unilaterally committed themselves to reducing their emissions of CO_2 by the year 2000 to the level of their emissions in 1990. This may sound good, and it is obviously better than allowing emissions levels to grow in a totally uncontrolled manner, as the United States and many other industrial nations are doing. The 1990 level of emissions, however, was making a net addition to the total every day, because it was far in excess of the capacity of the planet to recycle CO_2 without raising the surface temperature of the planet. A reduction to the 1990 level of emissions means *reducing the rate* at which we are adding to the atmospheric total to a rate below the current rate of addition, but it also means *continuing to add to the total*.

Stabilizing emissions at a level as high as the 1990 level will not stabilize temperature—it will continue the pressure to drive it up. In order to stabilize temperature, emissions must be reduced to a level at which the accumulated *concentration* of CO_2 in the atmosphere is stabilized. CO_2 must not be added by human processes faster than natural processes can handle it by means that do not raise the surface temperature. Natural processes will, of course, have to 'handle' whatever concentration of CO_2 we choose to produce, one way or another; some of those ways involve adjustments in parameters such as surface temperature that *we* will have a hard time handling. There is, therefore, nothing magic about the 1990 level of emissions. On the contrary, at that historically unprecedented level of emissions, the atmospheric concentration would continue to expand rapidly—it merely would not expand as quickly as it will at present levels or at the higher business-as-usual future levels now to be expected.

Emissions must be stabilized at a much lower level than the 1990 level, which means that emissions must be sharply reduced. The most authoritative scientific consensus (Houghton, Jenkins, and Ephraums, 1990: xviii, Table 2) said that in order to stabilize the atmospheric concentration of CO_2, emissions would have to be reduced below 1990 levels by more than 60 per cent! Even if this international scientific consensus somehow were a wild exaggeration and the reduction needed to be, say, a reduction of only 20 per cent from 1990 levels, we would still face a major challenge. Every day that we continue to add to the growing concentration, we increase the size of the reduction from current emissions necessary to stabilize the concentration at an acceptable total.

The need to reduce emissions, not merely to stabilize them at an already historically high level, is only part of the bad news for the industrial countries. The other part is that the CO_2 emissions of most countries that contain large percentages of the human population will be rising for some time. I believe that the emissions from these poor, economically less developed countries also ought to rise insofar as this rise is necessary to provide a minimally decent standard of living for their now impoverished people. This is, of course, already a (very weak) judgement about what is fair: namely, that those living in desperate poverty ought not to be required to restrain their emissions, thereby remaining in poverty, in order that those living in luxury should not have to restrain their emissions. Anyone who cannot see that it would be unfair to require sacrifices by the desperately poor in order to help the affluent avoid sacrifices will not find anything else said in this chapter convincing, because I rely throughout on a common sense of elementary fairness. Any strategy of maintaining affluence for some people by keeping other people at or below subsistence is, I take it, patently unfair because so extraordinarily unequal—intolerably unequal.

Be the fairness as it may, the poor countries of the globe are in fact not voluntarily going to refrain from taking the measures necessary to create a decent standard of living for themselves in order that the affluent can avoid discomfort. For instance, the Chinese government, presiding over more than 22 per cent of humanity, is not about to adopt an economic policy of no growth for the convenience of Europeans and North Americans already living much better than the vast majority of Chinese, whatever others think about fairness. Economic growth means growth in energy consumption, because economic activity uses energy. And growth in energy consumption, in the foreseeable future, means growth in CO_2 emissions.

In theory, economic growth could be fuelled entirely by forms of energy that produce no greenhouse gases (solar, wind, geothermal, nuclear—fission or fusion—and hydroelectric). In practice, these forms of energy are not now economically viable (which is not to say that none of them would be if public subsidies, including government-funded research and development, were restructured). China specifically has vast domestic coal reserves, *the* dirtiest fuel of all in CO_2 emissions, and no economically viable way in the short run of switching to completely clean technologies or importing the cleaner fossil fuels, such as natural gas, or even the cleaner technologies for burning its own coal, which do exist in wealthier countries. In May 1992 Chen Wang-xiang, general secretary of China's Electricity Council, said that coal-fired plants would account for 71 to 74.5 per cent of the 240,000 megawatts of generating capacity planned for China by the year 2000 (BNA, 1992). So, until other arrangements are made and financed, China will most likely be burning vast and rapidly increasing quantities of coal with, for the most part, neither the best available coal-burning technology nor the best energy technology overall. The only alternative China actually has with its current resources is to choose to restrain its economic growth, which it will surely not do, rightly or wrongly. (I think rightly.)

Fundamentally, then, the challenge of preventing additional avoidable global warming takes this shape: how does one reduce emissions for the world as a whole while accommodating increased emissions by some parts of the world? The only possible answer is: by *reducing* the emissions by one part of the world by an amount *greater than* the increase by the other parts that are increasing their emissions.

The battle to reduce total emissions should be fought on two fronts. First, the increase in emissions by the poor nations should be held to the minimum necessary for the economic development that they are entitled to. From the point of view of the rich nations, this would serve to minimize the increase that their own reductions must exceed. Nevertheless, the rich nations must, second, also reduce their own emissions somewhat, however small the increase in emissions by the poor, if the global total of emissions is to come down while the contribution of the poor nations to that total is rising. The smaller the increase in emissions necessary for the poor nations to rise out of poverty, the smaller the reduction in emissions necessary for the rich nations—environmentally sound development by the poor is in the interest of all.

Consequently, two complementary challenges must be met—and paid for—which is where the less obvious issues of justice come in.[1] First, the economic development of the poor nations must be as 'clean' as possible—maximally efficient in the specific sense of creating no unnecessary CO_2 emissions. Second, the CO_2 emissions of the wealthy nations must be reduced by more than the amount by which the emissions of the poor nations increase. The bills for both must be paid: someone must pay to make the economic development of the poor as clean as possible, and someone must pay to reduce the emissions of the wealthy. These are the two components of the first issue of justice: allocating the costs of prevention.

Allocating the costs of coping

No matter what we do for the sake of prevention from this moment forward, it is highly unlikely that all global warming can be prevented, for two reasons. First, what the atmospheric scientists call a 'commitment to warming' is already in place simply because of all the additional greenhouse gases that have been thrust into the atmosphere by human activities since around 1860. Today is already the morning after. We have done whatever we have done, and now its consequences, both those we understand and those we do not understand, will play themselves out, if not this month, some later month. Temperature at the surface level—at our level—may or may not already have begun to rise. But the best theoretical understanding of what would make it rise tells us that it will sooner or later rise because of what we have already done—and are unavoidably going to continue doing in the short and medium terms. Unless the theory is terribly wrong, the rise will begin sooner rather than later.

In the century and a quarter between the beginnings of the Industrial Revolution and 1993, and especially in the half century since World War II, the industrializing nations have pumped CO_2 into the atmosphere with galloping vigour. As of today the concentration has already ballooned.

Second, even if starting tomorrow morning everyone in the world made every exertion she could possibly be expected to make to avoid as much addition as possible to today's concentration, we would continue to add CO_2 much faster than it can be recycled without a rise in temperature for an indeterminate number of years to come. A sudden huge decline in the rate of addition to the total is not physically, not to mention economically, feasible. Further, needless to say, not everyone in the world is prepared to make every reasonable exertion. The 'leadership' of the United States, the world's largest injector of CO_2 into the planet's atmosphere, will not even commit itself to cap CO_2 emissions by 2000 at 1990 levels, as easy as that would be (and as little good as it by itself would do). Consequently, even a good-faith transition to sustainable levels of CO_2 emissions would make the problem of warming worse for quite a few years before it could begin to allow it to become better. Years of fiddling while our commitment to the warming of future generations expands will make their problem considerably worse still than it already has to be.

The second issue of justice, then, is: how should the costs of coping with the unprevented human consequences of global warming be allocated? The two thoughts that immediately spring to mind are, I believe, profoundly misguided; they are, crudely put, 'to each his own' and 'wait and see'. The first thought is: let each nation that suffers negative consequences deal with 'its own' problems, since this is how the world generally operates. The second is: since we cannot be sure what negative consequences will occur, it is only reasonable to wait and see what they are before becoming embroiled in arguments about who should pay for dealing with which of them. However sensible these two strategic suggestions may seem, I believe that they are quite wrong and that this issue of paying for coping is both far more immediate and much more complex than it seems. This brief overview is not the place to pursue the arguments in any depth, but I would like to telegraph why I think these two obvious-seeming solutions need at the very least to be argued for.

To each his own

Instantly adopting this solution depends upon assuming without question a highly debatable description of the nature of the problem, namely, as it was put just above, 'let each nation that suffers negative consequences deal with "its own" problems'. The fateful and contentious assumption here is that whatever problems arise within one's nation's territory are *its own*, in some sense that entails that it can and ought to deal with them on its own, with (only) its own resources. This assumption depends in turn upon both of two implicit and dubious premises.

First, it is taken for granted that every nation now has all its own resources under its control. Stating the same point negatively, one can say that it is assumed that no significant proportion of any nation's own resources are physically, legally, or in any other way outside its own control. This assumes, in effect, that the international distribution of wealth is perfectly just, requiring no adjustments whatsoever across national boundaries! To put it mildly, that the world is perfectly just as it is, is not entirely clear without further discussion. Major portions of the natural resources of many of the poorer nations are under the control of multinational firms operated from elsewhere. Many Third World states are crippled by burdens of international debt contracted for them, and then wasted, by illegitimate authoritarian governments. Thus, the assumption that the international distribution of wealth is entirely as it should be is hard to swallow.

Second is an entirely independent question that is also too quickly assumed to be closed: it is taken for granted that no responsibility for problems resulting within one nation's territory could fall upon another nation or upon other actors or institutions outside the territory. Tackling this question seriously means attempting to wrestle with slippery issues about the causation of global warming and about the connection, if any, between causal responsibility and moral responsibility, issues to be discussed more fully later. Once the issues are raised, however, it is certainly not a foregone conclusion, for instance, that coastal flooding in Bangladesh (or the total submersion of, for example, the Maldives and Vanuatu) would be entirely the responsibility of, in effect, its victims and not at least partly the responsibility of those who produced, or profited from, the greenhouse gases that led to the warming that made the ocean water expand and advance inland. On quite a few readings of the widely accepted principle of 'the polluter pays', those who caused the change in natural processes that resulted in the human harm would be expected to bear the costs of making the victims whole. Once again, I am not trying to settle the question here, but merely to establish that it is indeed open until the various arguments are heard and considered.

Wait and see

The other tactic that is supposed to be readily apparent and eminently sensible is: stay out of messy arguments about the allocation of responsibility for potential problems until we see which problems actually arise—we can then restrict our arguments to real problems and avoid imagined ones. Unfortunately, this too is less commonsensical than it may sound. To see why, one must step back and look at the whole picture.

The potential costs of any initiative to deal comprehensively with global warming can be divided into two separate accounts, corresponding to two possible components of the initiative. The first component, introduced in the previous section of this chapter, is the attempted prevention of as much warming as possible, the costs of which can be thought of as falling into the prevention account. The second component, briefly sketched in this section, is the attempted correction of, or adjustment to—what I have generally called 'coping with'—the damage done by the warming that for whatever reasons goes unprevented.

It may seem that if costs can be separated into prevention costs and coping costs, the two kinds of costs could then be allocated separately, and perhaps even according to unrelated principles. Indeed, the advice to wait and see about any coping problems assumes that precisely such independent handling is acceptable. It assumes in effect that prevention costs can be allocated—or that the principles according to which they will be allocated, once they are known, can be agreed upon—and prevention efforts put in motion, before the possibly unrelated principles for allocating coping costs need to be agreed upon. What is wrong with this picture of two basically independent operations is that what is either a reasonable or a fair allocation of the one set of costs may—I will argue, does—depend upon how the other set of costs is allocated. The respective principles for the two allocations must not merely not be unrelated but be complementary.

In particular, the allocation of the costs of prevention will directly affect the ability to cope later of those who abide by their agreed-upon allocation. To take an extreme case, suppose that what a nation was being asked to do for the sake of prevention could be expected to leave it much less able to cope with 'its own' unprevented problems, on its own, than it would be if it refused to contribute to the prevention efforts—or refused to contribute on the specific terms proposed—and instead invested all or some of whatever it might have contributed to prevention in its own preparations for its own coping. For example, suppose that in the end more of Shanghai could be saved from the actual eventual rise in sea level due to global warming if China simply began work immediately on an elaborate and massive, Dutch-style system of sea walls, dikes, canals, and sophisticated floodgates—a kind of Great Sea Wall of China—rather than spending its severely constrained resources on, say, purification technologies for its new coal-fuelled electricity generating plants and other prevention measures. From a strictly Chinese point of view, the Great Sea Wall might be preferable even if China's refusal to contribute to the prevention efforts resulted in a higher sea level at Shanghai than would result if the Chinese did cooperate with prevention (but then did not have time or resources to build the Sea Wall fast enough or high enough).

This fact that the same resources that might be contributed to a multilateral effort at prevention might alternatively be invested in a unilateral effort at coping raises two different questions, one primarily ethical and one primarily non-ethical (although these two questions are not unrelated either). First, would it be fair to expect cooperation with a multilateral initiative on prevention, given one particular allocation of those costs, if the costs of coping are to be allocated in a specific other way (which may or may not be cooperative)? Second, would it be reasonable for a nation to agree to the one set of terms, given the other set of terms—or, most relevantly, given that the other set of terms remained unspecified? Doing your part under one set now while the other set is up for grabs later leaves you vulnerable to the possibility of the second set's being stacked against you in spite of, or because of, your cooperation with the first set. It is because the fairness and the reasonableness of any way of allocating the costs of prevention depend partly upon the way of allocating the costs of coping that it is both unfair and unreasonable to propose that binding agreement should be reached now concerning prevention, while regarding coping we should wait and see.

The background allocation of resources and fair bargaining

This last point about potential vulnerability in bargaining about the coping terms, for those who have already complied with the prevention terms, is a specific instance of a general problem so fundamental that it lies beneath the surface of the more obvious questions, even though it constitutes a third issue of justice requiring explicit discussion. The outcome of bargaining among two or more parties, such as various nations, can be binding upon those parties that would have preferred a different outcome only if the bargaining situation satisfies minimal standards of fairness. An unfair process does not yield an outcome that anyone ought to feel bound to abide by if she can in fact do better. A process of bargaining about coping in which the positions of some parties were too weak precisely because they had invested so much of their resources in prevention would be unfair in the precise sense that those parties that had already benefited from the invested resources of the consequently weakened parties were exploiting that very weakness for further advantage in the terms on which coping would be handled.

In general, of course, if several parties (individuals, groups, or institutions) are in contact with each other and have conflicting preferences, they obviously would do well to talk with each other and simply work out some mutually acceptable arrangement. They do not need to have and apply a complete theory of justice before they can arrive at a limited plan of action. If parties are more or less equally situated, the method by which they should explore the terms on which different parties could agree upon a division of resources or sacrifices (or a process for allocating the resources or sacrifices) is actual direct bargaining. Other things being equal, it may be best if parties can simply work out among themselves the terms of any dealings they will have with each other.

Even lawyers, however, have the concept of an unconscionable agreement; and ordinary non-lawyers have no difficulty seeing that voluntarily entered agreements can have objectionable terms if some parties were subject, through excessive weakness, to undue influence by other parties. Parties can be unacceptably vulnerable to other parties in more than one way, naturally, but perhaps the clearest case is extreme inequality in initial positions. This means that morally acceptable bargains depend upon initial holdings that are not morally unacceptable—not, for one thing, so outrageously unequal that some parties are at the mercy of others.

Obviously this entails in turn that the recognition of acceptable bargaining presupposes knowledge of standards for fair shares, which are one kind of standard of justice. If we do not know whether the actual shares that parties currently hold are fair, we do not know whether any actual agreement they might reach would be morally unconscionable. The simple fact that they all agreed is never enough. The judgement that an outcome ought to be binding presupposes a judgement that the process that produced it was minimally fair. While this may not mean that they must have 'a complete theory of justice' before they can agree upon practical plans, it does mean that they need to know the relevant criteria for minimally fair shares of holdings before they can be confident that any plan they actually work out should in any way constrain those who might have preferred different plans.

If bargaining among nations about the terms on which they will cooperate to prevent global warming is to yield any outcome that can be morally binding on the nations who do not like it, the 'initial' holdings at the time of the bargaining must be fair. Similarly, the 'initial' holdings at the time of the bargaining about the terms on which they will cooperate to cope with the unprevented damage from global warming depends, once again, upon minimally fair shares at that point. Holdings at the point of bargaining over the arrangements for coping will have been influenced by the terms of the cooperation on prevention. Consequently, one requirement upon the terms for prevention is that they should not result in shares that would be unfair at the time that the terms of coping are to be negotiated. The best way to prevent unfair terms of coping would appear to be to negotiate both sets of terms at the same time and to design them to be complementary and fair taken together. This would deal with all the first three issues of justice at once. First, however, one needs to know the standard of fairness by which to judge. This is the third issue of justice.

Allocating emissions: transition and goal

The third kind of standard of justice is general but minimal: general in that it concerns all the resources and wealth that contribute to the distribution of bargaining strength and weakness, and minimal in that it specifies not thoroughly fair distributions but distributions not so unfair as to undermine the bargaining process. The fourth kind of standard is neither so general nor so minimal. It is far less general because its subject is not the international distribution of all wealth and resources, but the international distribution only of greenhouse gas emissions in particular. And rather than identifying a minimal standard, it identifies an ultimate goal: what distribution of emissions should we be trying to end up with? How should shares of the limited global total of emissions of a greenhouse gas such as CO_2 be allocated among nations and among individual humans? Once the efforts at prevention of avoidable warming are complete, and once the tasks of coping with unprevented harms are dealt with, how should the scarce capacity of the globe to recycle the net emissions be divided?

So far, of course, nations and firms have behaved as if each of them had an unlimited and unshakable entitlement to discharge any amount of greenhouse gases that it was convenient to release. Everyone has simply thrust greenhouse gases into the atmosphere at will. The danger of global warming requires that a ceiling—probably a progressively declining ceiling—be placed upon total net emissions. This total must somehow be shared among the nations and individuals of the world. By what process and according to what standards should the allocation be done?

I noted above the contrast between the minimal and general third kind of standard and this fourth challenge of specifying a particular (to greenhouse emissions) final goal. I should also indicate a contrast between this fourth issue and the first two. Both of the first two issues are about the allocation of costs: who pays for various undertakings (preventing warming and coping with unprevented warming)? The fourth issue is about the allocation of the emissions themselves: of the total emissions of CO_2 compatible with preventing global warming, what percentage may, say, China and India use—and, more fundamentally, by what standard do we decide? Crudely put, issues one and two are about money, and issue four is about CO_2. We need separate answers to Who pays? and to Who emits? because of the distinct possibility that one nation should, for any of a number of reasons, pay so that another nation can emit more. The right answer about emissions will not simply fall out of the right answer about costs, or vice versa.[2]

We will be trying to delineate a goal: a just pattern of allocation of something scarce and valuable, namely greenhouse gas emissions capacities. However, a transition period during which the pattern of allocation does not satisfy the ultimate standard may well be necessary because of political or economic obstacles to an immediate switch away from the status quo. For instance, current emissions of CO_2 are very nearly as unequal as they could possibly be: a few rich countries with small populations are generating the vast bulk of the emissions, while the majority of humanity, living in poor countries with large populations, produces less altogether than the rich minority. It seems reasonable to assume that, whatever exactly will be the content of the standard of justice for allocating emissions, the emissions should be divided somewhat more equally than they currently are. Especially if the total cannot be allowed to keep rising, or must even be reduced, the per capita emissions of the rich few will have to decline so that the per capita emissions of the poor majority can rise.

Nevertheless, members of the rich minority who do not care about justice will almost certainly veto any change they consider too great an infringement upon their comfort and convenience, and they may well have the power and wealth to enforce their veto. The choice at that point for people who are committed to justice might be between vainly trying to resist an almost certainly irresistible veto and temporarily acquiescing in a far-from-ideal but significant improvement over the status quo. In short, the question would be: which compromises, if any, are ethically tolerable? To answer this question responsibly, one needs guidelines for transitions as well as ultimate goals: not, however, guidelines for transitions instead of ultimate goals, but guidelines for transitions *in addition to* ultimate goals. For one central consideration in judging what is presented as a transitional move in the direction of a certain goal is the distance travelled toward the goal. The goal must have been specified in order for this assessment to be made.[3]

Two more kinds of questions

A principle of justice may specify to whom an allocation should go, from whom the allocation should come, or, most usefully, both. The distinction between the questions, from whom and to whom, would seem too obvious to be worth comment except that 'theories' of justice actually tend in this regard to be only half-theories. They tend, that is, to devote almost all their attention to the question 'to whom?' and to fail to tackle the challenges to the firm specification of the sources for the recommended transfers. This is one legitimate complaint practical people tend to have against such 'theories': 'You have shown me it would be nice if so-and-so received more, but you have not told me who is to keep less for that purpose—I cannot assess your proposal until I have heard the other half.'

Unfortunately, the answer to 'from whom?' does not flow automatically from all answers to 'to whom?'. Often a given specification of the recipients of transfers leaves open a wide variety of possible allocations of the responsibility for making the transfers. For instance, if the principle governing the allocation of certain transfers were, 'to those who had been severely injured by the pollution from the process', the potential sources of the transfers would include: those who were operating the process, the owners of the firm that authorized the process, the insurance company for the firm, the agency that was supposed to be regulating the process, society in general, only the direct beneficiaries of the process and no one else, and so on. Quite often proposals about justice are not so much wrong as too incomplete to be judged either right or wrong.

I have phrased the first four kinds of issues about justice, which arise from different aspects of the challenge of global warming, as, in effect, from-whom questions, precisely because this is the neglected side of the discussion of justice. What we are now noticing is simply that there is, in addition, always the question 'to whom?'. It is more likely that 'to whom?' will have an obvious answer than it is that 'from whom?' will, but it is always necessary to check. If we are discussing the costs of coping, for example, it might seem obvious that from whomsoever the transfers should come, they should go to those having the most difficulty coping. However, if the specification of the sources of the transfers is, 'those who caused the problem being coped with', then country A, which did in fact cause the problem in country X, might be expected to assist country X, and not country Y, even though country Y was having much more difficulty coping (but with problems that were not A's responsibility). Not much, unfortunately, is obvious, although I will try to show that a great deal is actually fairly simple, given common-sense principles of fairness.

One vital point that this abstract example of A, X, and Y illustrates is: answers to 'from whom?' and answers to 'to whom?' are interconnected. Once one has an answer to one question or the other, certain answers to the remaining question are inappropriate and, sometimes, another answer to the remaining question becomes the only one that really makes any sense. Often these logical connections are very helpful.

Two kinds of answers

We saw, in 'Four kinds of questions', that if one thinks hard enough about how the international community should respond to global warming, questions about justice arise unavoidably at four points:

1. Allocating the costs of prevention
2. Allocating the costs of coping
3. The background allocation of resources and fair bargaining
4. Allocating emissions: transition and goal.

And in 'Two more kinds of questions' we have just now observed that besides these more difficult questions about identifying the bearers of responsibility who should be the sources of any necessary transfers, there is always in principle, and often in practice, a further question in each case about the appropriate recipients of any transfers.

Before attempting to sort out specific proposed answers to this array of questions, it is helpful, I think, to notice that individual principles of justice for the assignment of responsibility fall into one or the other of two general kinds, which I will call fault-based principles and no-fault principles. A well-known fault-based principle is: 'the polluter pays'; and a widely accepted no-fault principle is 'payment according to ability to pay'. The principle of payment according to ability to pay is no-fault in the sense that alleged fault, putative guilt, and past misbehaviour in general are all completely irrelevant to the assignment of responsibility to pay. Those with the most should pay at the highest rate, but this is not because they have done wrong in acquiring what they own, even if they have in fact done wrong. The basis for the assignment of progressive rates of contribution, which are the kind of rates that follow from the principle of payment according to ability to pay, is not how wealth was acquired but simply how much is held.

In contrast, 'the polluter pays' principle is based precisely upon fault or causal responsibility. 'Why should I pay for the clean-up?' 'Because you created the problem that has to be cleaned up.' The kind of fault invoked here need not be a moralized kind—the fault need not be construed as moral guilt so much as simply a useful barometer or symptom to be used to assign the burden of payment to the source of the need for the payment. That is, one need not, in order to rely upon this principle, believe that polluters are wicked or even unethical in some milder sense (although one can also believe they are). The rationale for relying upon 'the polluter pays' could, in particular, be an entirely amoral argument about incentives: the polluter should pay because this assignment of clean-up burdens creates the strongest disincentive to pollute. Even so, this would be a fault-based principle in my sense of 'fault-based', which simply means that the inquiry into who should pay depends upon a factual inquiry into the origins of the problem. The moral responsibility for contributing to the solution of the problem is proportional to the causal responsibility for creating the problem. The pursuit of this proportionality can itself in turn have a moral basis (guilty parties deserve to pay) or an amoral basis (the best incentive structure makes polluters pay). The label 'fault-based' has the disadvantage that it may sound as if it must have a moral basis, which it may or may not have, as well as having a moral implication about who ought to pay, which it definitely does have.

An alternative label, which avoids this possible moralistic misunderstanding of 'fault-based', would have been to call this category of principles, not 'fault-based' but 'causal' or 'historical', since such principles make the assignment of responsibility for payment depend upon an accurate understanding of how the problem in question arose. This, however, has the greater disadvantage of suggesting as the natural label for what I call 'no-fault' principles, 'acausal', or 'ahistorical' principles. That would, I think, be more misleading still because it would make the no-fault principles sound much more ethereal and oblivious to the facts than they are. 'Payment according to ability to pay' does not call for an inquiry into the origins of the problem, but neither is it ahistorical or acausal. An historical analysis or a view about the dynamics of political economy might be a part of the rationale for an ability-to-pay principle, so it would be seriously misleading to label this principle 'ahistorical' or 'acausal' just because it does not depend upon a search for the villain in the not-necessarily-moralistic sense in which 'fault-based' principles do depend upon identifying the villain, that is, in the sense of who produced the problem. So I will stick with 'fault-based' for principles according to which the answer to 'from whom?' depends upon an inquiry into the question 'by whom was this problem caused?' and to 'no-fault' for principles according to which 'from whom?' can be answered on grounds other than an analysis of the production of the problem.

Principles for answering the second kind of question noted in 'Two more kinds of questions', 'to whom transfers should be made', also fall into the general categories of fault-based and no-fault. The principle 'make the victims whole' is ultimately fault-based in that the rightful recipients of required transfers are identified as specifically those who suffered from the faulty behaviour on the basis of which it will be decided from whom the transfers should come: on this principle, the transfers should come from those who caused the injury or harm and go to those who suffered the injury or harm. Indeed, one of the great advantages of fault-based principles is precisely that their cause-and-effect structure provides complementary answers to both questions: transfers go to those negatively affected, from those who negatively affected them. This specific principle, 'make the victims whole', embodies a perfectly ordinary view—and an especially clear one, since it also partly answers the third question, how much should be transferred, by indicating that the transfer should be at least enough to restore the victims to their condition prior to the infliction of the harm. The victims (to whom?) are to be 'made whole' (how much?—minimum amount, anyway) by those who left them less than whole (from whom?). This principle does not completely answer the question of 'how much?' because it leaves open the option that the victims are entitled to more than enough merely to restore them to their condition ex ante, that is, it leaves open the possibility of additional compensation.

An ordinary example of a kind of no-fault principle for answering to whom an allocation should go is: 'maintain an adequate minimum'. Naturally, the level of what was claimed to be the minimum would have to be specified and defended for this to be a usably concrete version of this kind of principle. It has the general advantage of all no-fault principles, however, in that no inquiry needs to be conducted into who was in fact injured, who injured them, how much they were injured, and to what extent their problems had other sources, and so forth. Transfers go to those below the minimum until they reach the minimum; then something else happens (for example, they are retrained for available jobs). Quite a bit of information is still needed to use such a no-fault principle, both to justify the original specification of the minimum level and to select those who are in fact below it. Yet this information is of different types from the information needed to apply a fault-based principle: one does not need an understanding of possibly highly complex systems of causal interactions and positive and negative feedbacks and/or lengthy chains of historical connections among potentially vast numbers of agents and multiple levels of analysis. The information needed to apply no-fault principles tends to be contemporaneous information about current functioning, which is often easier to obtain than the convincing analysis of fault needed for the use of a fault-based principle.

The evident disadvantage of a no-fault principle for specifying to whom transfers go is that it lacks the kind of naturally complementary identification of from whom the transfers should come that flows from the cause-and-effect structure of fault-based principles. In particular, it does not imply that the transfers should come from whomever caused those who are below the minimum to be below the minimum; in fact, it does not even assume that there is any clear answer or, for that matter, any meaningful question of the form 'who caused those below the minimum to be there?'. The consequence of the absence of the convenient complementary answers implied by fault-based principles is that with no-fault principles the answers to the 'to whom?' question and the 'from whom?' question must be argued for and established separately, not by a single argument such as arguments about fault. It might be, for example, that if the answer to the question 'to whom?' is 'those below the minimum', the answer to the question 'from whom?' may be 'those with the greatest ability to pay'. The point, however, is that the argument for using ability to pay to answer the one question and the argument for using maintenance of a minimum to answer the other question have to be two separate arguments.

COMPREHENSIVENESS VERSUS JUSTICE

With this framework in mind one can return to the choice between the recommendations of Stewart and Wiener, on the one hand, and Drennen, on the other. Stewart and Wiener make a kind of mistake that is often made by lawyers who take economics too seriously and equity not seriously enough. One of their chief arguments in favour of moving directly to a comprehensive treaty is that under a comprehensive treaty each nation could engage in what I will call homogenizing calculations of cost-effectiveness (Stewart and Wiener, 1992: 93–5). A major advantage of the comprehensive treaty is supposed to be that each nation could look at all uses of all GHGs and select the least-cost options. That is, one could begin the reduction of GHG emissions by eliminating the specific sources of specific gases the elimination of which would produce the smallest subtraction of economic value. The crucial feature of this approach to cost-effectiveness—the feature that leads me to call the approach 'homogenizing'—is that all gas sources (every source of every gas) are thrown into the same pot. Not a single distinction is made among gas sources, not even the distinction between essential and non-essential.

Now it may initially seem strange not to embrace a thoroughgoing least-cost-first approach, but I would like to try to argue not only that hesitation is not unreasonable but, further, that equity demands qualifications on that approach. First, I would like to explain my earlier slur against some economics and expand a bit on the worry about 'homogenization'. For standard economic analysis everything is a preference: the epicure's wish for a little more seasoning and the starving child's wish for a little water, the collector's wish for one more painting and the homeless person's wish for privacy and warmth, all are preferences. Quantitatively, they are different because some are backed up by a greater 'willingness to pay' than others, but qualitatively a preference is a preference. For a few purposes, perhaps, we might choose to treat preferences only quantitatively, in terms of willingness to pay. To choose, however, to discard all the qualitative distinctions built up during the evolution of human history is to deprive ourselves of a rich treasure of sophistication and subtlety. Some so-called preferences are vital, and some are frivolous. Some are needs, and some are mere wants (not needs). The satisfaction of some 'preferences' is essential for survival, or for human decency, and the satisfaction of others is inessential for either survival or decency.

Distinctions such as the one between needs and wants, or the one between the urgent and the trivial, are of course highly contested and messy, which is why we yearn for the simplicity provided by everything being a so-called preference, differing only in strength (willingness to pay). To ignore *these* distinctions, however, is to discard the most fundamental differences in kind that we understand. This is a general complaint against much mainstream economics. My specific complaint against Stewart and Wiener does not depend upon this stronger, more general one—I mention the general thesis because my specific thesis concerns a parallel form of clarity-abandoning homogenization.

To suggest simply that it is a good thing to calculate cost-effectiveness across all sources of all GHGs is to suggest that we ignore the fact that some sources are essential and even urgent for the fulfilment of vital needs and other sources are inessential or even frivolous. What if, as is surely in fact the case, some of the sources that it would cost least to eliminate are essential and reflect needs that are urgent to satisfy, while some of the sources that it would cost most to eliminate are inessential and reflect frivolous whims? What if, to be briefly concrete, the economic costs of abandoning rice paddies are less than the economic costs of reducing miles per gallon in luxury cars? Does it make no difference that some people need those rice paddies in order to feed their children, but no one needs a luxury car?!

It would be nuts not to follow the principle of least cost first as long as one was dealing with matters of comparable significance: to do otherwise would be to choose a more expensive means to an end that could be reached by a less expensive means—that *is* fundamentally irrational. While the elimination of *N* thousand hectares of rice paddy might well cost less in economic terms than the tightening of corporate average fuel efficiency (CAFE) standards enough to produce the same reduction in GHG emissions, however, the human consequences of reducing food production and of reducing inefficient combustion are far from comparable in their effects upon the quality of life— indeed, in the case of the food, upon the very possibility of life. These are not two different means to the very same end. The ends of the two different measures could be the same in the amount of GHG emissions they eliminate, but the ends are otherwise as different as reducing vital supplies of food and making luxury a tad more costly. Consequently, to apply a homogenizing form of cost-effectiveness calculation, as if the two measures differed only in how much they each cost to produce the same reduction in emissions, is seriously to distort reality. This kind of comprehensiveness obscures distinctions that are fundamental, most notably the distinction between necessities and luxuries.

The central point about equity is that it is not equitable to ask some people to surrender necessities so that other people can retain luxuries. It would be unfair to the point of being outrageous to ask that some (poor) people spend more on better feed for their ruminants in order to reduce methane emissions so that other (affluent) people do not have to pay more for steak from less crowded feedlots in order to reduce their methane and nitrous oxide emissions, even if less crowded feedlots for fattening luxury beef for the affluent world would cost considerably more than a better quality of feed grain for maintaining the subsistence herds of the poor.

It is of course a different story if *all* incremental costs for reducing emissions, wherever incurred, are to be allocated according to ability to pay. If the beef eaters will pay for the better feed grain for the subsistence herds of the poor with *additional* funds not already owed for some other purpose such as development assistance, it might, as far as the equity of the arrangements for reducing emissions goes, not be unfair to start with the least-cost measures. The least-cost measure paid for by those *most* able to pay is not at all the same as the least-cost measure paid for by those *least* able to pay. In terms of the framework laid out above, this would combine a no-fault answer to the question 'from whom?' (ability to pay), with a no-fault answer to the question 'to whom?' (maintenance of an adequate minimum).

If these two answers were fully justified—naturally, they require fuller argument—one or the other of two routes ought to be followed.[4] The homogenizing form of calculation of cost-effectiveness could be neutralized if it were accompanied by a firm commitment that costs are to be paid according to ability to pay and the actual establishment of mechanisms for enforcing the necessary transfers. Otherwise the costs ought to be partitioned—perhaps more than once but surely at least once—into costs that impinge upon necessities for the poor and costs that only impinge upon luxuries for the wealthy.

Thomas Drennen has suggested one type of such a partitioning, and he has based it upon two kinds of considerations: centrally upon the consideration of equity, but also upon the difficulties in measuring agricultural and other biological emissions of methane and verifying any mandated reductions. (It is ironic that the USA, which held up the control of nuclear weapons for years with exaggerated worries about verification, now wants to plunge ahead with the much messier matter of methane.) Drennen has arrived at his partitioning between what should be within the scope of the treaty, and what should not, by combining type of gas and type of use. The gases that Drennen would have under protocol control are CO_2 and methane, presumably because they are the largest contributors to global warming that are subject to human control. (Water vapour is larger but not controllable.) The uses that he would like to see reduced are the 'industrial-related' ones, not the agricultural ones. Drennen's strategy is to control non-biological anthropogenic emissions of CO_2 and methane. This is a much more sophisticated formulation than the much-discussed one that deals only with fossil-fuel CO_2. I would nevertheless suggest that a definition of the scope different from Drennen's, based even more directly on considerations of equity, would be preferable; although I also acknowledge that practical considerations may require that Drennen's pair, non-industrial-related CO_2 and non-industrial-related methane, still be used as proxies for, and as the nearest practical approximation of, the specifications that flow directly from attention to equity.

My main doubts—and they are relatively minor—are about the division into industrial-related and non-industrial-related. This division reflects quite well, as I understand them, the difficulties of measurement and verification of methane emissions. It is much easier to calculate (and change!) leaks from natural gas pipelines than to calculate (and change!) emissions from various varieties of rice and various species of ruminants. Precisely this division reflects less well, however, the concern about equity that Drennen and I share. Just as the methane emissions from beef feedlots are in service of the desires of the wealthy, many of the CO_2 emissions in China and India *could be*—I am not assuming that they all in fact are—in service of the needs of the poor. Some agricultural methane emissions are a luxury, and some industrial CO_2 emissions are a necessity. By the standard of equity we do not want to leave all the former uncontrolled, and control all the latter.

Now insofar as we really cannot measure, or even accurately estimate, biological emissions of methane, the lack of perfect fit with equity does not for now matter. Yet I am somewhat impressed by the contention by Stewart and Weiner that we should be able to arrive at accurate enough estimates for our purposes, especially if, as I contend, our purposes ought not to include the control of subsistence emissions. For the sake of scientific understanding we must eventually be able to measure all kinds of emissions, but the measurements of the kinds that we are not going to control can be rough in the beginning.

We should not have a homogenized—undifferentiated—market in emissions allowances in which the wealthy can buy up the allowances of the poor and leave the poor unable to satisfy even their basic needs for lack of emissions allowances. Drennen's partition between industrial-related and non-industrial-related may be the best approximation we can in practice make: most agricultural emissions are probably for subsistence and many industrial emissions are not. Better still, if it is practical, would be a finer partitioning that left the necessary industrial activities of the developing countries uncontrolled (not, of course, unmeasured) and brought the unnecessary agricultural services of the developed world, as well as their superfluous industrial activities, under the system of control.

If there is to be an international market in emissions allowances, the populations of poor regions could be allotted inalienable—unmarketable—allowances for whatever use they themselves consider best. Above the inalienable allowances, the market could work its magic, and the standard of cost-effectiveness could reign supreme. But the market for emissions allowances would not be fully comprehensive, as Stewart and Wiener recommend. The poor in the developing world would be guaranteed a certain quantity of protected emissions, which they could produce as they choose. This would allow them some measure of control over their lives rather than leaving their fates at the mercy of distant strangers.

NOTES

1. Less obvious, that is, than the issue whether the poor should have to sacrifice their own economic development so that the rich can maintain all their accustomed affluence. As already indicated, if someone honestly thought this demand could be fair, we would belong to such different worlds that I do not know what I could appeal to that we might have in common.

2. For imaginative and provocative suggestions about the final allocations of emissions themselves, see Agarwal and Narain (1991).
3. A serious attempt to deal with issues about a compromise transition is in Grubb and Sebenius's chapter, 'Participation, Allocation and Adaptability in International Tradeable Emission Permit Systems for Greenhouse Gas Control' in *Climate Change* (1992).
4. I have provided some relevant arguments in Shue (1992). Justice, or equity, is also discussed in several chapters in the UNCTAD publication, *Combating Global Warming: A Study on a Global System of Tradeable Carbon Emission Entitlements* (1992).

REFERENCES

Agarwal, Anil, and Sunita Narain (1991) *Global Warming in an Unequal World: A Case of Environmental Colonialism*. New Delhi: Centre for Science and Environment.

Bureau of National Affairs (BNA) (1992) 'Top Environmental Official Welcomes Summit Aid Pledges from Developed Nations—China', *International Environment Reporter: Current Reports* 15(July 1): 444.

Drennen, Thomas E. (1993) 'After Rio: Measuring the Effectiveness of the International Response', *Law & Policy* 15: 15–37.

Grubb, Michael, and James K. Sebenius (1992) 'Participation, Allocation and Adaptability in International Tradeable Emission Permit Systems for Greenhouse Gas Control', in *Climate Change: Designing A Tradeable Permit System*, by Organisation for Economic Co-operation and Development (OECD). Paris: OECD.

Houghton, J. T., G. J. Jenkins, and J. J. Ephraums (1990) *Climate Change: The IPCC Scientific Assessment*. Report by Working Group I. New York: Cambridge Univ. Press for the Intergovernmental Panel on Climate Change.

Shue, Henry (1992) 'The Unavoidability of Justice', in *The International Politics of the Environment: Actors, Interests, and Institutions*, edited by Andrew Hurrell and Benedict Kingsbury. Oxford: Clarendon Press.

Stewart, Richard B., and Jonathan B. Wiener (1992) 'The Comprehensive Approach to Global Climate Policy: Issues of Design and Practicality', *Arizona Journal of International and Comparative Law* 9: 83–113.

United Nations Conference on Trade and Development (UNCTAD) (1992) *Combating Global Warming: A Study on a Global System of Tradeable Carbon Emission Entitlements*. UNCTAD/RDP/DFP/1. New York: United Nations.

3

After you: may action by the rich be contingent upon action by the poor?*

Equity, or fairness, is at the heart of debates about possible international action to constrain global warming.[1] As two of the most important commentators on these negotiations, Michael Grubb and James Sebenius, have remarked:

> The allocation problem is unavoidable whatever control system is adopted, whether it be national emission targets, taxes, or other approaches. All systems that will have global impact will impose costs on participants, and must include some resource transfers if poorer countries are to take part... [I]n practice, some degree of equity considerations have appeared as a consistent feature in a large number of negotiated international environmental and natural resource regimes... Pure equity may well have a stronger role to play than in many agreements, in part because of the economic stakes involved combined with the need to gain widespread and very long term adherence to a control regime.[2]

By 'pure equity', Grubb and Sebenius mean equity taken seriously in its own right, equity per se, as distinguished from instrumental uses of equity for rhetorical and political purposes not intrinsically connected with equity.

The issues about equity, or fairness, can be distinguished into four practically related but analytically separable questions:

1. What is a fair allocation of the costs of preventing the global warming that is still avoidable?

---

* Copyright 1994, Henry Shue, as first published in *Indiana Journal of Global Legal Studies*, 1:2, pp. 343–66.

[1] Lawyers talk of 'equity'. Philosophers, and ordinary people, talk of 'fairness'. Belonging to at least one of the latter two categories, I will talk of 'fairness'. On the matters at hand, nothing that I can see turns on this terminology.

[2] Michael Grubb and James K. Sebenius, 'Participation, Allocation and Adaptability in International Tradeable Emission Permit Systems for Greenhouse Gas Control', in *Climate Change: Designing A Tradeable Permit System 193* (OECD Documents 1992). I have explored some implications of the impossibility in this case of skirting around the issues of fairness in Henry Shue, 'The unavoidability of justice', this volume.

2. What is a fair allocation of the costs of coping with the social consequences of the global warming that will not in fact be avoided?

3. What background allocation of wealth would allow international bargaining (about issues such as 1 and 2 to be a fair process)? And:

4. What is a fair allocation of emissions of greenhouse gases (over the long
term and during the transition to the long-term allocation)?[3]

It is a mistake to attempt to answer question 1 about the costs of prevention,
or mitigation, without simultaneously answering question 2 about the costs
of coping, or adaptation, because what is a fair allocation of the costs of
prevention depends in part upon what the allocation of the costs of coping
is going to be.[4]

In this chapter, I will discuss one aspect of the relation between the answer
to question 4 (long-term allocation of emissions) and question 1 (costs of
mitigation). If the current allocation of emissions is grossly inequitable, or
unfair, compared to an allocation that would be acceptable over the long term,
those with the inequitable current advantage ought to begin reducing emissions immediately at their own expense, and thereby contribute to the cost of
global warming prevention, irrespective of whether an international agreement exists regarding any of the four questions. In other words, I suggest that
one can get at least a partial answer to question 4 quite readily with implications about how to answer question 1.

COOPERATIVE STRATEGY BETWEEN
RICH AND POOR

What would be the quid pro quo in a cooperative strategy between rich
peoples and poor peoples to prevent, or at least to combat, global warming?[5]
I assume that the global total of greenhouse gas (GHG) emissions must, at the
very least, be prevented from continuing to rise. The most authoritative study
by leading scientists from several countries holds that the global total must be
sharply reduced.[6] Although I am thoroughly convinced that we face this far

---

[3] Henry Shue, 'Subsistence emissions and luxury emissions', this volume. The heart of this
analysis has also appeared as Henry Shue, 'Four Questions of Justice', in *Agricultural Dimensions
of Global Climate Change* 214 (Harry M. Kaiser and Thomas E. Drennen, eds, 1993).

[4] Shue, *supra* note 2.

[5] I will sometimes use 'peoples', along with 'nations', 'states', and 'countries', in order to signal
that I am not committed to any particular form of organization for large political societies and
certainly not to the contemporary nation state. Cf. John Rawls, 'The Law of Peoples', in *On
Human Rights* 41 (Stephen Shute and Susan Hurley, eds, 1993).

[6] *Climate Change: The IPCC Scientific Assessment* (J.T. Houghton et al., eds, 1990); the brief
'Policymakers' Summary' is also available in a separate and more colourful form as *Scientific*

greater challenge of scaling back the total emissions, any cooperative strategy faces a difficulty of the same shape, although not the same magnitude, even if we must only stop increasing the total. Even so, the task is daunting. While the global total of GHG emissions must stop growing, GHG emissions generated by the impoverished masses of the planet must grow if these individuals are to rise above the poverty in which generations have been trapped. The overall total of GHG emissions must cease growing while there is continued growth in one part of that total, the part generated by the poor. Arithmetically, this can be accomplished only if the only other part, the part generated by the rich, shrinks. In other words, even if the scientific consensus were wrong and the overall total did not need to shrink but needed only to be held constant, the contribution to the total by the rich would have to shrink by at least as much as the amount by which emissions by those rising out of poverty increase.

What is a reasonable quid pro quo? What may the rich and the poor ask of each other?

Contributions by the poor

Two dramatic requests that need to be made of the poor peoples are abundantly clear. First, the rich are clearly asking the poor to settle for levels of emissions per capita far below the current levels of the rich, that is, the rich are asking the poor to choose never to approach the levels that the rich currently produce. The poor must prepare to live with a level of economic activity compatible with per capita GHG emissions well below the present levels of the rich.[7] The planet simply could not tolerate a majority emitting GHGs at the per capita rate reached by today's rich minority, or anywhere near that rate. Even per capita rates well below those of the current rich minority, if produced by a majority of humanity, would send the global total emissions—which must, at the very least, be held constant—skyrocketing.

A unique sacrifice is, then, being asked of the poor majority of humanity: never before in recorded history have people ever *chosen* to live at an economic level both (a) much lower than levels previously attained by other people, and (b) lower than they themselves could sustain, for at least some time, with their own resources. The long-run unsustainability of higher emissions levels would

Assessment of Climate Change: The Policymakers' Summary of the Report of Working Group I to the Intergovernmental Panel on Climate Change (WMO/UNEP 1990). See also *Climate Change 1992: The Supplementary Report to the IPCC Scientific Assessment* (J.T. Houghton et al., eds, 1992).

[7] As I briefly indicate *infra*, emission levels could be kept low without keeping economic activity low provided only that we developed and switched to an energy technology that emits no GHGs, such as solar energy. I do not take this seriously in this discussion only because our politicians do not take it seriously in practice.

eventually reach everyone, including the poor, but they could temporarily enjoy the good times that the rich have been enjoying for decades. Is it humanly possible for whole peoples to choose less rather than more? We are asking people who have never enjoyed a plentiful, or even an economically adequate, life, to accept and help to implement a limit on the hopes they can have for their children's and grandchildren's economic welfare. They are asked to conspire in the imposition of limits on their own children's dreams.

Everything said thus far would be true even if there were now zero growth in human numbers, not the already frequent doubling of the global human population, which is becoming more frequent. The more humans there are, the lower the per capita emissions rate must be in order for the global total of emissions to hold constant. This is arithmetically evident. At some number of humans, the emissions rate per capita would become unbearably low.

The second request to be made of the poor, then, is energetic family planning. The rights of poor women to make their own choices must be respected, but all women and men must be shown reasons and provided with incentives to have small families. I hasten to add what I hope is generally understood by now: population growth is stimulated, not discouraged, by poverty. Starvation is not an effective, even were it a morally tolerable, method of population control over the long term. 'Energetic family planning' is not intended here as code for continuing to tolerate absolute poverty. On the contrary, even if there were no other sufficient reason to promote economic improvements,[8] promoting effective family planning would be reason enough to relieve crushing Third World debt and to take the other measures necessary for economic growth among the poorest.[9]

The second request, like the first, is unprecedented and extreme. Fertility rates among the rich decreased only when the prospects for each successive generation seemed to be better than for the one before. Parents in today's poor countries will, by contrast, be asked to have fewer children in a context in which economic improvement, though perceptible, is limited. At best, this is uncharted psychological terrain. Possibly, the rich will be asking parents in the poorest regions of the world to show a level of concern about the global environment unimaginable among today's rich. There must, we said first, be a limit on their children's dreams. There must also be a limit on the numbers of those very children. What are the rich prepared to offer as their part of the global effort?

---

[8] There are several others, including ethical ones.
[9] Cf. Nathan Keyfitz, 'Population Growth Can Prevent the Development that Would Slow Population Growth', in *Preserving the Global Environment* 39 (Jessica Tuchman Mathews, ed., 1991).

Contributions by the rich

The rich countries must emit less greenhouse gas. The rich countries must reduce their GHG emissions in order for the global total to remain constant, if the poorest billion are to be able to improve their lives. We must consume less and, probably, produce less. This too will obviously be unprecedented. It will take a genuine leader, a kind not now visible, who will, as must be done, promise people less.

There are two respects in which this overstates the challenge. First, there is undoubtedly a huge difference between standard of living, as measured by conventional economics, and quality of life. Much of what sustains a consumerist economy not only clogs our landfills but dulls our senses, clutters our minds, erodes our health, and fritters away our time and natural resources. We could clearly live much richer lives if we could be free from many of the gadgets, widgets, and other expensive junk that we sell to each other and then quickly discard. All that can be said for much of the stuff that expands the gross national product (GNP) is that making it, advertising it, distributing it, and discarding it all create jobs. Jobs are vital, so the rich need to create different kinds of jobs that, besides providing people with income, add to the quality of life.

A less wasteful way of life would surely reduce some aspects of 'standard of living' as it is currently (mis-)defined: there would certainly be fewer goods produced and fewer resources, probably including less energy, consumed. Insofar as less carbon-based energy was consumed, we would thereby reduce GHG emissions. A gallon of gasoline not purchased and burned means a smaller GNP (which has been considered bad) and fewer GHG emissions (which is, in fact, good). We could surely reduce GHG emissions while uncluttering and improving the quality of our lives. This would be doubly efficient: using less (quantity) to produce more (quality). Obviously, quality is complex. But many people would grant at least the negative thesis that quality of life would increase if the goods and services of our bizarre consumerist society, driven by artificial wants created out of whole cloth by television advertising, were reduced.

Second, more advanced energy technology would allow maintenance of current levels of economic activity with great reductions in GHG emissions. We could, if we chose, reduce the destructive emissions while maintaining, or even increasing, the economic activity, provided the energy sources were not carbon-based. All fossil fuel (coal, oil, and gas) is carbon-based. Burning it for energy releases carbon dioxide (CO_2), the single worst man-made GHG. As long as energy comes from burning fossil fuel and releasing CO_2, we face a wrenching choice between expanding the economy and reducing GHG emissions. Yet that choice is forced upon us only by our failure to invest enough in researching and developing alternative sources of energy and requiring internalization of the astronomical military, social, environmental, health, and other costs of reliance on a system of oil-moving tankers whose spills wreck the habitat and gas-burning automobiles whose fumes wreck the cities.[10]

If we switch to non-carbon-based energy, we may safely, as far as the danger of global warming is concerned, continue to increase levels of economic activity. We could, then, continue, in particular, to stimulate artificial wants and to satisfy them with expensive, expendable, energy-eating but employment-producing gadgets, if that was how we wanted to keep each other in jobs. In sum, if we would dethrone fossil fuels, we would not need to reduce economic activity. Our consumerist economy is sterile, without soul, and not worth saving if we are not capable of creating different kinds of jobs that add to the quality of life instead of detract from it. But that is no argument against developing energy technologies which would give us the choice. More importantly, the world's impoverished need higher levels of economic activity and millions more jobs, and it would be wonderful if this activity were created and sustained without GHG emissions, as it certainly could largely be.

Nevertheless, this argument is based on the assumption that current political leaders are incapable of breaking the fossil fuel dependency. Without effective leadership, GHG emissions can be reduced only by reducing energy usage. This would require reducing economic activity and GNP, once maximum efficiency has been attained through the elimination of sheer waste. This assumption is born out of political pessimism, for no leader in an industrialized country yet shows vision and courage to lead a revolution from entrenched energy habits, especially now that the fossil fuels of the former Soviet Union lie open to outside exploitation. If this assumption is, in fact, without merit, I will be very happy, and the problems discussed here will be considerably easier to handle.

FAIR TRANSITION

Issues concerning the fairness of alternative plans for transition to a world with a constant and sustainable total of GHG emissions and only moderate inequalities in global levels of emissions fall mostly on the side of the rich peoples. The poor have little flexibility and few choices.

---

[10] Solar energy is one of several sources with absolutely no GHG emissions; obviously this is not the place to consider which is precisely the best alternative, or package of alternatives, to fossil fuel, so I will simply refer negatively to non-carbon-based energy.

'Aggressive population control'

If it were possible to stabilize population size through an aggressive pro-gramme of population control, one might have to consider the moral merits. The quality of life of future generations might depend upon the population size at which human population is stabilized. In short, the stakes are high enough that one might at least consider even the violation of rights in order to improve the outcome for numberless succeeding generations. It may be worth mentioning very briefly why this melodramatic option is not open in the poor countries. For the more people among whom the global total of emissions must be divided, the smaller the per capita share. This makes the stakes in bringing the explosion of population under control very high. It might be thought that the urgency of the crisis dictates some kind of aggressive campaign of population control.

Recent experiences of the world's two most populous countries, China and India, are widely seen as testimony to the fact that 'aggressive population control' is an intellectual's fantasy and the kind of quick fix that virtually never works in practice. Men and women care profoundly about how many children they have, and the worse their economic condition, the more children they tend to want. Attempts by the state to force them to have fewer than they want meet either open rebellion, as in the case of Indira Gandhi's India, or evasion and subversion, as in the case of Deng's China. The aggressive Indian policy was openly abandoned; the Chinese one-child policy is quietly unravelling as the central government loses its grip on the levers that it used to employ to coerce or pressure would-be parents. Without thorough totalitarianism, parents must themselves want fewer children if they are to have fewer children. The desire for fewer children tends to come only with improving economic condi-tions (which, interestingly, are arising in China for some of the same reasons the state is losing its ability to enforce the one-child policy). Accordingly, while the demographics are far more complex than indicated, draconian measures, while perhaps fascinating for intellectuals to debate, are of no practical interest. Basically, birth rates will decline when quality of life improves, and not before.

For the rich, the fundamental question is: how rapidly, and at what price in economic dislocation for ourselves, should we reduce GHG emissions? How much time may we take in reaching the goal of a fair share of a sustainable total? What are the factors, ethical and non-ethical, that determine the an-swer? Within the ethical considerations, exactly what kind of imperative do we face? Here, we face real choices.

The no-regrets budget

The reduction in emissions has at least two stages. First is what might be called the no-regrets stage, the stage in which sheer waste of energy, the production of which emits GHGs, is reduced. Continued use of incandescent light bulbs, for example, simply wastes electricity, the production of which releases annually as much GHG as the burning of thousands of acres of tropical forests. New generation short fluorescent bulbs provide the same light for so much less electricity that the savings in electricity bills quickly pay for the new bulbs, even though they cost several times the price of incandescents.[11] Only ignorance, indifference, and vested interests in obsolete technology stand in the way of a great deal of no-cost—indeed, profitable—reduction in emissions.[12]

In Eastern Europe and the former Soviet Union the waste is still more colossal. Not all this environmentally destructive waste of energy could be eliminated at a net profit over a short time. Many instances require the wholesale replacement of large facilities by expensive new technology. Naturally, disputes are under way about the full extent of profitable reductions of the pure waste of energy. Whatever the precise extent of the no-regrets changes, there is a broad consensus that the opportunities for them are extensive and economically significant. Meanwhile, from an ethical point of view any failure to act in these instances is analogous to a refusal to rescue a drowning person by someone who would benefit from the exercise involved in the swim: it is fecklessness approaching perversity.

Thus, it is only after the initial stage of no-regrets reductions in emissions—primarily through reduction in waste of energy—that any net costs in conventional economic terms would be incurred. We must budget for the necessary changes, for example purchasing fluorescent bulbs. Yet this budget can be replenished from the savings its investment will generate in the short run. Call this the 'no-regrets' budget.

The mitigation budget

Only at a second stage come the true costs of mitigation. These are the costs of doing what we would have no other reason to do if we were not bound to resist global warming. Here is where the ethical considerations come to bear in major and non-obvious ways. Here is where the ethical issues about how costs are conceived make all the difference, provided one assumes that it is possible to cut emissions faster by spending more efficiently. How large any mitigation budget ought to be (another way of asking how rapid the transition ought to be from where we are now to where we ought to end up) depends on what kind of act, from an ethical point of view, a people would perform in resisting global

---

[11] See Thomas E. Drennen and Duane Chapman, 'Negotiating a Response to Climate Change: The Role of Biological Emissions', 10(3) *Contemporary Policy Issues* 49, 55–6 (July 1992).

[12] Alliance to Save Energy et al., *America's Energy Choices: Investing in a Strong Economy and a Clean Environment* (1991).

warming to any specific extent. Specifically, whether their action and its attendant cost would constitute their volunteering discretionary help or their ceasing wrongful harm. Most people seem to think that it is much worse to keep wrongfully harming others than it is not to start helping them.[13] Thus an obligation to eliminate certain emissions would be much stronger if the emissions were wrongfully harming others than it would be if the elimination of those emissions would merely be helpful to others.

A person can be wronged without being harmed (e.g. denied seating through racial discrimination on a plane which then crashes, leaving the wronged party much better off for having been wronged), but one cannot be harmed without being wronged. Simply being made worse off (e.g. driven out of business by a legal and fair competitor) does not constitute suffering harm. To be harmed means to be made worse off in some wrongful way. One clear example of being harmed is being made worse off through the violation of one of one's rights (e.g. denied seating through racial prejudice on a plane that would have delivered one to one's destination safely and on time). One can also be harmed by being made worse off through being deprived of one's fair share. Harms can be inflicted directly or indirectly, although, all other things being equal, the more indirect the infliction of the harm the less moral responsibility the source of the harm bears for it.

The next question is: does the rich's continual engagement in business as usual (emitting large quantities of GHGs) constitute an infliction of harm on anyone? Or does it constitute instead merely declining to pitch in to help with the problem of global warming? This query relates to the issue of the speed at which the rich should make the transition away from current practices; when one is actually engaged in harming others one ought to stop.

Obviously, one could consider whether past emissions of greenhouse gases were wrong. Yet I want to concentrate here on future emissions for two main reasons. First, if no good case can be made for treating future emissions as wrong, it will be all the clearer that past emissions cannot reasonably be treated as wrong. Past emissions involve additional issues. One example is the foreseeability of the seriousness of the risk of global warming.

Second, it is urgent that we know how to think about future emissions for the simple reason that we still control them and can still decide how much effort to make to reduce them. If it is wrong to release future emissions, it is much more imperative to stop them than if their release were merely costly but not genuinely wrong. Although it is important to settle whether our forebears did wrong to other people in releasing past emissions, the emissions themselves have either done their damage or are doing it in ways that we

---

[13] I am inclined to slice the conceptual pie at a different point from the line between not helping and harming, but since I am in a minority, it is more important to see what follows from the standard view held by most people.

cannot any longer affect. What would make future emissions wrong? They are wrong if they consume a grossly unfair proportion of a limited total. They are wrong if they make others worse off in a manner that is unfair.

NATURALLY LIMITED SUPPLIES

Supplies of things come in at least four rough categories. First, there are supplies that are entirely within human discretion (e.g. the money supply). If the relevant bureaucrats think more money is a good idea, then more money is, for better or worse, printed. The increase in the supply of money is by no means without costs, such as inflation, but if those with power decide there should be more, there is more.

Second, there are supplies that can be increased, not by bureaucratic fiat, but only through human investment of some combination of money, effort, and ingenuity, plus the 'cooperation' of the relevant natural or social forces. For example, the supply of paper at some future date could be increased by planting more trees now, or the supply of jobs at a future time could be increased by intelligent selection of economic policies now, provided a number of factors outside human control go well. These two kinds of supplies are more or less under human control, although the latter depends on contributing factors that are not.

Third, there are supplies that are, for all practical purposes, natural and beyond human control but unlimited (e.g. sunlight). Relative to imaginable human demand, the total supply is unlimited; the overall supply vastly exceeds any conceivable demand. Notoriously, it is turning out that fewer and fewer things actually belong in this category of unlimited natural supplies. Yet, while sunlight is limited in particular places at particular times, both naturally and as a result of smog, the aggregate amount available to the planet as a whole exceeds any imaginable need (which is one of the many appeals of solar energy).

Water, on the other hand, which used to seem abundant overall, is turning out to be chronically and severely limited. Even in the areas with relatively abundant and reliable rainfall, copious irrigation by massive agribusinesses, combined with profligate usage in burgeoning suburbs where brown lawns are viewed as evidence of criminal negligence, pumps down the groundwater levels much faster than normal rainfall can replenish them. Thus we confront an instance of the fourth kind of supply: supplies that are natural and limited and, for all practical purposes, cannot be increased by humans—by money, effort, and ingenuity, or by any other currently feasible means.

All this about supplies, man-made and natural, limited and unlimited, is perfectly obvious, if fairly crude. The only point for which I want to employ these categories is: dramatic implications result from the discovery that something that had generally been believed to be naturally unlimited is in fact naturally limited, that is, something that had been placed in the third category (correctly or incorrectly) belongs now, if not also then, in the fourth. A huge store of ethical considerations that are irrelevant to unlimited supplies 'lock in' when there turns out to be scarcity. Where there is scarcity, important ethical issues arise that do not arise amidst plenty. The fundamental fact is that it does not matter who gets what portion of an unlimited supply—it is not evident that 'portion' really makes any sense in that instance, any more than one can calculate a percentage of an infinite number. When, by contrast, there is scarcity, everything changes and who gets which portion becomes over-ridingly important. When the total supply is fixed and beyond human control, human shares become 'zero-sum': each bit that I get is a bit that you do not get, and vice versa. This means that I can only become better off by making someone else worse off. In economists' jargon, no Pareto improvements are possible.[14] As I emphasized above, making someone worse off need not entail harming her; one person has harmed another only if he has made her worse off in a manner that is wrong.

An earlier belief that the supply of something was natural and unlimited also need not have been false, at least as a belief relative to human demand at that earlier time. Water is a leading case. When human population and agricultural activity were sufficiently small, there was, for all practical pur-poses, an unlimited supply of water. Events may have overtaken him to a greater extent than he realized by the time that he said it, but Locke was certainly correct, at least about earlier periods, when he wrote about drinking from the stream while leaving 'enough and as good' behind for others.[15] Even after it came to be necessary to move large quantities of water for the satisfaction of human demand, via dams and pipelines, plenty was available if human ingenuity and investment could deliver it to the right place. But no more: some of the most vicious political battles in the USA today are over water in the Southwest and California.[16] The same is true around the world from the Narmada to the Yangtze. Relative to human demands for water, the supply of water is limited; the total natural supply cannot be changed by human action.

Odd as it may sound, an intellectual revolution is occurring on the subject of man-made wastes, although the word is spreading much too slowly. I am old enough to remember having been unashamed as a youth to have the job of

---

[14] A Pareto improvement is a change that makes at least one party better off without making any party worse off.

[15] *John Locke, Two Treatises of Government*, ch. V, paras 27 and 33 (Peter Laslett ed., 1960).

[16] See Peter H. Gleick, *Water in Crisis* (1993); Richard Conniff, 'California: Desert in Disguise', *National Geographic*, November 1993, Special Edition, at 38–53.

gathering up the household box of tin cans and carrying it back into the woods where I dumped the cans into what we called 'the sinkhole', where the ground had conveniently opened up a depression. At the time it seemed as if everyone who had a woods must have had a sinkhole for their tin cans. I did not realize that something later to be named 'solid waste' would become a national problem, nor that 'trash collection' would become 'waste management'. By now, most well-informed people understand the basic point about solid waste, although the international traffic in toxic wastes, which are dumped in the poor countries by rich countries with about as much concern or care as I discarded my tin cans in the woods, is being kept as quiet as possible by the rich.[17]

The invisible non-solid wastes that seem to be drifting off harmlessly into the sky and on into the endless universe are another story. Where something visible to the human eye, such as bad urban smog, results, people notice, with or without taking action. Yet even reasonably well-informed ordinary people have not internalized sufficiently what they may know in the abstract about emissions of greenhouse gases into that great sewer, the sky: there are natural limits, too, on what the sky can swallow up (without effects, such as the surface temperature rising). As Garrett Hardin, with whom I profoundly disagree on the policy implications, is correct to keep insisting: 'One can never do merely one thing.'[18] The sky is not a great open pit with the planet Pluto at the bottom; our own planet's atmosphere is a net that catches many things, and greenhouse gases are among those things. Without the greenhouse effect, the earth would of course be uninhabitably cold.

There are two different points here we still fail to comprehend. The underlying one is that, however imperceptible gases such as CO_2 may be to the naked human eye, these gases are not nothing. There are limits to the net quantities of these gases that the atmosphere can handle without making adjustments, such as surface temperature increases, that are quite uncongenial

[17] The Clinton administration is opposing the movement to create a total ban on the shipment of toxic waste produced in rich countries to poor countries; the USA wants to continue shipping its toxics to poor countries whose governments provide 'assurances' that they are handling them properly. The ban is proposed under the now very weak Basel Convention on the Transboundary Movements of Hazardous Wastes and Their Disposal. See 'Basel Convention Working Group Says Progress Made on Waste Disposal Guidelines', 16(12) *International Environment Reporter: Current Reports* 431–2 (16 June 1993). The Clinton administration's position appears to be flatly in contradiction of Principle 14, *Rio Declaration on Environment and Development*: 'States should effectively cooperate to discourage or prevent the relocation and transfer to other States of any activities and substances that cause severe environmental degrad-ation or are found to be harmful to human health.' I have discussed some of the general issues underlying such practices in Henry Shue, 'Exporting Hazards', 91 *Ethics* 579, 579–606 (1981) reprinted with revisions in *Boundaries: National Autonomy and Its Limits*, 107–45 (Peter G. Brown and Henry Shue, eds, 1981).

[18] Garrett Hardin, *Living Within Limits*, 199 (1993). Cf. David Feeny et al., 'The Tragedy of the Commons: Twenty-Two Years Later', 18 *Human Ecology* 1, 1–19 (1990).

to humans and many other species of plants and animals. Like a malign Noah we will carry many other species along with us to their extinction. Fully appreciating this is for many people (except the scientists who study the dynamics at work) still a matter of a psychological incorporation, or emotional internalization, of something we already know only in the abstract, akin to a headline on CNN so devoid of context as to be virtually meaningless except as an answer to a trivia question.

Much more intellectually revolutionary, and far less appreciated or internalized by our economic and political leadership, is the ethical point that these facts make applicable to us now. Since there are unchangeable limits on the atmosphere's capacity to absorb net emissions (without increases in surface temperature), there is a limited supply of absorptive capacity. The capacity of the planetary atmosphere for gaseous emissions, no less than the capacity of the local landfill for solid waste, is limited. This capacity is valuable, indeed vital, but scarce. Who gets how much of it, therefore, is a basic ethical issue. In particular, it raises all the questions about justice that are raised by the choice of a process for allocating any scarce natural resource, especially a resource essential for minimal economic welfare or subsistence.

Three qualifications that are scientifically important should be noted, although they in no way undercut the normative significance of the realization of the limited capacity of the atmosphere for gases. First, it is not the case that there is some natural limit on the total capacity of the planet to deal with GHGs by all possible means. At least for naturally occurring GHGs such as CO_2, as opposed to human concoctions such as the chloroflurocarbons (CFCs), there are naturally occurring sinks as well as natural sources. If one increases the supply of sinks (by, for example, planting new forests), one can increase the supply of the gas without increasing the net load upon the atmosphere. In this case the new terrestrial sinks deal with the additional CO_2 at the surface of the planet. Surface recycling of CO_2 has been going on for millennia. All that coal and oil that we have recently been gobbling up has, indeed, sequestered phenomenal amounts of CO_2, keeping it safely out of circulation. This is why it is such a horrific problem that over mere decades we are, by racing through the coal and oil, thrusting carbon back into the atmosphere that had been removed from it for millennia. It is net emissions to the upper atmosphere on which there are limits. If increases in the supply of any gas are matched by increases in the supply of its sinks, everything is handled at the surface and there is no net increase in atmospheric levels. We are, of course, not adding sinks remotely as fast as we are pouring carbon into the atmosphere by burning oil and coal. On the contrary, we are not making a net addition of sinks at all. We are both vastly increasing the supplies of CO_2 and simultaneously significantly decreasing the sinks for CO_2 (e.g. cutting down more forests than we plant). So we are compounding net atmospheric levels. It is difficult to imagine how we could be doing more to upset the carbon balance if we tried.[19]

Second, a technical point simply needs to be acknowledged. Evidently, as the concentration of CO_2, in particular, in the atmosphere increases, the capacity of the oceans to sequester CO_2 also increases. The increase in the ocean's capacity to absorb, however, by no means matches the increase in atmospheric concentration, nor is it unlimited.[20] Hence, the same scientists who point out that one should note this variable oceanic sequestering for full accuracy also emphasize that one should take no great comfort from it. So, for all practical purposes, the limit of net atmospheric emissions that can be handled without increases in surface temperature is fixed.

Third, besides the oceans there are other little-understood feedbacks, with clouds being probably the most important. Some cloud activity provides a negative feedback to global warming. No one fully knows the overall effect of water vapour at various heights above the surface. Nevertheless, it is not inconceivable that some increases in global warming will, for example, unleash water vapour from the surface into the air. This will produce a negative feedback to the warming (e.g. by cloud blockage of sunlight) which will actually increase the net amount of CO_2 the atmosphere could handle without further warming. One can only speculate about the recursive feedbacks that would then occur.

Still, my fundamental point about limits and scarcity holds. Whatever exactly the capacity for net atmospheric CO_2 (and other GHGs) is, that capacity will be limited by natural processes not decided by humans. The absorptive capacity may be scarce at a somewhat higher absolute amount, or at a somewhat lower absolute amount, but *it will be scarce*. That is, there will be much less than humans would like, at least as long as the politically entrenched

---

[19] For a straightforward and short overview pointing to more technical discussions, see Irving M. Mintzer, 'Living in a Warming World', in *Confronting Climate Change: Risks, Implications and Responses*, 1–13 (Irving M. Mintzer ed., 1992). For a more passionate but no less well-informed, sadly beautiful, and book-length treatment of the profound human significance of human-induced climate change, see Bill McKibben, *The End of Nature* (1990).

[20] Some had hoped that the 'fugitive carbon', the unaccounted-for difference between calculations of sources and calculations of sinks, would turn out to be in the oceans, proving them to be a larger sink than previously thought. It now seems more likely that the sources have been overstated through overestimates of the carbon released by deforestation, most notably in Brazil. See David Skole and Compton Tucker, 'Tropical Deforestation and Habitat Fragmentation in the Amazon: Satellite Data from 1978 to 1988', 260 *Science* 1905, 1909 (1993). See also R.T. Watson et al., 'Sources and Sinks', in *Climate Change 1992: The Supplementary Report to the IPCC Scientific Assessment*, 25, 25–46 (J.T. Houghton et al., eds, 1992).
That the CO_2 released by deforestation in the Third World was being overestimated—in particular, by World Resources Institute (WRI) in Washington—was the central *scientific* thesis in Anil Agarwal and Sunita Narain, *Global Warming in an Unequal World: A Case of Environmental Colonialism* (1991). Their *political* thesis, that Third World deforestation was being intentionally exaggerated in order to blame more of global warming on the Third World, seems to me to be groundless. Their *ethical* thesis I take very seriously. For subsequent analysis by WRI, see World Resources Institute, 1992–93 *Report* 118–20, 348, 352–3 (1992).

dependence on fossil fuel is perpetuated. How the total amount of emissions is divided is important, and ethically charged, without anyone's understanding the extraordinarily complex ricocheting atmospheric phenomena that will determine the absolute level of the total. Only if all the natural feedbacks taken together turned out luckily to be strongly negative (against surface warming) and so negative that they always stayed ahead of the pressure of endlessly increasing human emissions of GHGs, would the problem of scarce absorptive capacity disappear. However, there is no reason to count on that being the case.

Accordingly, I accept the following:

1. Global warming is dangerous enough that it ought to be minimized, at least until the costs of doing so in the quality of human life and the enjoyment of basic rights, become excessive.

2. Global warming can be minimized only if a constant global total of GHG emissions is sustained (i.e. the total global GHG emissions must soon stop rising). Or:

2'. Global warming can be minimized only if a constant global total of GHG emissions much smaller than the global total in 1990 is reached and then sustained (i.e. total global GHG emissions must soon be reduced significantly below the 1990 level).

For those who like the servings on their plate divided into facts and values, (1) is my basic relevant value; and (2) and (2') are my basic relevant facts. The difference between (2) and (2') is a scientific issue. I rely, as I already mentioned, upon only the far weaker (2), although I am in fact persuaded, by the breadth of the scientific consensus behind it, of (2').[21] A naturally limited planetary capacity to absorb CO_2 (and other GHGs) and a consequent shortage of emissions capacity relative to human demand need to be handled in some civilized manner. How should that scarce capacity be allocated among humans?

PRE-AGREEMENT STANDARDS OF FAIRNESS

Before we take up this question, it may be worth discussing its origins. We had turned to the question of what I would like to call the true 'mitigation budget',

---

[21] The difference between (2) and (2') does not matter here because either way we have a naturally limited total of a valuable resource, permissible emissions of GHGs, that must be allocated across humanity. The difference between (2) and (2') matters greatly as one spells out distributive principles more fully and concretely because, generally speaking, the more severe the scarcity, the more radical the principles appropriate. On the scientific questions, see the sources cited in note 6 above.

as distinguished from the no-regrets budget that could take measures to eliminate waste with no long-term costs. Unlike the no-regrets budget, the true mitigation budget involves expenditures directed solely at global warming. In principle, there is a total annual mitigation budget for the globe, the overall amount that would be reasonable to spend each year to control GHG emissions (that is, either (2) to stop total global GHG emissions from rising or (2′) to reduce total global GHG emissions to a lower sustainable level).[22] It is far more likely that enough will be spent actually to do the job of controlling emissions if there is agreement about how the totals are to be shared across nations. In any case, extensive international negotiations have in fact been under way to formulate an agreement allocating the costs.[23] For obvious reasons, only a few nations will take expensive measures without assurance that other nations will do their part as well.

The original question was: how should the total costs of mitigation be shared? By what principles should portions of the total be assigned among various nations (or other parties)? My suggestion earlier was that it would make a great deal of difference if some nations were, at the start of the mitigation process, already doing something wrong. The purpose of the intervening discussion about naturally limited supplies was to provide the background to make judgements about nations doing wrong. In what sense could a nation already be doing something wrong before any agreement had been reached about who should pay for what? Once there is an international agreement specifying how much of the total costs of mitigation each nation should bear, a nation obviously could do wrong by failing to provide its share of the costs. This would be a wrong specifically in the form of an unfairness: an agreement would specify fair shares of the total cost, and then a nation would refuse to bear its share while other nations paid their shares.[24] But how could a nation

---

[22] Needless to say, all kinds of issues arise about how to arrive at annual totals, most notably issues about allocations across generations. Simply in order to try to deal with one impossibly difficult problem at a time, I am restricting my attention here to issues about how to share the costs (to be borne at any one time) across nations. Obviously the international and the intergenerational are interconnected, and one will almost certainly need later to revise whatever will have been said about one as soon as one can see what to say about the other.

[23] See generally, Daniel Bodansky, 'The United Nations Framework Convention on Climate Change', 18 *Yale J. Int'l L.* 451 (1993).

[24] I do not believe that one is bound to do one's share only if everyone else is doing hers. If the shares are indeed fair, and if the undertaking in question does not depend for its success on unanimous cooperation, one is, I think, normally bound to do one's own share in spite of unfair non-performance by others. This does not mean, of course, that one should make fruitless contributions merely because they were part of an original plan that is subsequently coming unstuck. Presumably helping to control global warming is less like helping to construct an arch than a wall: an arch will collapse if not every stone is put into place—so trying somehow to stick one's own stone into place in the absence of others is pointless—but a wall that is not as tall as it was supposed to be may stand nevertheless and may do some good—so one should add the stone one promised even in the face of betrayal by others. David Hume used the metaphor to contrast benevolence and justice as he understood them. See *David Hume, Enquiries Concerning Human*

be doing anything wrong before any agreement had been made specifying who was to pay for what?

In fact, we do not generally believe that one is bound to do only what one has explicitly and voluntarily agreed to do. On the contrary, we regularly judge agreements to be fair and unfair, which reflects the fact that we take some elemental principles of fairness to be more fundamental than explicit agreements and to include standards that agreements themselves must satisfy in order to be binding.[25] This means that there are elemental moral standards that laws, treaties, and other human agreements must satisfy in order to deserve compliance—in order to be morally as well as legally binding. In the extreme case, civil disobedience may be considered necessary when what is legally binding is believed to violate more fundamental standards that are morally binding.

Since these standards might be in some sense prior to human conventions, one is tempted to call them 'natural'.[26] That is a mistake. That conceptualization creates unnecessary problems, introduces false issues, and invites misleading comparisons. It is enough for present purposes to notice that there is a standard human practice of assessing agreements as fair and unfair, which it is very difficult to imagine giving up or wanting to give up.[27] Regardless of whether this reflects anything 'natural', it certainly reflects something very deep, which ordinary people respect and are not about to abandon (nor is there any reason why they should). This is enough. So I will speak of pre-agreement standards and post-agreement standards. The latter are based upon particular agreements. The former are used to judge which agreements to take seriously, and include standards of fairness.

EXCESS AND ENCROACHMENT

Thus, the suggestion is this: prior to any agreement about how the costs of mitigation ought to be shared, a nation might be doing something wrong—as

---

Understanding and Concerning the Principles of Morals, 305–6 (P.H. Nidditch, ed., 3rd edn 1975), quoted and discussed in Brian Barry, *Theories of Justice: A Treatise on Social Justice*, 150–2 (1989). See also Brian Barry, 'Can States be Moral? International Morality and the Compliance Problem', in *Liberty and Justice: Essays in Political Theory*, 171–4 (1991).

[25] Even the Western social-contract theorists, who probably assigned about as exaggerated a role in human life to voluntary agreement as one conceivably could, all without exception assumed 'natural rights' that were more fundamental than agreements and contracts and could not be violated by any contract that anyone could be bound to keep.

[26] See Joseph Boyle, 'Natural Law and International Ethics', in *Traditions of International Ethics*, 112–35 (Terry Nardin and David R. Mapel, eds, 1992).

[27] We also assess agreements as voluntary and involuntary, which reflects another kind of standard that is more fundamental than any particular agreement. Many have thought this reflects a 'natural' right to liberty.

judged by a pre-agreement standard of fairness. It could be using more than its fair share of the naturally limited supply of capacity to absorb emissions, thus producing more than its share of emissions. I will explain this against the background provided by the preceding discussion of naturally limited supplies.

It is essential to notice that my focus of attention has moved from shares of mitigation costs to shares of GHG emissions, from dollars to gases. The proposed connection between gases and dollars is this: a country has no right to emit gases in excess of its pre-agreement fair share. Not only does one have no right to produce those emissions, but it is wrong to produce them because, in using up a scarce and valuable capacity, a country is unfairly impinging upon the fair shares of others to the extent that it is exceeding its own share. Consequently, even prior to any agreement about sharing mitigation costs, a country ought to eliminate the emissions that go beyond its entitlement. The costs of cutting back those emissions and the costs of stopping unfair wronging ought to be paid no matter what (even if there should be no agreement to undertake other mitigation of global warming). Most particularly, a country ought not to use the threat of refusing to cut back the excess emissions as leverage in the bargaining over the terms of any cooperative mitigation effort. A threat to refuse to stop doing what is wrong is simple bullying in any case.

The preceding paragraph highlights two separate theses. One is that it makes sense to talk about fair shares of emissions quite apart from international agreements—violating pre-agreement fairness is wrong. The other is that ceasing to commit wrongs is not an acceptable part of the quid pro quo of the bargaining about further mitigation. That this kind of bullying is unacceptable within any process of bargaining that purports to be fair is readily apparent, but it is a somewhat separate point from the fact that not all judgements of fairness depend upon prior agreements.

Three aspects of the nature of pre-agreement unfairness merit discussion. First, the kind of wrong involved in exceeding one's share of allowable emissions has nothing to do with intention or foreseeability. One can wrong others without intending to and without having been able to foresee that one would. Naturally, it is worse still to commit any given wrong with foresight or intention than to do the same thing without them, but the absence of intention and foreseeability by no means eliminates the wrongful character of many acts. As a general point, this is uncontroversial; it is simply worth noticing.

Second, if the first point were not true, it would not affect present and future emissions. Once we know that we are exceeding our share, we are certainly not any longer doing so either unintentionally or without foresight. On the contrary, we are continuing to exceed our share in full knowledge that this is what we are doing and only because we do not choose to stop. Hence, future emissions become the key to reducing GHG levels. Although I do not think that lack of intention or foresight carry much weight, they are simply irrelevant to future emissions. Now we know: future harms are foreseen.

Third, and by far most important, the seriousness of this wrong derives from the fact that anyone's excess comes out of other people's shares. It is one thing simply 'to make a pig of oneself' in a context in which one's own gluttony has no effects on others. While simple gluttony is at least a minor vice, in isolation from bad effects on others it might not be a major one. The point is that in GHG emissions there is no surplus from which gluttons can indulge themselves. The supply of emissions-absorbing capacity is too small relative to the demand, but cannot be increased, and therefore is zero-sum. Because the total is, for all practical purposes, fixed, anyone's *excess encroaches* upon someone else's share. It deprives them of something they badly need. Excess consumption of emissions-absorption capacity is wrong because it makes others seriously worse off by being unfair to them.

BY WHAT STANDARD UNFAIR?

Noticing that there is, in general, such a thing as pre-agreement fairness, which most would readily acknowledge, is different from discovering what in particular is fair and unfair in specific contexts. One could grant all that has just been said about how wrong it would be to exceed one's share of emissions, where the capacity to absorb them was scarce and valuable, and still not know what counts as one's share or on what grounds that is decided. What is a fair share? What is the decision that it is fair based upon? Our case is, I believe, one of those not uncommon ones in which specifying the precise location of a boundary—in this case, between fair and unfair shares of a limited total of emissions—is theoretically challenging and fascinating, but judging in actual practice whether some parties have crossed the line is as easy as can be. When a ball lands near the line, it is challenging and interesting to discuss what counts as the edge of the line. When the ball lands among the spectators, the ball is out. In the case of GHG emissions by the rich industrialized countries, we know that the ball is out of bounds even though we can imagine closer calls that might arise in future and might leave us then to scratch our heads. For now, we know all that we need to know in order to judge and to act.

Sources vary on the precise details, but the general shape of things could not possibly be clearer. CO_2 is both the most important GHG produced by human activity and the only GHG for which there are anything like accurate estimates. Thomas E. Drennen has calculated that the industrialized countries, with 15.7 per cent of the global population, emit 48.5 per cent of the carbon, while the developing countries (not including China), with 51.9 per cent of the population, emit 14.9 per cent of the carbon. China accounts for 23.5 per cent of the people and 10.3 per cent of the carbon, while the Commonwealth of Independent States and Eastern Europe have 8.8 per cent of the people and 26.2 per cent of the carbon.[28] Thus, 'citizens of the industrialized world are responsible for emitting 11.9 tons of CO_2 per capita per year, ten times more than their counterparts in developing countries (1.1 tons)'.[29] Thomas Drennen implicitly makes what I take to be the absolutely correct judgement that it is individual persons who are our ultimate concern. In addition to per capita measures, there are measures of emissions per square mile, per unit of GNP, and various other variables, but these other measures seem to me to be less central. It is the needs of individual human lives that finally matter.

When one adds in the facts, first, that the total emissions must be held constant and, second, that many of those whose emissions are only one-tenth the emissions of others are living in extreme poverty which they can exit only via economic development that would sharply increase their emissions per capita, it becomes clear that some very powerful justification would be needed to show that the status quo is acceptable. One need not be an egalitarian to conclude that it is unacceptable that a small minority should live in economic conditions 900 per cent better than the conditions in which the majority of humans (51.9 per cent) live. Of course much of the industrialized emissions are the result of sheer waste of energy, not superior quality of life, so perhaps real quality of life is only 800 per cent or 700 per cent of the quality of life of the world's majority. Would it be morally tolerable for a minority to live 100 per cent better than the global majority? That is, at least, worth discussing. Is it morally tolerable for a minority to live 900 per cent better than the global majority when the shares in economic welfare are zero-sum? I cannot imagine any plausible reasons for continuing to maintain arrangements that produce such radical inequalities in the prospects for different human lives. Only after we have reduced the inequalities to levels at which we are living only, say, five or six times as well as the majority of our fellows need we seriously worry whether we are overdoing it and earnestly search for the precise line between unacceptable and acceptable inequalities.

THREE ALTERNATIVES

In sum, I see only three choices. First, reduce GHG emissions at the expense of the industrialized peoples through investment in research and development to produce economically feasible non-carbon-based energy, which might

---

[28] Thomas E. Drennen, 'Economic Development and Climate Change: Analyzing the International Response', 142 (1993) (unpublished PhD dissertation, Cornell University).
[29] Thomas E. Drennen, 'Economic Development and Climate Change', at 8.

improve the quality of life and might not reduce the standard of living. Second, reduce the GHG emissions at the expense of the industrialized peoples through reduction in their energy-consuming economic activity, which will certainly reduce the standard of living and may or may not reduce the quality of life. Or third, continue the annual increases in GHGs pumped into the upper atmosphere as the rich get richer and the poor become more numerous and/or possibly slightly less poor.

Either of the first two choices would require considerable initiative and imagination. Since the third requires only business as usual, I sadly expect the third, which I have argued here is grossly unfair and which a consensus of scientists is arguing is possibly catastrophic. 'Too sophisticated to burn books, we burn the planet.'[30] And only a few of us briefly warm ourselves even from that.

---

[30] Bill McKibben, *The Age of Missing Information*, 85 (1992).

4

Avoidable necessity: global warming, international fairness, and alternative energy*

FOUR QUESTIONS OF FAIRNESS

A wide international consensus of scientists is convinced that this planet faces serious danger from changes in temperature and other aspects of weather to which agricultural systems probably cannot adjust with the speed that would be necessary in order to avoid serious disruptions in food supplies for millions of vulnerable humans.[1] In order to reduce the magnitude of this danger, which is generally referred to as 'global warming' (in spite of the fact that temperature is only one of its many aspects), it would be necessary to have comprehensive international cooperation, cooperation that might not need to be literally universal across countries but would certainly need to include both the richest industrialized countries, whose emissions of 'greenhouse gases' are the main current sources of global warming, and the most populous less industrialized countries, whose emissions will be the main future sources of warming if they try to follow the model of industrialization pioneered by the now rich. If this scientific consensus is even roughly correct, one of the most important issues facing humanity is: on what terms can rich and poor, industrialized and less industrialized, cooperate in dealing with this common global threat?

Unfortunately, we face here not simply one question about fairness, but at least four questions, the answers to which must in the end nevertheless be coordinated. For the sake of a clear analysis, we need initially to see that at least four different issues of fairness must be handled by any cooperative international arrangement. For the sake of effective action, we then need ultimately to tie the

* 'Avoidable Necessity: Global Warming, International Fairness, and Alternative Energy', in *Theory and Practice*, NOMOS XXXVII, edited by Ian Shapiro and Judith Wagner DeCew (New York: NYU Press, 1995), pp. 239–64. This chapter has endnotes rather than footnotes.

answers to the four questions together into a single coherent package. At greater length elsewhere I have explained why I believe that all the following four questions must be answered before we can know how to proceed fairly:

1. What is a fair allocation of the costs of preventing the global warming that is still avoidable?

2. What is a fair allocation of the costs of coping with the social consequences of the global warming that will not in fact be avoided?

3. What background allocation of wealth would allow international bargaining (about issues such as 1 and 2) to be a fair process? And:

4. What is a fair allocation of emissions of greenhouse gases (A) over the long term and (B) during the transition to the long-term allocation?[2]

In this chapter I will try to begin to answer question 4A, by discussing what the allocation of emissions should ultimately become; I will not tackle the other crucial, and of course practically more immediate, issue of how rapid should be the transition from the current allocation to a fair one.[3]

IDEAL BEFORE NON-IDEAL?

This is a messy chapter in at least the following sense: in addition to recognizably normative discussion about fairness and inequality, essential reference is also made not only to various elements of the scientific consensus about the problem of global warming but also, in the defence of practical judgements among alternative solutions, to supposed aspects of human psychology and even speculations about US politics. Those influenced by the magisterial work of John Rawls, among others, may well feel that this is a muddled manner in which to proceed, especially perhaps on issues about fairness to which Rawls has contributed so much. In particular, it might be thought wiser first to establish what the ideal arrangements would be and then gradually to factor in the relevant psychological, political, and other empirical elements needed to decide what to do in the thoroughly non-ideal situation that everyone, Rawls included, is well aware we face.

A firm believer in the principle that the proof of the pudding is in the eating thereof, I will not offer a general defence of my method in advance and in the abstract but will leave readers to judge its persuasiveness and fruitfulness in the concrete. I would, however, like to make one preliminary critical comment about the much neater-sounding thought that one should engage in ideal theory first and only then move, compromising as one goes, into non-ideal theory. The critical comment is that, at least in Rawls's original formulation, the line between ideal theory and non-ideal theory is indeterminate.

Famously, the construction of the argument by Rawls in *A Theory of Justice* has four stages: original position, constitutional convention, legislative stage, and application of rules to particular cases.[4] In the original position use is made of general truths, but not of knowledge about particular contingencies; in stages two through four the particular contingencies are gradually fed back in and—it would seem to me—some of the general truths fall out (since we do not in reality know all the general theoretical truths that in the original position we are hypothetically assumed to know). The four-stage sequence gives us 'a schema for sorting out the complications that must be faced'; at each stage 'the veil of ignorance is partially lifted'.[5] Thus, at, for instance, the second stage (of the constitutional convention),

> in addition to an understanding of the principles of social theory, they now know the relevant general facts about their society, that is, its natural circumstances and resources, its level of economic advance and political culture, and so on.[6]

And Rawls shortly adds:

> I imagine then a division of labor between stages in which each deals with different questions of social justice. This division roughly corresponds to the two parts of the basic structure. The first principle of equal liberty is the primary standard for the constitutional convention . . . The second principle comes into play at the [third] stage of the legislature.[7]

To put it unkindly, we have just been told that the constitution is to be written with no attention to its economic and social implications, since neither the relevant (second) principle nor the relevant information—'the full range of general economic and social facts'—becomes available until the constitution is finished and the third (legislative) stage begins. Many reasons have been given during the succeeding two decades of debate why this picture of the relation of constitution and legislation—resting upon a sharp division between liberties, on the one hand, and everything else, on the other—is profoundly misguided.[8] But it does at least appear to be clear.

Not so. At *every* one of the four stages of construction, including the original position, one may conduct two kinds of inquiry, ideal and non-ideal; and within non-ideal inquiry two kinds of compromise principles are needed:

> One consists of the principles for governing adjustments to natural limitations and historical contingencies, and the other of principles for meeting injustice.[9]

Most references to Rawlsian non-ideal theory that I have encountered make it sound as if compromise is allowed only to deal with partial compliance, the second of the two kinds of compromise expected in non-ideal contexts. That Rawls is not in the sentence quoted simply overindulging in the creation of typologies by mentioning a different, first kind of non-ideal theory is indicated by his having just previously made precisely the same distinction on the issue of supreme importance to him, restrictions on liberty:

> There is, however, a further distinction that must be made between two kinds of circumstances that justify or excuse a restriction of liberty. First a restriction can derive from the natural limitations and accidents of human life, or from historical and social contingencies . . . For example, even in a well-ordered society under favorable circumstances . . . the principle of participation is restricted in extent. These constraints issue from the more or less permanent conditions of political life . . . In the second kind of case, injustice already exists, either in social arrangements or in the conduct of individuals. The question here is what is the just way to answer injustice.[10]

Here too, Rawls distinguishes the subsequently much more widely discussed (second) case of partial compliance, or departures from justice, from the (first) case of natural limitations and historical and social contingencies.

Now, these remarks by Rawls raise several kinds of important issues that I cannot pursue, and I will call attention to only one. For purposes even of *ideal* theory at the constitutional (second) stage of construction, people know 'the relevant general facts about their society, that is, its natural circumstances and resources, its level of economic advance and political culture, and so on'. For purposes of *non-ideal* theory at the same constitutional (second) stage, restrictions specifically in liberty 'can derive from the natural limitations and accidents of human life, or from historical and social contingencies'. I submit that there is no clear commonsensical answer and no clear answer within Rawlsian theory to the question: where is the line between 'the relevant general facts about their society' (ideal theory) and 'historical and social contingencies' (non-ideal theory)? That means that there is no systematic way to know which information is appropriately invoked in arguments within ideal theory and which is appropriately invoked in arguments within the first of the two kinds of non-ideal theory. One is left, I think, to muddle through, judging relevance as one goes. It will turn out below, in the case at hand, that empirical judgements about what is in fact avoidable and what is not avoidable are integral to the argument for a standard of fairness and that judgements about what is politically and ideologically palatable are crucial to the choice of what specifically to do.

GLOBAL WARMING: NOT YOUR USUAL PROBLEM

Global warming, which is almost certainly not yet occurring, will be produced here on the surface where people live by the accumulation of unprecedented amounts of greenhouse gases in the atmosphere a few miles above. A greenhouse gas just is a gas that to some significant degree holds heat on the planet by reflecting heat back from the atmosphere toward the surface. Water vapour, which humans cannot control, is a major greenhouse gas, as are CFCs, which we could control (and which we invented), and to some extent are controlling. By far the most important greenhouse gas produced by humans is carbon dioxide. All fossil fuels—coal, oil (including of course the gasoline refined from oil), and natural gas—are carbon-based, and burning any fossil fuel for energy thrusts carbon dioxide into the air. Carbon that had been, as the scientists say, 'sequestered'—taken out of circulation—for thousands of millennia in coal, gas, and oil is now suddenly being flooded in torrents into the atmosphere as carbon dioxide. It would be extraordinary indeed if this sudden, massive, and increasing rush of carbon dioxide into the upper atmosphere had no effect here on the surface. And there is indeed plenty of reason to think that, roughly, once the concentration of carbon dioxide doubles, which will not take much longer at current rates, the surface temperature of the earth will rise enough to affect the volume of seawater and thus the location of high tide (water expands with rises in temperature), the top velocity of hurricane winds (the maximum speed rises with the temperature of the underlying ocean water), and many other aspects of the weather vital to agriculture and human food supplies, especially the supplies to the poorest people.[11]

Nothing that is done by way of the atmospheric mechanism behind global warming can have any effect here on the surface until many years after it is done. The delay between cause and effect in this case spans decades. This is an utterly critical, and quite unusual, fact about the greenhouse effect. What has to be changed in order to affect global warming is the accumulated total of gases at certain levels of the atmosphere, but what atmospheric scientists call the 'residence times' of the critical gases are, in the case of carbon dioxide, several generations—about a century—and even in the case of one of the shortest-lived ones, methane, a full generation (fifteen years). Needless to say, atmospheric chemistry is exceedingly complex and lively (far beyond anything I know about): no gas particle simply floats in place for one hundred years and then evaporates. Nevertheless, generally speaking it would be the case that if, *per impossibile*, we could avoid adding any carbon dioxide to the earth's atmosphere for an entire year, the atmosphere would at the end of the year contain all the carbon dioxide it had contained at the beginning except for the last of the carbon dioxide emitted around 1895, which would finally have played itself out at that location. Most of the vastly greater quantities of carbon emitted in the unprecedented burst of industrial activity since World War II, most notably, is continuing to be active in one way or another, especially since more is emitted every year than was emitted the year before. The carbon dioxide emitted in 1995 will mostly still be very active in, say, 2050. Consequently, we could by beginning now significantly affect the temperature and weather in 2050, but there is next to nothing that we could do now to change the temperature in 2000. That die has been cast. Global warming is the ultimate case of the prevention being easier than the cure.

Consequently, the argument that we should not take any action until we have observed global warming is a little like the argument that one should not stop smoking until one has seen tumours in one's lungs. It is admittedly possible for one to smoke all one's life and not get lung cancer. Consequently, it is true that one might give up smoking 'for nothing' in the sense that one might never have developed cancer (or any of the other serious diseases that are provoked and promoted by smoking) even if one had never given up the smoking. To be fully analogous, however, the cigarette smoke would have to remain trapped in one's lungs for up to, say, fifty years after one stopped smoking so that if one wanted to be free of continuing effects of the smoke at age sixty-five, one had to stop smoking when one was fifteen. Some aspects of nutrition are exactly like this: a child whose nutrition up to about age two does not permit adequate bone development will suffer from fifty-five to sixty in ways that cannot then be undone (one of many reasons why the levels of childhood malnutrition that we routinely tolerate in the Third World are criminal).

In fact, insofar as those harmed by the effects of any given emissions are not the members of the emitting generation but are instead the members of succeeding generations, the analogy with smoking would become still closer if it were the case that the 'passive' smoke from the cigarettes of one generation collected and remained in the lungs of their grandchildren to do damage there decades after its emission.

Consequently, we face what might be called quasi-irreversibility. One of the reasons why the extinction of plant and animal species is such a serious matter is that extinction is strictly irreversible; one cannot bring a species back once it is gone. It is, by contrast, perhaps not the case that once global warming has been set in motion it literally cannot be reversed, although we should not glibly assume that no reverberating cycle of positive feedbacks, making the warming worse and worse, will be unleashed beyond some unknown threshold.[12] For the same reason, namely that no one knows nearly enough, we cannot be certain that negative feedbacks would not correct any warming that occurs; higher temperatures might, for example, lead to additional water vapour in the atmosphere in the form of clouds that would block enough incoming sunlight to restore the temperature to its former level, and so forth. Because clouds, in particular, are so little understood, we really do not know what to expect: 'Most of the variability between global climate models can be traced to differences in cloud radiative feedback.'[13] While this wonderfully curative response to unprecedented human activity is not inconceivable, it is important to note that we lack adequate empirical grounds for counting on it and that, if it were to occur, it would be entirely outside human control and would have many other effects as well, not necessarily all positive from a human point of view. Basically we are blithely unleashing planetary forces that we do not understand and that we have no reason to believe we can tame.

LESS ECONOMIC ACTIVITY OR DIFFERENT ENERGY SOURCES?

The main challenge, then, is to reduce new emissions of carbon dioxide to a level that will allow the atmospheric concentration to stabilize at a quantity that is not too much higher than the quantity we are accustomed to living with and for which agriculture has evolved. Obviously, there are two broad possibilities. First, without necessarily using any less energy, and therefore without necessarily reducing our standard of living, we could use energy from sources other than fossil fuels; possibilities include nuclear (fusion as well as fission), hydroelectric, geothermal, wind, passive solar, and photovoltaic. Remarkably little interest has been shown—and a pitifully small proportion of the gigantic public subsidies for energy has been spent—on energy sources other than fossil fuels and nuclear fission by the current political structure.

The second possibility is to continue using fossil fuel as our predominant source of energy but reduce our consumption of energy, which will sooner or later mean reducing our standard of living.[14] This is complicated since 'standard of living', as conventionally measured, and quality of life are, I think, two different matters. Every time someone wastes a tank of gasoline in pointless driving, the GNP goes up, along with the carbon dioxide. To some extent, reductions in standard of living could improve quality of life. On the other hand, advanced medical technology is possible only on the basis of considerable societal wealth. In addition, people do need jobs in order to support themselves, so net reductions in standard of living must generate jobs, which is quite possible (if, for example, production becomes less energy intensive and more labour intensive). I have neither the space here nor the competence to pursue this controversial (and ethically charged) but important matter. For simplicity, then, I will talk as if reductions in standard of living are always a problem, as they are now widely perceived to be and are always treated by conventional economists as being.

Basically, then, we can either switch energy sources, away from fossil fuels, or we can reduce energy consumption and thereby reduce standards of living. Since our political leaders have so far steadfastly ignored the former option of changing our source of energy, I will in most of the rest of this chapter explore the latter option, returning to reconsider alternative energy sources only at the end in the light of how the main argument develops. For now and through most of the remaining discussion, then, I will assume (1) that if we are going to reduce carbon emissions, we will have to make sacrifices in our standard of living through reducing economic activity. As I indicated in the previous paragraph, I will also assume (2) that making sacrifices in our standard of living is bad. Making sacrifices in our standard of living, I am thus assuming in company with the vast majority, would mean making sacrifices in our self-interest.

Even so, there is one other complication that is most fortunate for us in the rich industrialized countries. Much of the carbon emissions by the rich countries are from energy that is entirely wasted. These emissions are the product of sheer inefficiency. (Not to mention that carbon-emitting fossil fuel is used when non-carbon-based fuel could easily have been substituted.) We are failing to make what have become popularly known in the public policy community as 'no-regrets' emissions reductions. A politically broad-based and authoritative consortium of public interest groups said: 'Our conclusion: whether one simply wants to minimize costs to consumers, or to mitigate global warming, vigorous adoption of energy-efficiency measures and accelerated use of renewable energy sources make sense... Current government policies and the marketplace are structured in a way that encourages the wasteful use of fossil fuels, not the efficient use of all available energy sources.'[15] Only appalling indifference to the fate of future generations, our own and other people's, allows this part of the rich country emissions. We could actually save money for ourselves while improving conditions for the future. Our remarkable failure to do so reflects political incompetence and social irrationality of a high order, in addition to flagrant disregard of the environment itself and of future generations as well as a craven degree of subservience to the entrenched interests of fossil fuel firms. In these respects, all that we have to do is to stop shooting ourselves in the foot with bullets that ricochet and harm others as well. This part of the current behaviour of the rich is utterly feckless.

The 'no-regrets' reductions in carbon emissions could actually enable us in the rich countries to raise our living standards if we invested the savings from the increased energy efficiency into improvements for ourselves. Not everything that needs to be done to minimize global warming can be done at a profit, of course. Additional measures would actually require reductions in the living standards of the rich. Presently we will move on to the emissions reductions that would reduce our living standards, that are definitely contrary to at least our immediate self-interest (continuing to ignore any divergence between standard of living and quality of life). The costs to rich countries of these reductions would represent a genuine contribution by us to the cost of mitigating global warming. Where true sacrifices must be made, questions of fairness arise about who should make them.

CURRENT CARBON EMISSIONS

Who ought in fairness to make which sacrifices for the sake of combating global warming depends upon several factors. For a final decision I believe that one needs the answers to at least the first and second, as well as the fourth, of the questions listed at the beginning of the chapter; one would not in practice need the answer to the third if the parties were willing to decide the terms of cooperation on principle rather than to engage in no-holds-barred bargaining based on current power. Certainly the answer to the fourth question—What is a fair allocation of emissions of greenhouse gases?—is crucial because, other things equal, the allocation of the costs of resisting global warming ought to move the allocation of emissions in the direction of a fair allocation.[16] That is, if the current allocation is unfair because some countries' emissions now exceed their fair share, and if some emissions need to be cut in order to prevent or reduce global warming, then, other things equal, the emissions to eliminate are the ones that are already excessive, the ones that go beyond a fair share.[17] So we need to know what a fair share would be in the end. But first we should take note of actual shares.

How do the current carbon emissions of the rich and poor countries compare? The electrifying fact about current rich country carbon emissions is that they constitute almost half the emissions by all humanity, while the residents of the rich countries are only about 16 per cent of the human population: 15.7 per cent of the people are producing 48.5 per cent of the carbon emissions.[18] Thomas E. Drennen, who made the calculations just cited, also puts the comparison in per capita terms: 'Citizens of the industrialized world are responsible for emitting 11.9 tons of CO_2 per capita per year, ten times more than their counterparts in developing countries (1.1 tons).'[19]

Drennen's ratio of 10:1 is an extremely conservative (from the point of view of the rich) calculation. A ratio of 8,150:1, for US to Indian citizens, in *net* emissions of *all* greenhouse gases has been calculated by Anil Agarwal and Sunita Narain of the Centre for Science and Environment in New Delhi.[20] Obviously these two ratios are not comparable: the particular figures I am citing from Drennen concern carbon dioxide only, not all greenhouse gases, and they are gross, not net. I am persuaded by Drennen's arguments that figures on biological emissions, most notably methane emissions, are not firm enough to be relied upon yet and that for this and other reasons we should in the immediate future concentrate upon carbon dioxide.[21] There is also no doubt that carbon dioxide is the single most important anthropogenic greenhouse gas, whether or not one should focus on it exclusively.[22] I am, in addition, dubious about the nationalistic allocation of sources and sinks that underlies this particular result drawn from Agarwal and Narain. In any case, a ratio of 10:1 may already cry out sufficiently loudly for correction before one even contemplates allegedly still more unequal ratios, especially when one considers the role of carbon emissions in everyone's life.

EQUAL MINIMUM OF THE ESSENTIALS

The making of per capita calculations implicitly assumes that equal shares per capita are at the very least a relevant benchmark, if not a moral requirement. I would like now to examine a little more fully the question of per capita equality in carbon emissions. Some discussions seem to assume—although perhaps no one has actually put it quite so crudely as I now will—that if 'all men are created equal', then each is entitled to an equal share of emissions. That, however, by no means follows. This is another case in which the specific facts about something, namely carbon emissions, are highly relevant to how they should be viewed morally, namely whether it is or is not the case that they are among the things that ought to be distributed equally. We need to know what role carbon emissions play in life as it is actually lived today.

That all persons are fundamentally equal has to imply that they have important equal entitlements, but it need not imply that they have equal entitlements in everything or even in most things. Some things should not, or cannot, be distributed centrally, and consequently cannot be guaranteed any particular kind of distribution. Some things would lose their value if they were equally distributed. And so forth. Belief in fundamental human equality still leaves open what shape, if indeed any particular shape, the distributions of many things should have.[23] So any position on the distribution of something as particular as greenhouse gas emissions requires specific, factually informed argument.

And it is argument that is entirely dependent on distinctive features of the stage of economic history in which we find ourselves. It happens to be the case that a single form of industrial society has spread throughout the globe and that this dominant form of industrialization is overwhelmingly dependent upon energy obtained from burning fossil fuels, which of course emits carbon dioxide in huge amounts. The poor and powerless confront industrialized or industrializing societies in which engaging in enough economic activity to support oneself and one's family requires the emission of carbon. In the fossil fuel dependent economies into which poor people are born, people can survive only through activities that generate carbon. Significant carbon emissions are a necessity of life, given the energy sources made available to the poor by those who control energy policy. For as long as these emissions are a necessity of life—for as long as those who control the economy, and energy policy in particular, choose to rely heavily on fossil fuel—everyone ought to be allowed at least an equal minimum amount of emissions sufficient for at least a decent life. Several points about this need to be noted.

First, while it may be obvious, it is also important that an equal minimum is not full equality. It is not my contention that there ought to be no departures from equality in possession of necessities of life—I am not suggesting that there is an objection to anyone's having more of one of the necessities of life than anyone else has when the total supply of the commodity or capacity in question is more than adequate for everyone's needs.

Second, I am not asserting, nor am I denying, the perfectly general thesis that if something is a necessity of life, every person is entitled to at least an equal minimum amount of it. Considerable discussion, in which I have participated, has occurred in recent years at that level of generality; here I want to concentrate on a more specific case.[24]

Third, the critical special feature of the case of carbon emissions is that its becoming zero-sum, or more precisely its mattering that it would become zero-sum, is the result of social choices not made by the poor. Whenever it is decided that, because of the threat of global warming, the global total of annual carbon emissions must be capped, carbon emissions become zero-sum. Once the total is frozen, every emission produced by someone is an emission that cannot be produced by anyone else. This only matters because the rich and powerful have allowed fossil fuels to become so inordinately important to all economies. I will attempt now to explain this crucial third point more fully.

MAKING A ZERO-SUM CAPACITY ESSENTIAL TO LIFE

A strong international scientific consensus holds that the global total of annual emissions must not merely be capped but must be reduced; the original report of the Intergovernmental Panel on Climate Change considered that in order to stabilize the atmospheric concentration of carbon dioxide, annual emissions would have to be cut more than 60 per cent below 1990 levels![25] Thus, if the politicians ever act upon the available scientific knowledge, we will have a *shrinking zero-sum:* a timetable for annual reductions in the global total. Then, not only will it be the case that every emission produced by one source is an emission that may not be produced by another, but some of the emissions permitted in a given year will have to be forgone by everybody in the subsequent year. The annual total would have to be reduced in order to prevent expansion of the accumulated concentration in the atmosphere. The relation between the annual emissions and the atmospheric concentration is just like the relation between the budget deficit and the national debt (and will, heaven help us, take similar leadership and courage to be dealt with). The national debt continues to *rise* even in years in which the budget deficit is *reduced* because even a reduced deficit adds to the total debt; its being reduced merely means that it does not add as much as it otherwise would have added. Reduced annual carbon emissions similarly continue to add to the atmospheric concentration as long as the annual emissions exceed the earth's capacity to recycle them before they reach the atmosphere; therefore, we can reduce total global emissions and still continue to expand the atmospheric concentration (but less rapidly than before) if we do not reduce emissions to a sustainable level.

Moreover, even if we were to reduce current emissions to a sustainable level (comparable to a balanced budget), that would still do nothing to *reduce* the atmospheric concentration (comparable to the national debt). The national debt does not go away simply because one balances some or all annual budgets; neither does the atmospheric concentration.[26] Worse, the longer one waits to attain annual balance, the higher the level at which the national debt/atmospheric concentration stabilizes. Global warming is caused, not directly by the level of annual emissions, but by the level of the atmospheric concentration. Stabilization at, for example, a doubled concentration of carbon dioxide can be expected to mean warmer temperatures (worse hurricanes and the rest) for a long time. Worse still, the longer our 'leaders' wait to deal with the problem, (1) the greater the concentration of greenhouse gases in the atmosphere will be (since it is now ballooning every year because the annual total every year is greater than the sustainable amount) and (2) the larger the human population—the 'emitters'—will be (population too will stabilize at a higher absolute number the longer it takes to stabilize it).

In spite of what I have just noted, in order to keep the assumptions of my argument as weak as possible, I will appeal merely to the fact that total tolerable emissions will, if ever we get organized to deal with global warming, be zero-sum. This alone will already produce extraordinary stringency, un-precedented except possibly during wartime. If, as I have noted in the last few paragraphs, we must in fact have a total that is a shrinking zero-sum, the stringency will, I fear, be more than humans can bear. On the assumption that the shape of the problem, although not its urgency, is the same with a simple zero-sum as with a shrinking zero-sum, I will appeal in subsequent argument here only to the former.

Now, whatever may be the case in general about whether every human being is entitled to an equal minimum share of all necessities for life, I submit that every human being is definitely entitled to an equal minimum share of an essential zero-sum capacity such as carbon emissions, and all the more so if the capacity was made essential for the powerless by decisions by the powerful.

Consider again the difference between what makes carbon emissions essential—what makes them a necessity of life—and what makes them zero-sum. Carbon emissions are currently *essential* because those who control energy policy have failed to fund the research and development of sources of energy alternative to fossil fuel (that are safe, affordable, and so forth).[27] The day may come when the bulk of carbon emissions becomes thoroughly unnecessary through the availability of safe, affordable alternatives, but this is certainly out of the hands of the poor people of the Third World. We in the countries that dominate the global energy regime present them with a world in which one can live above medieval subsistence standards only by means of carbon emissions. Carbon emissions must become *zero-sum* because the surface temperature of the planet will rise with effects harmful to human beings unless the concentration of greenhouse gases, especially carbon dioxide, is (1) stabilized at (2) a level that is not too much higher than pre-industrial levels (obviously no one is claiming to know exactly where this level is).

There is a sense in which both characteristics, the necessity and the zero-sum character, would be the result of human choice. Carbon emissions will not be zero-sum unless (and until) a decision is made by the international community to specify and enforce a global cap. Yet it is seriously misleading, I think, to stress the extent of choice about making emissions zero-sum. If the international consensus of scientists is correct, we have no choice, in the usual sense of that phrase, that is, we have no good alternative. We could go ahead and allow carbon emissions to continue to compound decade after decade as we have in fact done since the beginning of the Industrial Revolution, although faster still as the billions in the Third World industrialize using fossil fuel, and just let the temperature rise (the hurricanes intensify, the sea level rises, and so forth), attempting to cope with the effects directly and efficiently after they occur rather than attempting to prevent or at least reduce them before we are certain of their nature or extent. Such a wait-and-see approach has its advocates among economists.[28]

This is not the place for a full-scale discussion of adaptation versus mitigation. Simply in summary, I believe that the strategy that counsels wait and see what happens and then adapt to it, using all the wealth we will by then have accumulated by not having been earlier panicked into premature and wasteful mitigation strategies, fails to give sufficient weight to, among other things, (1) the vulnerability of the poorest to even small 'temporary' changes in food supplies, (2) the slowness with which agriculture can adapt to climate change (not only temperature itself, but length of season, amount of rainfall, timing of rainfall, evaporation rates, etc.), and (3) the extinction of species of plants and animals that could not adapt rapidly enough.[29] It might be that wealthy people could, at great expense, adapt rapidly enough after changes in climate had occurred, but poor people and many species of plants and animals could not. Consequently, I would judge the *zero-sum* character of future carbon emissions to be basically the *unavoidable* response to the facts about complex natural mechanisms that humans are unable fundamentally to modify.[30]

Carbon emissions are essential, by contrast to zero-sum, only as long as we choose to rely heavily on fossil fuels. Indeed, if gasoline for automobiles were priced to internalize the phenomenal environmental and health effects of combustion-engine-driven transportation, solar energy would instantly become overwhelmingly competitive in the market. Problems, such as what to do about batteries for electric cars that could be recharged by the sun but that would not power cars for trips as long as we sometimes like to take, would be solved in profitable ways in a flash if gasoline and gasoline-powered car prices reflected even a small fraction of the environmental destruction they are wreaking. An adequate gas tax could have revolutionary technological effects. Consequently, I would judge the *essential* character of future carbon emissions to be an easily *avoidable* social choice resulting from vested economic interests in fossil fuels and combustion engines and from political failure to have the courage to confront the interests and to have the imagination to construct the alternatives.

Further, it is only because we choose to travel behind combustion engines and to generate so much of our electricity with coal, oil, and gas that carbon emissions must be capped and become zero-sum. If total demand for carbon emissions fell well below the recyclable level because so much of our activity was powered by other energy sources, usage would probably no longer be zero-sum either. That is, if they stopped being essential, they would probably soon stop being zero-sum as well because demand would fall so far below any cap that would have needed to be enforced.

The significance of the discussion of the essential and therefore zero-sum character of carbon emissions is this: it is technological choices by the rich and powerful, most notably the choice to remain dependent upon fossil fuels, that make carbon emissions a zero-sum matter. It is of course a natural fact that atmospheric concentrations beyond a certain level will produce rises in surface temperature of a certain number of degrees (and, I have suggested without full argument, that the resulting effects could be successfully adapted to only by people with levels of wealth most people are unlikely to command at the time). This fact has human significance, however, only because we, with the ability to influence energy policy, make carbon emissions essential to life by choosing not to pursue alternative energy sources aggressively. Therefore, carbon emissions will become zero-sum if we choose to resist unlimited global warming, but they will become zero-sum only because (and as long as) we choose to rely upon fossil fuel. If, on top of all this, we ourselves used up so much of the allowable total of carbon emissions that the poor of the world did not have available the minimum amount needed for a decent life, we would have (1) created a politico-economic system in which carbon emissions were zero-sum and then (2) used so much of the total that there was not enough left for the survival of others. We would, in other words, first have created the scarcity and then consumed so much that even minimal amounts did not remain for others. That would be truly unconscionable, and it is precisely what we are doing. We, with influence over energy policy, have made a world in which everyone can survive only if each restricts himself or herself to a fair share of something we have allowed to become essential to life, carbon emissions, and now we are proceeding to consume vastly more than our share, leaving grossly inadequate amounts for others.

SHRINKING OUR ECONOMY?

Presidents of the United States tend to propose solving our problems by 'growing' our economy. The problem of global warming cannot be solved by 'growing' any economy fuelled by fossil fuels, unless some now unforeseen technology for recapturing the carbon is created and made economically feasible very rapidly. Carbon emissions must be reduced; the only way to do that is to use less carbon-based fuel, as long as there is no recapture technology in place. The only way to do that, as long as the economy is overwhelmingly dependent upon fossil fuels, is to shrink the economy. As long as fossil fuels are essential, choosing to resist global warming means their usage becomes zero-sum. If we are not to deprive the poor and powerless of the planet of an essential of life (carbon emissions) by using vastly more than our share of it, we ought to reduce our usage enough to shrink it to our share of the constricted total for the globe. This means shrinking our energy usage, and thereby our economy, vastly. These are the required reductions in the living standards of the rich to which I referred earlier.

No way! Right or wrong, we will not do it! No US president is likely to suggest it. If a president were politically foolhardy enough to suggest shrinking the economy, the US populace would drive him or her out of office, even if he or she could correctly say that the number of jobs would not shrink. We will disrupt the climate for everyone before we will willingly accept a lower standard of living for ourselves. Fairness to 'foreigners' simply does not today move enough people to sacrifice their own interest. I would guess that few would deny this judgement about the practical persuasiveness of the kind of argument I have given, its theoretical merits aside. Explaining the roots of the unshakable attachment to 'economic progress' would be much more challenging. Perhaps the ideology of modernity that structures the thought of our era contains so deep a commitment to endless economic growth that attempts to challenge the growth automatically disqualify the challenger from being taken seriously.[31] No one in the mainstream advocates limits to growth because as soon as someone advocates them, he or she is no longer in the mainstream; only Thoreauvian cranks think that we should not 'grow' the economy. One could hope that arguments about the difference between standard of living and quality of life, which I have left aside, can have some bite here, but I am not optimistic.

What then? I leave aside recapture technology, which would allow us to continue our love affair with gas and oil but reduce the carbon emissions anyway by extracting them before they are released into the air. Journalistic accounts report some research under way, mostly in Japan, but I do not know of any reason to be hopeful that such a technology will in fact be developed soon enough to keep atmospheric concentrations from becoming high enough to cause temperature rises at surface level. Recapture aside, we have three options: (1) go ahead and produce global warming, thereby avoiding any decision to make carbon emissions zero-sum; (2) reduce our economic activity and standard of living, thereby living within our fair share of a zero-sum emissions total; or (3) develop non-carbon-based energy sources, thereby making carbon emissions non-essential and making any zero-sum total set to resist global warming irrelevant to most people's lives, thereby defusing this issue of fairness. I believe the first option is dangerous and irresponsible toward future generations, although I have not tried to argue for that here but have simply assumed it, along with most of the scientists who study the phenomena and their likely social effects. I believe the second option is politically and psychologically impossible; I have not argued for this either, but I think that, however much I wish otherwise, an examination of the ideology of our time would support my political pessimism.

That leaves the third option as the only practical hope. If gas and oil were taxed heavily enough to incorporate the costs of their harmful effects, solar technology would immediately be competitive.[32] Other non-carbon-based energy technologies should be vigorously researched as well. Economic growth could continue, undisturbed by concern about carbon. Even so, this option means taking on the oil companies and the automobile companies, and millions of affluent consumers, who have given a new meaning to 'autocracy' (or could quiet, electric-powered vehicles somehow still give the sense of power now derived from roaring combusion engines?). Is this too politically and psychologically impossible? If so, we rich, and our children, and their children, appear to be headed for a destabilized climate; the children of the poor seem headed for 'temporary' disruptions in food supplies. If even narrow political change concentrated upon the energy regime is impossible, climate change is likely, hastened and exacerbated by extreme unfairness.[33]

NOTES

1. See J. T. Houghton, G. J. Jenkins, and J. J. Ephraums, eds, *Climate Change: The IPCC Scientific Assessment* (New York: Cambridge University Press for the Inter-governmental Panel on Climate Change (WMO/UNEP), 1990); the brief 'Policy-makers' Summary' is also available in a separate and more colourful form as *Scientific Assessment of Climate Change: The Policymakers' Summary of the Report of Working Group I to the Intergovernmental Panel on Climate Change* (Geneva: Intergovernmental Panel on Climate Change (WMO/UNEP), 1990). Also see J. T. Houghton, B. A. Callander, and S. K. Varney, eds, *Climate Change 1992: The Supplementary Report to the IPCC Scientific Assessment* (New York: Cambridge University Press for the Intergovernmental Panel on Climate Change, 1992).
2. Henry Shue, 'Subsistence emissions and luxury emissions', this volume, 48 ff. The heart of this analysis has also appeared as Henry Shue, 'Four Questions of Justice', in *Agricultural Dimensions of Global Climate Change*, ed. Harry Kaiser and Thomas Drennen (Delray Beach, FLA: St. Lucie Press, 1993), 214–28.
3. I have already attempted to show that the answer to question 4(A) has strong implications for the answer to question 1—see Henry Shue, 'After you: may action by the rich be contingent upon action by the poor?', this volume—and am beginning to consider question 4(B)—see 'Environmental change and the varieties of justice', this volume. In addition, I have argued that the answers to questions 1 and 2 are mutually dependent—see 'The unavoidability of justice', this volume.
4. John Rawls, *A Theory of Justice* (Cambridge: Belknap Press of Harvard University Press, 1971). Obviously too important to be taken up here is the question of the nature and the significance of the methodological differences displayed by John Rawls, *Political Liberalism*, The John Dewey Essays in Philosophy, 4 (New York: Columbia University Press, 1993). The methodology of Rawls's later work is the subject of Frank I. Michelman, 'On Regulating Practices with Theories Drawn from Them: A Case of Justice as Fairness', in *Theory and Practice*, NOMOS XXXVII, edited by Ian Shapiro and Judith Wagner DeCew (New York: NYU Press, 1995).
5. Rawls, *A Theory of Justice*, 196, 197.
6. Rawls, *A Theory of Justice*, 197. It is the reasoning that would be done in the constitutional convention that is the primary focus of Michelman, 'On Regulating Practices with Theories Drawn from Them', in *Theory and Practice*, NOMOS XXXVII, edited by Ian Shapiro and Judith Wagner DeCew (New York: NYU Press, 1995).
7. Rawls, *A Theory of Justice*, 199.
8. A perceptive early critique of the sharp split between two clusters of primary goods is Norman Daniels, 'Equal Liberty and Unequal Worth of Liberty', in *Reading Rawls: Critical Studies on Rawls' 'A Theory of Justice'*, ed. with a new Preface by Norman Daniels, Stanford Series in Philosophy (Stanford, CA: Stanford University Press, 1989), 253–81. For a later more comprehensive argument see Thomas W. Pogge, *Realizing Rawls* (Ithaca, NY: Cornell University Press, 1989), 109–60.

9. Rawls, *A Theory of Justice*, 246.

10. Rawls, *A Theory of Justice*, 244–5.

11. See Cynthia Rosenzweig and Martin L. Parry, 'Potential Impacts of Climate Change on World Food Supply: A Summary of a Recent International Study', in *Agricultural Dimensions of Global Climate Change*, ed. Harry Kaiser and Thomas E. Drennen (Delray Beach, FLA: St. Lucie Press, 1993), 87–116; and, for fuller detail, C. M. Rosenzweig et al., *Climate Change and World Food Supply* (Oxford: Environmental Change Unit, 1993). Temperature changes are expected to be least at the equator and greater as one moves toward the poles. It is often speculated that, therefore, people in the tropics will suffer least. Rosenzweig and Parry indicate that this is completely unfounded and that the impoverished residents of the tropics, who live closest to the margin, will probably suffer most. On the significance of the study, which was funded by the EPA, see 'Warming Will Hurt Poor Nations Most', *Science News* 142:8 (22 August 1992): 116. Also see M. Lal, ed., *Global Warming: Concern for Tomorrow* (New Delhi: Tata McGraw–Hill, 1993).

12. Or that a significant rise would not unleash a precipitous fall in temperatures— precedents for sharp reversals have now been documented. See Greenland Ice-core Project (GRIP) Members, 'Climate Instability during the Last Interglacial Period Recorded in the GRIP Ice Core', *Nature* 364 (15 July 1993): 203–7. These scientists warn that the stability of climate during the years of human civilization so far is extraordinary: 'Given the history of the last 150 kyr, the past 8 kyr has been strangely stable' (207). They also note that 'mode switches' from warming to cooling 'may be completed in as little as 1–2 decades and can become latched for anything between 70 yr and 5 kyr' (207). The science correspondent for the *New York Times* observed: 'The data are likely to bolster concern that future changes in climate might not be spread over many centuries, allowing farmers to adjust to altered growing conditions and coastal cities to deal with rising sea levels, for example'—see Walter Sullivan, 'Study of Greenland Ice Finds Rapid Change in Past Climate', *New York Times* (15 July 1993): A1 and B9.

13. Martin I. Hoffert, 'Climate Sensitivity, Climate Feedbacks and Policy Implications', in *Confronting Climate Change: Risks, Implications and Responses*, ed. Irving M. Mintzer (New York: Cambridge University Press for Stockholm Environment Institute, 1992), 42.

14. Obviously, I am speaking in general terms about very complicated matters. For example, in 1980 the United States ranked eleventh in the world in GNP per capita (at $11,360) but was second only to Canada in 1979 energy consumption per capita (at 308 MBtu), while Switzerland ranked third (behind only the United Arab Emirates and Kuwait, which just sit and rake in monopoly profits on oil) in GNP per capita (at $16,440) with energy consumption of only 128 MBtu per capita—Duane Chapman, *Energy Resources and Energy Corporations* (Ithaca, NY: Cornell University Press, 1983), 281, Table 14–1.

15. Alliance to Save Energy et al., *America's Energy Choices: Investing in a Strong Economy and a Clean Environment* (Cambridge, MA: Union of Concerned Scientists, 1991), 2.

16. Other things that might not be equal are the answers to the other questions, about the costs of prevention as such and the costs of mitigation as such. I, in fact, think that *all* the answers to *all* the questions line up in the same direction, but the arguments have to be made one at a time—see my other chapters on this subject cited above. I hope to bring all these arguments together under the working title, *Compound Injustice*.

17. The opposite, more efficient strategy is: make the lowest-cost reductions in emissions first, irrespective of whose emissions they are. The now-popular strategy of 'joint implementation' is claimed by some proponents to combine fairness and efficiency by making lower-cost reductions in emissions in poor countries (for efficiency) and having them paid for by rich countries (for fairness) with funds that would otherwise have produced smaller, more expensive reductions in their own emissions. This strategy is in danger, however, of freezing international inequalities in standards of living into place. I believe it is essential to take into account the human function of the activities that produce the emissions—see 'Subsistence emissions and luxury emissions', this volume; and Onno Kuik, Paul Peters, and Nico Schrijver, eds., *Joint Implementation to Curb Climate Change* (Boston: Kluwer Academic Publishers, 1994).

18. Thomas E. Drennen, 'Economic Development and Climate Change: Analyzing the International Response' (PhD dissertation, Cornell University, 1993), 142 (Table 6.5).

19. Drennen, 'Economic Development and Climate Change', 8.

20. Anil Agarwal and Sunita Narain, *Global Warming in an Unequal World: A Case of Environmental Colonialism* (New Delhi: Centre for Science and Environment, 1991), 18. This publication often appears in bibliographies as an example of 'egalitarianism', but neither its empirical calculations nor its ethical theses are often examined. Also see Gilberto C. Gallopin et al., 'Global Impoverishment, Sustainable Development and the Environment', *International Social Science Journal* 121 (1989): 375–97; Alvaro Soto, 'The Global Environment: A Southern Perspective', *International Journal* 47 (Autumn 1992): 679–705; and J. K. Parikh, 'IPCC Strategies Unfair to the South', *Nature* 360 (December 1992): 507–8.

21. Drennen, *Economic Development and Climate Change*, ch. 4, 'The Role of Biological Emissions'; an earlier version appeared as Thomas E. Drennen and Duane Chapman, 'Negotiating a Response to Climate Change: The Role of Biological Emissions', *Contemporary Policy Issues* 10:3 (July 1992): 49–58.

22. I have argued for a slightly different position from Drennen's in 'Subsistence emissions and luxury emissions', this volume.

23. Michael Walzer has argued eloquently for the diversity of goods and the diversity of distributive arrangements in *Spheres of Justice: A Defense of Pluralism and Equality* (New York: Basic Books, 1983). I believe that the diversity across goods is clearer than the diversity across societies for the same good.

24. It is noteworthy that Rawls now explicitly acknowledges fundamental economic rights—see John Rawls, 'The Law of Peoples', in *On Human Rights*, ed. Stephen Shute and Susan Hurley (New York: Basic Books, 1993), 62.

25. J. T. Houghton et al., eds, *Climate Change: The IPCC Scientific Assessment*, xviii, table 2.

26. There is the qualification that the atmospheric residence time of every gas is finite, so that gradual natural diminutions occur. However, the atmospheric residence time of CO_2 is on the order of a century, so the real value of the national debt may well be reduced by inflation faster than the accumulation of CO_2 will be reduced by natural attrition.

27. Even conservative business groups now realize that the neglect of research and development on alternative energy has been excessive: 'CED believes that federal support for renewable energy R&D should be given a higher priority than it was during the 1980s'—Committee for Economic Development, Research and Policy Committee, *What Price Clean Air? A Market Approach to Energy and Environmental Policy* (New York: Committee for Economic Development, 1993), 80–1.

28. See Wilfred Beckerman, 'Global Warming and International Action: An Economic Perspective', in *The International Politics of the Environment*, ed. Andrew Hurrell and Benedict Kingsbury (Oxford: Oxford University Press, 1992), ch.10.

29. On the food supplies, see Rosenzweig and Parry, 'Potential Impacts of Climate Change on World Food Supply', in Kaiser and Drennen.

30. If, contrary to what I have just argued, this were a relatively unconstrained social choice, my argument that the powerless are at the mercy of discretionary choices by the powerful would be doubly strong.

31. A sophisticated explanation along these lines is currently being worked out by Chris Reus-Smit; see his paper, 'The Normative Structure of International Society and the Justice of International Environmental Accords', prepared for a conference on 'Global Environmental Change and Social Justice', Peace Studies Program, Cornell University, photocopy.

32. The most positive recent development is a 'conceptual breakthrough' in solar technology made at the University of New South Wales—Matthew L. Wald, 'New Design Could Make Solar Cells Competitive', *New York Times*, 14 June 1994, C9.

33. A highly accessible account of the scientific issues, written for laypersons, has been published by the chairman of the Science Working group of the IPCC. See John Houghton, *Global Warming: The Complete Briefing* (Elgin, Ill.: Lion Publishing, 1994).

5

Equity in an international agreement on climate change*

EQUITY UNDER A GLOBAL EMISSIONS CEILING

Fundamental fairness

What diplomats and lawyers tend to call 'equity' incorporates important aspects of what ordinary people everywhere call 'fairness'. The concept of fairness is neither Eastern nor Western, neither Northern nor Southern, but completely universal. People everywhere understand what it means to ask whether an arrangement is fair or is instead biased toward some parties over other parties. If you own the land but I supply the labour, or you own the seed but we own the ox, or you are old but we are young, or you are female but I am male, or you have an education and I do not, or you worked long and hard but I was lazy—in situation after situation it makes perfectly good sense to ask whether a particular division of something among two or more parties is fair to all the parties, in light of this or that difference between the parties. All people understand the question, even where they have been taught not to ask it. Should the educated be paid more than the uneducated (and if so, how much)? Should hard workers be paid more than the lazy (and if so, how much)? What about the educated but lazy *versus* the uneducated but hard working? What would be fair? Or, as the lawyers and diplomats would put it, which arrangement would be equitable?

Ultimate scarcity

Adoption of a global ceiling on greenhouse gas (GHG) emissions will radically transform the international context for fairness. Three levels need to be noted:

* From icipe Science Press, *Equity and Social Considerations Related to Climate Change*: Papers presented at the IPCC Working Group III Workshop on Equity and Social Considerations related to Climate Change, Nairobi, Kenya, 18–22 July 1994. World Meteorological Organisation/United Nations Environment Programme. 1st Edition. Copyright © 1995 by Secretariat of Nairobi Workshop/Kenya Meteorological Department. Reprinted with permission from icipe Science Press, Nairobi.

science, policy, and equity, or fairness. The scientific investigations reported by the IPCC provide a solid basis for the conclusion that there is in fact a *natural limit* on the GHG emissions that this planet can process without rises in surface temperature (and other complex and mostly unwelcome climate change). If the science is correct, the natural limit exists whether we choose to acknowledge it or not. If, at the level of policy, we acknowledge the natural limit and attempt to protect ourselves against the unwanted effects of exceeding the limit, we will act under the Framework Convention on Climate Change to impose on GHG emissions a *global ceiling*. While the natural limit on the earth's capacity to handle anthropogenic additions in excess of natural emissions is a given, the global ceiling would be a political limit that we imposed upon ourselves. If we wish to protect ourselves against the untoward (for us) natural effects of limitless increases in anthropogenic GHG emissions, we may have no good alternative to the adoption of a global ceiling. Nevertheless, the adoption of a global ceiling is a political choice of policy, and we are responsible for the effects of the policies we adopt, most of all the effects on those powerless to influence our choices.

A political decision to adopt a global ceiling on GHG emissions has implications for equity that are far more radical than has so far been recognized. A serious decision to deal with the natural limit on the planet's capacity to dispose of GHG emissions by imposing a political limit on the emissions produced by humans totally transforms the international situation. The reason is simple: imposition of an emissions ceiling makes emissions, as the economists like to say, *zero-sum*. For equity this change has powerful implications. First, we should review the meaning of 'zero-sum', which is widely understood; then we can see the significance of making a quantity zero-sum, which is so far less widely understood.

A total is zero-sum if more of it for anyone means less of it for everyone else. A total that is zero-sum is a total that cannot be enlarged. If there is +3 for you, there must be −3 for the rest of us; the sum is always zero: +3 and −3 are zero. Consequently, each time that some of whatever is in question is used by someone, less of it is left for all the others. Establishing a global ceiling on GHG emissions makes GHG emissions zero-sum. Since the total cannot be increased, because a ceiling has been placed on the total, each unit of emissions used up by one party will be a unit not available for use by all the other parties. This much, the definition of 'zero-sum', is obvious and undeniable.

What is the significance for equity of a policy that makes emissions zero-sum? Once a total is zero-sum, anyone who consumes more than her own share harms others whose share she thereby takes away. Overconsumption is encroachment.[1] In the consumption of anything the total of which is to be

---

[1] I previously argued that in this new context '*excess encroaches*'; see 'After you: may action by the rich be contingent upon action by the poor?', this volume, 84–6.

kept zero-sum, there is no such thing as 'a little harmless overindulgence'. Once one has used up one's own share, continuing to consume is always using up someone else's share, provided only that the case is one in which there are fair shares.[2] One is doing harm to the person whose share one is consuming, because one is depriving the other person of the only share she could have had (without her in turn depriving some third person). Overconsumption—that is, use of more than one's share—always wrongs someone once a total has become zero-sum. Overconsumption is no longer, if ever it was, a victimless crime. Those for whom the supply of something important is zero-sum face the ultimate scarcity: one cannot exceed one's share without doing wrong. In a zero-sum context, other people are always affected by consumption in excess of one's own share. There is no slack, in fact—or in equity.

It is also undeniable, I think, that creating the ultimate scarcity of a zero-sum total has the significance for equity just indicated. However, this fundamental implication for equity may not be obvious and so it may be worth illustrating it in a different case. Suppose I very much enjoy dining and I can afford to buy as much food as I like. Yet there are many malnourished, hungry people in the world. Whatever money I spend on food for myself that I enjoy but do not need to eat is money that I could instead have spent providing desperately needed food to some number of persons who do not have enough. Now, people disagree about the ethics of a case like this. Some people believe that once you have enough for yourself, you ought then to assist others in obtaining what they need, before you go on to enjoy luxuries, even if their need is in no degree your fault. These people believe that it is selfish to indulge yourself when others lack necessities. Other people, by contrast, think that as long as you are minding your own business and not doing others any harm, you are not morally required to provide any positive assistance to others that you do not choose to provide (although it is certainly admirable to choose voluntarily to help in ways not required).

The ethical debate just sketched is about the extent to which any one human being is required to help other human beings in need. The answers normally defended range across a wide spectrum between the one extreme, no help is required (although it is praiseworthy to volunteer it anyway), to the other extreme, as much help as possible is required once one's own needs are satisfied. It would be irrelevant to the subject at hand to pursue this issue, although it is widely discussed and important in its own right. I have mentioned this other debate here strictly in order to note one of the crucial features of the circumstances it assumes, namely it is entirely about whether to offer help to someone for whose plight one bears no causal responsibility. The question there is: how much responsibility does one have to help someone

---

[2] I explain below why questions of fairness arise in the case of GHG emissions, although I think the reasons are actually quite obvious.

whom one has not harmed? Ought I to help a complete stranger to whom I have done no harm?

It is a very different matter if I have in fact wronged the person whose plight is under consideration—if that person's plight was caused by harm that I did. The question of whether I ought now to help someone whose need for this help results from harm that I myself inflicted is radically different from the question of whether I ought to help a stranger whom I have never harmed. And the reason that the situation is so different when harm has been done is that one of the most basic principles of equity in every culture about which I know—and I hope any significant counter-examples will be brought to my attention—is: Do no harm. One may or may not be expected to help in this or that context, but one is always expected not to harm (but for exceptional overriding circumstances). Consequently, the obligation to restore those whom one has harmed is acknowledged even by those who reject any general obligation to help strangers. Whatever one's obligation to help people with whom one has no previous connection, one virtually always ought to 'make whole', insofar as possible, anyone whom one has harmed. And this is because one ought even more fundamentally to do no harm in the first place.

Now it should be clearer why the adoption of a global ceiling on GHG emissions so radically transforms the international context for equity. In the absence of a ceiling on total emissions, some party might out of goodwill volunteer to refrain from excessive emissions as a gesture of helpfulness to other parties.[3] Or it might not. With the ceiling in place, a party that does not refrain from excessive emissions—meaning emissions in excess of its own fair share of the total—is, far from simply not volunteering to help, inflicting harm. Using up emissions that in equity belong to another party constitutes the doing of harm to that other party. People have different views about when, if ever, one is obligated to help strangers, but I have not yet encountered anyone who was prepared seriously to defend the view that one is free to inflict harm on strangers.

After the adoption of a global ceiling, volunteering to help will consist only in volunteering to consume less than one's equitable share. Consuming no more than one's share will have become subject to what may be the most fundamental principle of equity of all: Do no harm. The adoption of a ceiling on total emissions moves the consumption of more than one's share of allowable emissions into a new category of equity, the category of rock-bottom prohibited wrong.

---

[3] I am assuming, merely for the sake of argument, that there are no other independent grounds for thinking that emissions already ought to be restricted. That is far from clear, but I simply want to concentrate on one line of argument at a time.

Fair shares

If a party's exceeding its fair share will be a fundamental wrong, it becomes exceedingly important how fair shares are determined. But isn't the question of the appropriate standards for allocating fair shares one of those eternal enigmas, endlessly debated but never settled? No, it is not. I next want to suggest that the problem of specifying fair shares is actually much easier than it may seem. First, we should quickly remind ourselves of one purely contingent but for now unavoidable feature of the world economy.

It happens that at present a considerable quantity of GHG emissions in the form of CO_2 is a necessity for any decent human life. This does not need to be the case and presumably will not always be the case. CO_2 emissions are as vital as they are now only because the particular form of industrialization created by the now-industrial societies rests upon a fossil fuel energy regime. As long as industrial activity is dependent upon fossil fuel, which is of course carbon-based, the industrial activity necessary for a modestly adequate standard of living will generate large quantities of CO_2, a GHG with an especially long atmospheric residence time. Of the people in the world, more than 99.9999 per cent have had absolutely no voice in the adoption or the dogged retention of the energy regime of fossil fuel into which they are simply born. Whenever we can advance beyond dependency upon fossil fuel, anthropogenic CO_2 emissions may become of little consequence and even less interest. Meanwhile, the current global energy regime makes CO_2 emissions central and vital.

Given a global ceiling on total GHG emissions, how do we calculate national shares? Equity in shares begins at the bottom—with those whose actual shares are smallest. (Those who already have large shares do not normally need to appeal to others to treat them fairly.) The first concern of equity is with the minimum share. Every national share of the global total must be at least enough for every person in the nation to lead at least a minimally decent life.[4] Why?

Obviously a philosophically adequate theoretical explanation would require considerably more scope than a single chapter and would venture beyond present concerns. Brief remarks can, however, be made at two levels of abstraction. At the more abstract level it is evident that the notion of human equality is meaningless unless it entails that there is something—and of course not necessarily only one thing—to which all human beings are equally entitled. If an assertion of human equality did not mean that something must be kept

---

[4] I want to emphasize that I am now discussing only the *minimum*. I am not implying that no one is entitled to anything but the minimum in saying that everyone is indeed entitled to the minimum. I have discussed this at greater length in *Basic Rights* (Princeton University Press, Princeton, NJ, 1980).

equal among human beings, it is not apparent what it could mean. It would presumably be empty, misleading rhetoric.

At a more concrete level, then, the question becomes: which thing(s) must be guaranteed equally to all humans if we believe in human equality? Obviously many values central to human life are possible candidates for guaranteed equal distribution: opportunities (of various kinds), legal due process, wealth, health, liberties (of various kinds), cultural autonomy, economic development, healthy environment, physical safety, and much else. On the other hand, in many of these cases there are at least plausible reasons why it is not the best arrangement to have political institutions that centrally maintain an equal distribution of the value in question, however important it may be in itself. If one is interested not in beautiful phrases but in serious policy, one must look in the case of each of these values in detail at the advantages and disadvantages of specific institutions that could guarantee an equal distribution of the value in question. My own view, as I am sure is already obvious, is that some of the items listed should be guaranteed an equal distribution and some should not. Which so far tells you nothing, or almost nothing.

Once it is acknowledged, as I have just done, that even some important human values ought not to, or need not, be guaranteed an equal distribution, it becomes especially important to be careful about which views are, and which are not, categorized as 'egalitarian'. Most politically thoughtful people already have strong feelings for or against 'egalitarianism', so that they react strongly—one way or another—to any view so labelled. Naturally, no one—and certainly not I—owns a copyright on the term 'egalitarian' and people are free to define it as they think best, within reason. But because the term is so emotionally charged, and because politically important differences are in danger of being overlooked, I am simply urging caution. I would suggest reserving *egalitarian* for views specifying that equality ought to be enforced at levels well above minimum levels. The view that everyone is entitled to the same basic minimum, by contrast, is naturally referred to as an endorsement of a *guaranteed minimum*.

For example, the view that everyone has a right to literacy is embracing what I would call a guaranteed minimum of education. The view that everyone has a right to a university education is egalitarian. Similarly, endorsing free preventive health services (such as immunizations against communicable diseases) for everyone is endorsing a guaranteed minimum, but advocating free surgery for everyone who could benefit from it is advocating egalitarianism. There is a sense in which literacy and a baccalaureate are different points on the same spectrum—education—and differ only in degree: one is less, and the other is more, education. Yet they are points that are so far apart that in most contexts it would be wildly misleading not to distinguish them. Similarly for vaccinations and operations. Accordingly, while it may not strictly speaking be an error to call an advocate of universal rights to literacy and immunizations an 'egalitarian', that practice simply leaves one with no good name for the advocate of open universities and full national health services, which are what I would call full-bloodedly egalitarian measures.

Now, as I am sure has already become clear, I do not think that it is accurate or helpful to call those advancing the view that every person is entitled to a minimum level of GHG emissions 'egalitarians', irrespective of whether calling them that would make them more or less appealing to any particular other person. At this time, a human being is born into a world in which, because of the kind of industrial system that—for better or for worse—we in fact have, it is impossible to lead a decent life without benefiting from a minimum level of CO_2 emissions. CO_2 emissions are at present a vital necessity. As I have already mentioned, this could change. If we moved from a fossil fuel regime to a solar regime, CO_2 emissions would become inessential. Meanwhile, they remain essential.

CO_2 emissions are, then, a necessity and, moreover, for everyone except those who actually control energy policies, an imposed necessity. No one, except the few who purposefully choose to cling to the fossil fuel regime against less primitive technologies, chooses to lead a life that requires CO_2 emissions rather than a life that does not; all the ways of life available to the vast majority of humanity within the current energy regime require CO_2 emissions. This is the technological world every person inherits at birth. The dominance of this particular energy technology is one good reason why everyone ought to be guaranteed at least the minimum level of CO_2 emissions that the technology makes necessary to a decent life.

In sum, the argument actually has three steps. First, a commitment to human equality, which is not to be empty talk, requires that some aspects of life be equal for all persons. The question becomes: which specifically? Second, the most plausible candidates in general to be the subjects of guaranteed equalities are necessities of life in circumstances in which people are unable to secure them for themselves. Necessities that are often impossible for people to provide for themselves include food (when, for example, there is disruption of agriculture by natural disaster or warfare), literacy, and vaccines. Many of us would argue that all such basic material necessities must be guaranteed equally to all who are powerless to obtain them for themselves—one simply displays contempt for the life of the helpless who are excluded from such social arrangements.[5] Third, the position is stronger still in the case of what can be called avoidable necessities: what are in fact at this time necessary but would not be necessary if we chose to pursue different technology.[6] CO_2 emissions are a necessity of life only until we advance beyond fossil fuel energy

---

[5] See Shue, *Basic Rights*, ch. 5.
[6] See 'Avoidable necessity: global warming. international fairness, and alternative energy', in this volume.

technology. The recent Australian 'conceptual breakthrough' in solar technology makes photovoltaic, for example, an even more promising alternative than it already was.[7] Our technological choice of energy source has powerful implications for equity. If we make it impossible for people to lead a decent life without CO_2 emissions, we must in equity not actively prevent them from producing those minimum emissions.

Double imposition

And that is precisely what we would do if, given our real current technological circumstances, we imposed a global ceiling on CO_2 emissions without protecting at least a minimum level of CO_2 emissions for every human being. A global ceiling on emissions at a stage in economic history when we are dependent upon fossil fuel technology, unaccompanied by a firm guarantee for every person of a minimum level of emissions, would constitute a double imposition. On the one hand, we would impose a continuing fossil fuel regime, when we could be moving to alternative regimes, thereby imposing the necessity of CO_2 emissions, and, on the other hand, we would impose a limit on emissions that did not allow everyone to have this necessity, if the shares of the limited total were not equitable. Here, as everywhere, equitable means: protection for an adequate minimum. An equitable emissions regime must, if it includes a limit upon the global total, guarantee a universal right to a minimum share of emissions. Otherwise, the limited total will be completely used up by a relative few. Minimum shares need protection against encroachment by overconsumers.

If ever there were, for example, the kind of system of tradeable emissions permits that economists like so much, the trading system would have to rest upon a foundation of inalienable—non-tradeable—permits securely guaranteeing a minimum level of emissions to every human being.[8] To fail to arrange for rights to minimum inalienable emissions would permit the infliction of serious harm—as explained above in 'Ultimate scarcity'—upon the people whose equitable share would as a result of that lack of protection be taken away from them through being used by others. In fact, whatever the particular scheme aimed at promoting efficiency—tradeable permits or other arrangements—the one feature demanded by equity in any international agreement imposing a global ceiling is protection for minimum shares.

---

[7]   See Matthew L. Wald, 'New Design Could Make Solar Cells Competitive', *New York Times* 14 June 1994, C9. The story reports the work of Prof. Martin A. Green at the Center for Photovoltaic Devices and Systems, University of New South Wales.

[8]   I have explained the necessity of underlying inalienable rights to minimum emissions more fully in 'Subsistence emissions and luxury emissions', this volume.

AN ALTERNATIVE TO INTERNATIONAL
AGREEMENT?

There may be no completely good solution to global warming, which is a serious and unprecedented problem. Even in the cases of simpler and less important problems, one often must choose the least bad option from among a bad lot because there is no good option. This means that options must almost always be assessed comparatively. Having seen a little of what equity can tell us about what international agreements ought to include, we might do well next to see what consideration of equity can tell us about the leading alternative, which seems to be something called 'joint implementation'.

Two preliminary clarifications. First, I say 'something called "joint implementation"' because it is my impression that exceedingly, and importantly, different schemes are being called, as I will henceforth for brevity style it, JI. I shall have much more to say about this presently. Second, what I am treating as two options, international agreement and JI, could be combined as one: an international agreement to allow and encourage specific forms of JI. Such agreed-upon JI would be a type of regulated, or at least guided, JI. When I refer to JI as an alternative to international agreement, I mean unregulated types of JI pursued ad hoc outside any guiding framework specified by international agreement. Indeed, my main point about JI will be that various activities that might by their proponents be called 'JI' differ so much from each other that it is essential that forms that deal effectively and equitably with global warming be encouraged and others be discouraged.

Conflicting conceptions

Since one of the primary issues about JI is how to specify it and which undertakings to treat as examples of it, it may clarify matters to begin with three ideal types; 'ideal', not in the sense of supremely good, but in the sense of supremely simple. To maintain simplicity I shall in each of the three instances assume arbitrarily that electricity generation is the enterprise in question, although obviously many different kinds of activities might fit the three patterns I am about to sketch. The one essential element of all JI—the element that is the basis for calling what is done 'joint'—is that an activity is performed in nation #1 by a firm that in some sense, which needs to be further specified, 'belongs' to nation #2. Extremely different purposes can, however, be pursued.

Type 1: *Less-cost reduction*. Prior to the JI, an electricity generating facility in nation #1 was producing M megawatts of electricity by emitting E tons of carbon. The JI consists of a firm from nation #2 building a new electricity generating facility in nation #1 that produces N megawatts of electricity

($N > M$) by emitting D tons of carbon ($D < E$). The firm from nation #2 has enabled nation #1 to produce *more* electricity than before with *fewer* emissions. The firm has obviously delivered a superior technology. Whose property the technology is—whether it is sold, given, licensed, or what, and on what terms—is a supremely important issue, and it too is an issue of equity. However, issues about equity in technology transfer are relatively familiar and thoroughly aired, and JI does not raise them in any unique way, so I will not devote space to those equitable issues here.

JI_1, as I will call this, has two critical features. First, it *reduces* GHG emissions and thereby contributes positively to the mitigation of global warming. Second, it contributes positively to the standard of living in nation #1. Strictly speaking, it only directly increases the amount of electricity available; and a nation can certainly waste electricity in such a way that the standard of living of its people is not raised, or its population growth can cancel out any potential per capita increases. However, these also are familiar and separable problems that I will not pursue here. My ideal types are designed to highlight other issues more integral to global warming.

The third feature that JI_1, has, by assumption, is that the firm from nation #2 can produce the reduction in emissions in nation #1 at less cost than any firm could produce a reduction of the same magnitude in nation #2. This must be added by assumption because the simple fact that a profit-seeking firm chose to conduct its activity in nation #1 rather than in nation #2 could have many explanations other than the costs being lower in nation #1—perhaps some source (inside or outside nation #1) was willing to pay to have the work done in nation #1 and no one was willing to pay to have it done in nation #2 even if nation #2 was not already generally using the technology in question. One could of course vary the case by assuming not merely less cost but globally least cost; again, there is obviously even less reason to assume this would happen automatically in an unregulated market.

Type 2: *Source-sink equivalence.* Something quite different is also referred to as 'joint implementation'. Here, prior to the JI, an electricity generating facility in nation #2 was producing R megawatts of electricity by emitting J tons of carbon. The JI consists of a firm from nation #2 engaging in two activities: (a) building a new electricity generating facility within nation #2 that produces S megawatts of electricity ($S > R$) by emitting K tons of carbon ($K > J$) and (b) introducing new carbon sinks into nation #1 that absorb ($K - J$) tons of carbon. The firm in nation #2 (a) has enabled nation #2 to produce *more* electricity than before but with *greater* emissions, while it also (b) has produced additional sinks (in nation #1) sufficient to eliminate any net increase in emissions. Altogether, the firm in nation #2 has enabled nation #2 to produce *more* electricity than before with the same net emissions, which is certainly an increase in one kind of efficiency.

JI_2, as I will call this, has two critical features, as did JI_1, but the features are very different. First, it leaves GHG emissions *unchanged* and thereby contributes nothing positive to the mitigation of global warming. Second, it contributes positively to the standard of living in nation #2. Theoretically, it could contribute to the standard of living in nation #1 as well; if the carbon sink consisted of new forests and the most productive use of land in nation #1 happened to be for additional forest, nation #1 might benefit as well. This leaves, however, many unanswered questions, such as how the same land can both provide the maximum sustainable yield of lumber (to benefit nation #1) and provide sufficient carbon sinks (to enable nation #2 to avoid making GHG emissions worse by its increased production of electricity). One cannot help wondering how often everyone on both sides will be so fortunate as to find that the best carbon sinks from the perspective of nation #2 will just happen to be the most productive use of land from the perspective of nation #1. Obviously, if the firm paid nation #1 enough for the right to use the land as a sink, that rent could make this the most productive use of that land. Equally obviously, few poor nations have the leverage to strike such a fair deal. In any case, I will not stipulate whether the standard of living in nation #1 does or does not rise. The second feature of JI_2 is simply that the standard of living in nation #2 does rise.

Type 3: *Less emission increase.* Something else quite different yet is also being anointed as 'joint implementation'. Now, prior to the JI, an electricity generating facility in nation #1 was producing M megawatts of electricity by emitting E tons of carbon (exactly as in JI_1). But in this instance the JI consists of a firm from nation #2 building a new electricity generating facility in nation #1 that produces Q megawatts of electricity $(Q > M)$ by emitting F tons of carbon $(F > E)$. The firm from nation #2 has enabled nation #1 to produce *more* electricity than before but with *greater* emissions. It is actually indeterminate from what I have said so far whether the new technology is more efficient—qualitatively better—or simply bigger in capacity. For the sake of the argument, however, assume that it is more efficient: Q/F is a better ratio than M/E.

Call this JI_3. Its two critical features are the following. First, it *increases* GHG emissions and thereby interferes with the mitigation of global warming. Second, it contributes positively to the standard of living in nation #1.

The critical features of these three ideal types—obviously there are others—can be summarized as follows:

Three types of joint implementation

	Emissions effects	Most likely rise in standard of living
JI_1	reduction	poor nation
JI_2	unchanged	rich nation
JI_3	increase	poor nation

If nothing else, I hope that it is now evident that very different enterprises are being denominated 'joint implementation' and that it is vital for reasonable discussion to determine at least the very small amount of information that I have built into my three ideal types before one tries to say whether something with this now-trendy name is good or bad.

Dubious equity

More important, however, are the profound questions about equity that are raised by merely these three types of so-called JI—and above all, by their interactions if they are not coordinated under a framework agreement. A short chapter does not provide scope for an adequate exploration of the full complexity of any of the implicated dimensions of equity. If serious analysis of the equity of 'joint implementation' does not begin immediately, however, the headlong rush of practice will far outstrip careful reflection, not to mention intelligent control for the sake of the global environment.

Consider JI_3, to take only one example. Bearing in mind that the fundamental purpose of this whole exercise is to establish (at least) a global ceiling, one might initially think that JI_3, because it increases GHG emissions, ought to be discouraged in favour of JI_1, which reduces GHG emissions. Yet it is fairly obvious that whether an instance of JI_3 ought to be discouraged or not (a) cannot be determined in the abstract and (b) turns on equity. Whether an instance of JI_3 in nation X is on the whole good or bad depends upon whether ex ante the people of nation X have their fair share of the global total of emissions. If they do not have their fair share, instances of JI_3 may be exactly what they need. If they already have more than their share, JI_3 may simply contribute toward worsening global warming while making the world more inequitable.

In general, whether a specific nation should or should not increase its GHG emissions depends entirely upon whether its current share of GHG emissions is greater or less than its fair share. This simple finding, which I believe is difficult to controvert, has three powerful implications about JI as an alternative to international agreement, that is, about freelance, unguided JI:

1. We need an agreed standard of equity before 'joint implementation' is undertaken. Only *after* a standard of equity has been specified will it be possible to know whether a proposed instance of JI that would produce a net increase in emissions—such as JI_3—is acceptable or unacceptable. Otherwise, the cause of preventing global warming will be damaged for no good reason by some JI, namely any JI that increases the share of a nation whose share is already too great, given the global ceiling and the GHG emission needs (as long as we continue with a fossil fuel regime) of

everyone else. It is irresponsible to tolerate everything that anyone wants to call 'joint implementation'.

2. We need effective mechanisms to encourage helpful JI and to discourage harmful JI, as assessed by the standard of equity.

3. We need effective mechanisms to produce compensating reductions in GHG emissions for all justified increases in GHG emissions, that is, for all increases in GHG emissions in nations that now have less than their fair share of the global total. On the one hand, for the sake of equity it is important that JI go forward in nations whose emissions are now below their reasonable share of the global total, at least where the increases will improve the lives of their own people.[9] On the other hand, for the sake of establishing a global ceiling it is important that net emissions not increase. Net increase can be avoided only if any increases in nations that are entitled in equity to greater emissions are cancelled out by reductions elsewhere, namely nations where current emissions exceed a fair share of the global total. To be responsible, any plans for JI that will increase emissions somewhere must also include a specification of where and how emissions will be decreased by the same amount.

In sum, equity requires that 'joint implementation' not be accepted as a substitute for an international agreement that includes the three features listed. 'Joint implementation' can be acceptable only within the framework of an international body to implement it. Uncontrolled JI—JI outside an agreed framework—is liable to deal with the difficult choices we face by taking the ever-popular route of avoiding painful redistribution by enlarging the pie. With a larger pie, those who now have less than enough can have more without taking anything away from those with more than their share. Unfortunately, our fundamental goal of a global ceiling on GHG emissions means that there can be no enlargement of the emissions pie. And that means that painful redistribution is the only conceivable route to greater equity, as long as we cling to the fossil fuel regime that makes large GHG emissions essential to economic health. Decentralized and uncoordinated instances of 'joint implementation' are incapable of solving the problem.

A wild wave of market fundamentalism is surging around the globe. 'The market' is being fervently preached as the solution to every problem. Market mechanisms do some jobs extremely well, and it is folly not to use them where they are best means to the end. Equitably preventing global warming is a

---

[9] Emissions increases on the territory of a nation obviously need not improve the welfare of the citizens of that nation. There is great danger that what is described as a firm from nation #2 building a new facility for nation #1 is simply nation #2 producing additional emissions for its own purposes from a platform in another nation's territory. These accounting problems arise equally for all schemes that involve national assignments of emissions, including schemes that ignore equity.

special—indeed, an extraordinary—case, and this mission cannot be accomplished without prior political choices to structure the incentives that market actors encounter. If we simply unleash firms to engage in whatever they choose to call 'joint implementation', the result will be neither a global ceiling on total emissions nor an equitable distribution of the shares of whatever total there is. These are not goals to which one can get by purely market means. Incentives to maintain a global ceiling and incentives to distribute it equitably cannot arise spontaneously—they need to be put in place by international agreement. Otherwise, we will fail to prevent global warming, much less to accomplish it with equity.

CONCLUSION

The fundamental focus of equity is the protection of an adequate minimum for those unable to protect their own. This chapter briefly touches upon equity under two options: international agreement upon an emissions ceiling and unregulated 'joint implementation'. The adoption of a global ceiling creates an urgent need for an adequate minimum share to be guaranteed to those who cannot take it for themselves. The acceptability of projects of 'joint implementation', of which three highly divergent types are sketched, depends upon whether they increase the emissions of those who now have less than their fair share or those who already have more, as judged by equity. Consequently, 'joint implementation' should not proceed without agreed background standards of equity.

6

Environmental change and the varieties of justice*

PREFERENCE AND FAIRNESS

The various policy communities, including those focused on the environment, tend to accept existing human preferences uncritically when making social choices. Policy analysts seem to do everything but challenge these preferences. They adopt them as if they were a baby left on their doorstep.[1] If, for instance, many people in metropolitan areas happen to prefer travelling to work in private automobiles rather than in commuter trains, policy analysts will judge the health of the economy by the ease with which these people obtain and operate the cars they want. Yet there is no good reason to treat the preferences held at any one time, including the present time, as if they were somehow foundational and immune to assessment when we are setting long-term public policy. Indeed, total passivity toward all current human preferences is hopelessly misguided, especially in the face of environmental change caused by global warming. We must decide which preferences to modify in this particular case, however difficult it may be to bring about such change either in ourselves or in others.[2]

---

* Reprinted, Henry Shue, 'Environmental Change and the Varieties of Justice', in *Earthly Goods: Environmental Change and Social Justice*, edited by Fen Osler Hampson and Judith Reppy pp. 9–29. Copyright 1996, Ithaca, N.Y.: Cornell University Press, used by permission of the publisher Cornell University Press.

[1] This point has been powerfully established over the years by Mark Sagoff. See his *Economy of the Earth: Philosophy, Law, and the Environment* (Cambridge: Cambridge University Press, 1988), especially ch. 5, 'Values and Preferences'.

[2] It is possible that Americans are more ready to consider change than my worst fears suggest. 'People are struggling with deep ambivalence about their own values . . . Watch television for a day and you will get a clear picture of what Americans supposedly want in life: new cars, a big house, stylish clothes, the latest gadgets—and of course, fresh breath. Yet when Americans describe what they are looking for in life, their aspirations rarely center on material goods.' *Yearning for Balance: Views of Americans on Consumption, Materialism, and the Environment* (Bethesda, MD: Harwood Group, 1995), p. 14. Four focus groups are reported on.

Most preferences entail a claim upon resources.[3] The preference for oper-ating private automobiles, for example, entails claims upon, among many other things, the energy and the raw materials needed to make the necessary steel, the gasoline needed to operate the combustion engine, the hospitals needed to treat the injuries from the accidents, and the capacity of the planet somehow to deal with the tons of carbon dioxide emitted annually by each car, along with all the other troublesome—and toxic—emissions. Because the satis-faction of a preference ordinarily demands the consumption of resources (and in this case, the creation of emissions), to calculate from existing preferences is in effect to postulate that all (and only) current preferences are created equal. One is tacitly assuming without explanation that no matter how divergent the respective 'calls' of two different preferences upon a society's or the planet's resources, the preferences are, at least prima facie, equally reasonable and equally urgent.

This is manifestly implausible. One person's desire for an additional jar of caviar is not equal in urgency to another person's need for an additional bowl of black beans. One state's intention to spend three hundred billion on military preparations is not equal to another's need to spend three billion on infrastruc-tural improvements. Yet if we fail to question the relative value of such competing preferences, then we ask only, Does the party in question in fact command enough credit to borrow the necessary sum? We fail to ask, Is this a preference that deserves to be honoured, or even tolerated, by the rest of us (including, presumably, future generations who will in due course have preferences of their own)? We fail to ask, Will the satisfaction of this preference consume limited resources that should be used differently—or protected from use, either now or forever?

Whenever I, my state, or my group wants something, this usually means that we propose to consume whatever social wealth, natural resources, polit-ical attention or indulgence, and individual time and energy are needed to produce it. No matter how productive the project, it will still require resources that could have been used many other ways, or preserved in their current state and not consumed for human purposes at all. Everyone understands that investment and consumption have opportunity costs, but there is more at stake here than the obvious costs. Why should just *these* preferences control *these* resources? Whose preferences control policy? Who pays the costs? Who receives the benefits? Whose costs and benefits 'count', and whose do not? *Every* calculation presupposes such a division between those internal to the group and those external to it.[4] Those whose preferences count, according to

---

[3] This is lucidly noted by Will Kymlicka, *Liberalism, Community, and Culture* (New York: Oxford University Press, 1989), pp. 37–8. Obviously the limiting case is a preference for a service that can be performed without consuming any resource or expending any marketed energy.

[4] This is true whether the calculation concerns some kind of subjective preferences or some kind of objective costs not calculated on the basis of preferences. I leave the much-discussed

the method of social choice used, control the resources. One cannot think seriously about any environmental policy without thinking seriously about which preferences should be taken into account in setting the policy. 'Ours now' may seem the obvious answer to us now, but it will not seem obvious to the people of distant places and distant times.

To decide whose preferences count, we must first examine our grounds for the selection of preferences. To assume that any party—individual, group, or institution—is entitled to satisfy all and only the preferences that it is in a position to satisfy without *in future* engaging in force or fraud is to assume that no reasonable questions can be raised about past or present holdings of resources. Iain Wallace and David B. Knight make such an assumption near the beginning of Chapter 4 in *Earthly Goods: Environmental Change and Social Justice*. After defining the net primary organic productivity (NPP) of a regional ecosystem as 'essentially its capacity to convert incoming solar energy...into terrestrial vegetation, on which "higher" lifeforms ultimately depend', Wallace and Knight conclude with an ethical judgement:

> By this measure, a tropical rain forest is intrinsically more productive (has a higher mean NPP) than a boreal forest, a high-altitude grassland, or a hot or cold (tundra) desert, for instance. Other things being equal, a 'traditional' or preindustrial society situated in an ecosystem of higher NPP will enjoy a higher sustained material standard of living than one situated in an ecosystem of lower NPP, as a direct result of the underlying nonisotropic natural conditions. Economic *inequality* between societies in these circumstances is not, prima facie, an *inequity*.

This seems to assume, in general, that human history counts for nothing and, in particular, that all present territorial holdings are equally beyond challenge—that each people has precisely the territory and resources it ought to have—however much force, fraud, or environmental havoc lies behind the current locations of borders, resources, and persons. Stated clearly, these assumptions are difficult to credit. That many children, through no fault of their own, are born into regions that simply cannot sustain a high standard of living no matter how hard people try, whereas others, through no merit of their own, are born into regions that do sustain a high standard of living throughout life is prima facie an inequity, an extreme inequality in life prospects that cries out for either change or justification. Why should the life prospects of two otherwise similar infants be so radically different for reasons entirely beyond their control?

---

subjective/objective problem aside in favour of the less noticed internal/external problem, which comes up either way.

The only alternative to denying the significance of the contingencies, crimes, and accidents in the history of the world so far is to construct some kind of reasonable standards (other than the mere fact of possession) for the assessment of current holdings of territory and other resources. This involves the formulation of what philosophers call principles of distributive justice. If it is incredible that the status quo must simply be accepted as uniformly justifiable whatever the divergences in the undeserved fates of individuals, we must try to establish some criteria for distinguishing the outrageous from the tolerable. We must try to make sense of our conviction that not everything is equally acceptable by spelling out some standards of acceptability in the division of resources, including most especially the initial allocation from which our markets and other distributive processes must begin. And we have to do this before we can know *whose* actual preferences, and *which* of them, to count.

Still, how much do we need to know about justice in order to make environmental policy? If several parties (individuals or groups) have conflicting preferences, they would do well to work out some mutually acceptable arrangement. Yet don't they need to have a set of principles of justice before they can arrive at a limited plan of action?

If parties are more or less equally situated, they should bargain directly. Other things being equal, it may be best if parties can work out among themselves the terms of any dealings they will have with each other. Even lawyers, however, have the concept of an unconscionable agreement; and non-lawyers have no difficulty seeing that an agreement can have objectionable terms if one party is subjected to some kind of undue influence. Parties can be unacceptably vulnerable to other parties in more than one way, naturally, but perhaps the clearest case is extreme inequality in initial positions. This common-sense understanding means that even morally acceptable bargains depend upon initial holdings that are not morally unacceptable.

Obviously to recognize acceptable bargaining we must have some knowledge, at least, of the standards for fair shares. If we do not know whether the actual shares that parties currently hold are fair, we cannot know whether any actual agreement they reach is morally conscionable. The simple fact that they all agreed is never enough. They don't have to have a complete theory of justice before they can agree upon practical plans, but they do need to know the relevant criteria for fair shares of holdings before they can be confident that any plan they actually work out should in any way constrain those who might have preferred different plans. As Chris Reus-Smit forcefully reminds us in Chapter 5 of *Earthly Goods: Environmental Change and Social Justice*, there are deep issues of justice in addition to issues about fair shares, but here I can sketch only some of the latter.

FAIRNESS AND GLOBAL WARMING

It is time to concentrate on a concrete instance of environmental change: climate change, or 'global warming'. In general, issues of fairness fall along multiple continua: (1) fairness specific to the problem at hand versus general background fairness; (2) fault-based standards versus no-fault standards; and (3) standards for transitions (or extrications) versus final standards. In sum, conceivably either the fairness of background conditions or the fairness in handling a specific problem might be judged by either fault-based or no-fault standards that are taken to be either transitional or final. Since each of these three kinds of considerations are presumably continuous, comments on fairness can be located in many different three-dimensionally specified conceptual locations. Quite a few disputes consist, not over different answers to the same question, but over questions not obviously different. Here we need to see the precise shape that various issues of fairness take in the case of a particular kind of environmental change, global warming.[5]

The basic issue of fairness characteristic of global warming seems to be: How should responsibility for solving the problem of global warming be divided, given how responsibility for creating the problem is in fact divided? And given how the benefits of the activities that produce the problem have in fact been distributed? The most important contributors to global warming are gases produced by the industrial activities that have made the rich rich. How much should the poor contribute to the solution of a problem caused by activities from which they have not benefited nearly as much? The division of causal responsibility for global warming and the division of benefits from the activities causing global warming are conceptually distinct (even if in fact they tend to fall together) but both are specific to the problem at hand.

A little further reflection suggests that 'the problem' is at least two distinct problems: (1) What is a fair allocation of the costs of avoiding however much additional warming still can, and reasonably should be, avoided? And (2) What is a fair allocation of the costs of social adjustments to however much warming is not in fact avoided? The answers may be similar; I have tried elsewhere to show that they are profoundly interconnected.[6] Yet no matter how interconnected, these two questions about global warming are still distinct.

---

[5] A comprehensive view of the issues specific to global warming, splendidly integrating economic and ethical considerations, appeared since this writing: Michael Grubb, 'Seeking Fair Weather: Ethics and the International Debate on Climate Change', *International Affairs* 71 (1995): 463–96. Also see Matthew Paterson, 'International Justice and Global Warming', in *The Ethical Dimensions of Global Change*, ed. Barry Holden (London: Macmillan, 1996).

[6] 'The unavoidability of justice', this volume.

One might also pursue a *general* inquiry: How should responsibility for solving the problem of global warming be divided in light of the background inequalities in wealth and power that are the present bitter fruit of centuries of colonialism, imperialism, unequal development, war, greed, stupidity, or whatever exactly one thinks are the main features of the history of the international political economy? One might think that background injustices not directly connected to the origins or fruits of global warming itself dwarf in significance any injustices that are. Or the opposite—that is part of what would have to be considered in analysing the problem in depth, as I would try to do if I were here not merely outlining the normative agenda.

I establish elsewhere that the gross inequalities among nations in wealth and power, and therefore in bargaining leverage, will undercut the moral legitimacy of the outcomes of actual bargaining among nations on the matter of global warming.[7] Actually, as long ago as Stockholm 1972, the rich countries conceded the principle that the poorer countries are entitled to special arrangements that they are in no bargaining position to demand; one understanding of this was reaffirmed in the terms of the Montreal Protocol (1987) on ozone destroyers and again in Principle 6 of the Rio Declaration on Environment and Development (1992), which begins: 'The special situation and needs of developing countries, particularly the least developed and those most environmentally vulnerable, shall be given special priority.'[8]

Some standards of fairness refer to, and therefore depend upon an accurate analysis of, the respective causal roles already played by the various parties.[9] Notably, portions of Principle 7 of the Rio Declaration contain the following rather delicate (and exceedingly vague) statements: 'In view of the different contributions to global environmental degradation, States have common but differentiated responsibilities. The developed countries acknowledge the responsibility that they bear in the international pursuit of sustainable development in view of the pressures their societies place on the global environment.'

On the other hand, a different standard of fairness refers strictly to features of the existing situation without any reference to their origins or development. Again, the remainder of Principle 7 of the Rio Declaration states 'and [in view of]

---

[7] 'The unavoidability of justice', this volume. For a contrasting approach, see Adam L. Ironstone, 'From "Cooperator's Loss" to Cooperative Gain: Negotiating Greenhouse Gas Abatement: Note', *Yale Law Journal* 102 (1993): 2143–74.

[8] The Rio Declaration is widely reprinted; one source is *Agenda 21: The United Nations Programme of Action from Rio* (Final Texts of *Agenda 21*, Rio Declaration, and Statement of Forest Principles, E.93.1.11) (New York: Department of Public Information, United Nations, 1993), pp. 9–11. A valuable commentary on possible precedents in the Montreal Protocol is Jason M. Patlis, 'The Multilateral Fund of the Montreal Protocol: A Prototype for Financial Mechanisms in Protecting the Global Environment', *Cornell International Law Journal* 25 (1992): pp. 181–230.

[9] As just indicated, the causal role investigated could be either specific to global warming or general to the world economy.

the technologies and financial resources they command'. Such payment according to ability to pay, for example, is a no-fault standard. It in no way presupposes or implies that the better-off did (or did not do) anything wrong in the process of becoming better off. This rationale for assigning heavier burdens to the better-off refers forward, not backward; it concerns how well the better-off can bear the heavier burdens, not how they came to be better off.

The fault-based 'polluter pays principle' (already established within the OECD and embraced explicitly, with qualifications, in Rio Declaration, Principle 16), by contrast, assigns heavier burdens to those who have contributed more heavily to the problem (regardless of ability to bear the burdens). Obviously one notable way by which parties come to have greater ability to pay is by having externalized the costs of their own gains: by not paying for the effects of their own pollution. In such cases, 'ability to pay' and 'polluter pays' may select the same parties, but that is a contingent conjunction in the applications of principles that are conceptually distinct. These two principles may assign different degrees of responsibility to any one party, making the Rio Declaration ambiguous in principle, which may or may not matter in practice.

The application of a no-fault standard depends upon the facts about the current situation—such as the extent of inequalities in standards of living and technological resources—whereas the application of a fault-based standard depends upon facts about stretches of the past, often rather long and wide stretches. Consequently, attempts to apply fault-based standards are virtually guaranteed to become embroiled in more or less irresolvable controversy about historical explanations. Yet never to attempt to assess fault is to act as if the world began yesterday.

Philosophers and political theorists have written little about what might be called principles for transitions, either principles of the fairness of transitions or principles of any other kind directly applicable to how change is brought about. Normally, we are offered an ultimate ideal and, in effect, wished good luck in figuring out how to reach it. I think this is lazy and irresponsible, but transitions are hard, intellectually as well as practically. Consider 'extrication ethics': here the central question is not simply 'How do we manage to reach the ideal?' but 'How do we manage to reach the ideal given that we are already in a swamp, as the saying goes, up to our ass in alligators?'[10] Since I am inclined to think the swamp general and the alligators numerous, I tend to think that extrication ethics largely constitutes transition ethics—but no matter.

---

[10] I take the term 'extrication ethics' from Tony Coady, who has written one of the few helpful discussions of the subject of which I am aware. See C. A. J. Coady, 'Escaping from the Bomb: Immoral Deterrence and the Problem of Extrication', in *Nuclear Deterrence and Moral Restraint*, ed. Henry Shue, Cambridge Studies in Philosophy and Public Policy (New York: Cambridge University Press, 1989), ch. 4, especially pp. 193–225.

Wherever we are, we face questions about how to get to where we are supposed to be, as well as questions about where exactly that is.

One of my teachers used to talk about what he called the pacifist's fallacy. Since I think pacifism is quite a reasonable position—although I do not hold it—this strikes me as an unfortunate name. Nonetheless, the underlying point is important. And the point is that even if we know beyond all doubt that the ideal society is one without violence, it simply does not follow that we should start being non-violent now and count on the power of our example to bring others around to our way of thinking. It may be that when we consider exactly where we are starting from—for example, a 'swamp' of injustice, unemployment, exploitation, and looting—we need to begin by applying state violence to end popular violence, or popular violence to end state violence. (Or the pacifist may actually be correct that violence from either side will only lead sooner or later to more violence, and that we would do best to start being non-violent right away.)

The point is just that even if the pacifist should turn out ultimately to have been correct, she will need to have given some kind of means/end analysis in order to show that she is correct.[11] The rationale for any transition/extrication ethic needs to be spelled out well beyond a statement of the ideal. The transition ethic *may* be the same as the ideal ethic: sometimes the best route to the ideal is to set an uncompromising example of the ideal, right here in the swamp. Since, in general, much human activity is self-defeating, we cannot, however, simply assume that good examples will routinely inspire emulation, whatever the circumstances.

In sum, then, the minimum list of essential, and unavoidable, questions about distributive justice in the case of global warming contains at least the following:

1. What is a fair allocation of the costs of preventing the global warming that is still avoidable?

2. What is a fair allocation of the costs of coping with the social consequences of the global warming that will not in fact be avoided?

3. What background allocation of wealth would allow international bargaining—about issues such as (1) and (2)—to be a fair process?

4. What is a fair allocation of emissions of greenhouse gases (a) over the long term and (b) during the transition to the long-term allocation?[12]

---

[11] Philosophers will be quick to note that this assumes that the agent is to some extent responsible to advance, or at the very least, not to retard, the ideal and not simply to embody it, whatever the consequences. I am sailing over deep waters here but hope not to have to plunge into them.

[12] These questions are explored at greater length in 'Subsistence emissions and luxury emissions', this volume. The heart of this analysis has also appeared as Henry Shue, 'Four Questions of Justice', in *Agricultural Dimensions of Global Climate Change*, ed. Harry Kaiser

At least these four issues need to be explored in order to develop reasonable policies. In the remainder of this chapter I struggle a bit more only with 4(b), which I find the hardest to grasp in the case of global warming, as an illustration of what might be involved in taking fairness seriously in one instance of environmental change.

COMING FULL CIRCLE

The underlying question remains: Who counts? It seems to me—although nothing about this is entirely clear—that the answer has to be, roughly, everyone whose fundamental interests are seriously affected. There are fewer and fewer tolerable environmental externalities, because there is no one 'out there' on whom the significant externalities may be inflicted. The internal/external line must be drawn, if it can be drawn at all, where the actual serious consequences end—the consequences that affect fundamental interests. *Whose* fundamental interests? Anyone's that are seriously affected. I call this the naturalistic approach: one follows the effects of the activity in question at least as far as they continue to impinge significantly on human beings (and on ecological elements that play an irreplaceable role in human life).[13] If the activity is, say, setting the discount rate for US banks, and the effects include jobs in Mexico, then, assuming that jobs are vital resources for living any kind of acceptable life, and that resources for living individual lives are the interest at stake, the justice of a particular discount rate and/or a particular process for setting discount rates depends in part on its effects on Mexican jobs—not just on its effects on US jobs.

The alternative approach can be called voluntaristic, because it depends upon the implicit contention that one can somehow renounce responsibility for the significant effects one is in fact having upon some people. Keeping the

---

and Thomas Drennen (Delray Beach, FLA: St. Lucie Press, 1993), pp. 214–28. Responding to the version of this analysis into four conceptually separable questions in *Law & Policy*, Michael Grubb has emphasized the depth of the economic ties between allocation of emissions (4) and allocation of the costs of abatement (1). See Grubb, 'Seeking Fair Weather', especially pp. 483–8. If a nation's allotment of permissible emissions is well below what would have been its business-as-usual level (not to mention its current level), the nation will need, other things being equal, to spend large amounts on abatement, either its own or other nations', perhaps through 'joint implementation'. Allocating the costs of prevention in order to produce an allocation of emissions that had been chosen independently on grounds of fairness would of course be one no-fault allocation of prevention costs. The answer to question 1 would be derived from an answer to question 4(a).

[13] As fewer and fewer people remain on the external side of the line, the tenability of the internal/external line itself is jeopardized, as was noted by an anonymous reviewer, who also emphasized the ecological elements irreplaceable in human life.

same example, one would say, on the voluntaristic approach, 'Yes, I realize that this is having significant negative effects on Mexican jobs, but my social contract is exclusively with US workers and so I do not take effects on Mexican jobs into account.' This might be part of a theory about the obligations of citizens to one another or of a theory about the obligations of a government to the governed. Its crucial feature is that effects on some people, the outsiders, may be disregarded—or discounted.

Discounting outsiders is, nowadays, more plausible than totally ignoring them. Perhaps it would not be acceptable to give the outsiders' interests no weight at all, but they are to be given significantly less weight than the interests of insiders. What seems crucial (and strange) about this—and the reason why I call it voluntaristic—is the assumption that it is up to the insiders to *decide* how much, or whether, to weigh the interests of outsiders in their balance. I have only the responsibilities I choose to take upon myself.[14] We have only the responsibilities we choose to acknowledge. What spectacular assertions of autonomy! Could anything be less compatible with an understanding of ecological interdependencies?

What if I begin to smoke a cigar during a meeting, and people begin to complain. But the complaining people are all women, and I say: 'I decided long ago not to acknowledge any responsibility toward women'? What if we build a factory that discharges pollutants into the Rhine in Switzerland, and people begin to complain. But the complaining people are all Dutch, and we say: 'We decided long ago not to acknowledge any responsibility toward non-Swiss'? What if we build an economy that lays debts upon future generations, and (self-appointed spokes)people begin to complain. But the people complained on behalf of are all non-existent, and we say: 'We decided long ago not to acknowledge any responsibility toward non-existent people'?

Whatever else can be said about these three cases (and the differences among them), it is preposterous to think that I, or a self-defining 'we', can simply decide as we please whether we fancy taking on responsibilities, construed as optional, toward people defined by us as others or outsiders, when their lives are being affected by ours. Voluntarism does not plausibly extend this far.

Moral responsibility for correcting injustice must, to a considerable extent, track causal responsibility. This ties it to what I call fault-based standards; this is a very specific way in which the topics of relevant communities and justice interlock conceptually. When my pursuit of my interests affects the resources available to others for their pursuit of their interests, I cannot *simply* declare those others out of bounds: 'Sorry, but you are the wrong sex, or nationality, or

---

[14] This level of would-be voluntarism has been subjected to a strong critique in Nancy J. Hirschmann, *Rethinking Obligation: A Feminist Method for Political Theory* (Ithaca, NY: Cornell University Press, 1992).

generation to count under my system.' It may be highly relevant whether I interfered with your pursuit of your interests intentionally, or accidentally, or unknowingly, or unforeseeably, or how much harm my own interests would have suffered if I had chosen to protect yours, and so on. But it is quite irrelevant that I would simply prefer not to count you and your kind at all— that cannot be simply for me to say, especially if shortages or harms I produced, however unintentional and unforeseen, deeply affect the livability and richness of the only life you can expect to have. The voluntaristic view is like the renunciation of paternity: renounce it as one may, the objective fact of one's effect on the world remains. One can deny responsibility, but one cannot remove it.

What, then, should we be doing in the short term (in the initial transition away from where we are now) specifically about global warming?

TRANSITION/EXTRICATION

Change is rarely instantaneous, and we must, therefore, recognize in most cases some kind of transition period between wherever we are now and where we ought to be. But there are transitions and transitions—it matters very much how long a transition takes, and it especially matters whether some people live their entire lives and die before the transition is completed. If there is anything to which every human being has a right, it is no good for the people whose lives have meanwhile come and gone that arrangements are gradually being made so that eventually everyone will enjoy what those now dead were also entitled to have enjoyed but did not.

Let me focus on two rough categories of ethical standards, which I will call minimum standards and short-range standards. These are not traditional labels, and there is nothing sacred about these particular names for the categories, which are meant to be one extreme and the middle of a continuous scale. Obviously the other extreme on this continuum could be called long-range standards. The next item in that direction, beyond long-range standards and off the ethical charts, is something like high-minded hypocrisy: empty talk about what would be 'great' without any sense of obligation to act. Thus by 'long-range standards' I do not mean dreamy rhetoric; I mean genuine standards that we are bound to meet as soon as we reasonably can, but which are in fact going to take a while to satisfy even if we pursue them with appropriate vigour. For example, many of us believe that we ought to eliminate racist and sexist attitudes as well as racist and sexist behaviour. But no matter how aggressively we attack racist and sexist behaviour, there is no way that we can in the short run eliminate the attitudes that lie behind the behaviour. Besides eliminating the behaviour, we can also criticize and pour contempt upon the attitudes, but that will not destroy the attitudes, in ourselves or in others. This persistence of the attitudes is definitely not a minor problem that we can happily live with. The trouble is that the only way to try to eliminate the recalcitrant attitudes immediately would be to engage in some kind of 'moral re-education' that would look a lot like brainwashing, and that would be either ineffective or highly undesirable in light of some of our other firmly held ethical standards. So long-range standards specify not places to which 'you cannot get from here' but places we cannot go straight to right away.

I have already by implication given an example of what I take to be a short-range standard. There are reasons why we cannot suddenly make all men respect women—why we cannot change all our attitudes—but we can make men stop harassing women. I am not of course predicting that we can in fact be 100 per cent successful—nothing is 100 per cent successful—but we can seriously and soberly adopt a policy that says, not 'We want to get sexual harassment down to tolerable levels' (whatever 'tolerable levels' might mean), but 'We intend to stamp out sexual harassment.' Even this cannot be done tomorrow, of course, which is why it is a short-range and not an 'instantaneous' standard. Laws will have to be passed against harassment, regulations will have to be issued defining it, victims will have to find the courage to make accusations even though they know they will be considered troublemakers, accused men will have to have fair hearings, convicted men will have to be punished, and word will have to get around that harassing women is just not much fun any more. This will all take some time, but it need not take an awful lot of time, particularly if the attitudes also begin to change, so that the men do not just figure that it is not worth the hassle any more but come to believe that women deserve better.

Beyond short-range standards lie what I call 'minimum standards'. Since the other categories are called 'long-range' and 'short-range', it would have been natural to call this one 'immediate', and by 'minimum standards' I do mean standards that must be satisfied as quickly as is humanly possible. These minimum standards must always be fully satisfied. In principle, no transition period, however brief, is acceptable. As always, we will probably not in fact succeed in effecting such standards immediately, but we should try, sincerely and seriously. In my analogies of racism and sexism, now we are talking about lynchings and rapes. If we said to someone, 'Look, lynching—or rape—is just unacceptable, and we want it stopped,' and he said, 'Right—I've got a plan, and we will reduce lynchings to 90 per cent in one year, to 80 per cent after two years, and so on until in a decade we will have come close to wiping out lynching,' I think our response would be, 'You don't seem to understand about the seriousness of lynching—we don't want no lynchings by the year 2006—we want no lynchings now.'

The practical difference may be best seen in budgetary terms. Suppose (unrealistically) that it turned out that for every 10 per cent increase in the enforcement budget of the FBI you could cut lynchings by 10 per cent, and someone proposed that the budget should go up 10 per cent per year for ten years, I think the natural response would be: 'No—we want the budget for this increased 100 per cent immediately because we can't think of anything worse than lynching that merits the money more.'

The budgetary analogy suggests another useful way to mark the differences among long-range standards, short-range standards, and minimum standards, namely the amount of sacrifice one should be willing to make in order to satisfy the standard. Once again, this is a continuum, not a series of bright lines, but the differences are significant. For a long-range standard one is serious and is making some effort, but one is also working on many other fronts, some of which are more urgent. For a short-range standard one is making serious investments of time, money, and energy, although a very few things are more urgent still. Those 'few things (that) are more urgent still' are the minimum standards, which have higher priority than anything else, with the possible exception of any necessary preconditions of them all, such as the prevention of the extinction of the human race. At least some of the requirements of justice are, I think, minimum standards.

Unfortunately, one cannot specify what we should do in the short run without having some answer to where we should arrive in the long run; one needs to know the objective before one can concretely consider whether it may be compromised. In terms of my earlier analysis of four questions of distributive justice, one cannot answer 4(b) independently of 4(a). Discussions so far of the allocation of greenhouse gas emissions that would be fair over the long run have tended toward the simplistic extremes. At the one extreme, those with rich-country interests in mind tend to assume that current emissions constitute some kind of unquestionable baseline of entitlement (from which we should simply start trading emissions permits), as if the great Hegelian fantasy had spontaneously materialized: the real is rational and just, and the rational and just is real. That there is some type of property right in current levels of emissions is an intelligible thesis, but one that is far from obvious (and, I argue, far from true).

At the other extreme, those with poor-country needs in mind tend to assume that if one grants that all human beings are equal in dignity and worth, it immediately follows that all human beings have a right to equal greenhouse gas emissions, which amounts, roughly, to equal levels of economic activity. Anil Agarwal and Sunita Narain famously endorsed equal per capita emissions in their much-cited (but, I suspect, less read) 1991 monograph; and Dr Atiq Rahman, director of the Bangladesh Center for Advanced Studies, said the same in August 1993 before a hearing of the US House of Representatives.[15] While I believe these advocates of equal emissions are much

[15] See Anil Agarwal and Sunita Narain, *Global Warming in an Unequal World: A Case of Environmental Colonialism* (New Delhi: Centre for Science and Environment, 1991), and

closer to the truth than the defenders of current emissions, I do not think that either polar position can be adequately supported.[16] This turns out to be a fascinating but highly complex issue, which I cannot begin to analyse here. Elsewhere I have defended the view that the universal human right is not to equal emissions *simpliciter* but to an equal minimum; if ever the minimum were universally guaranteed, departures from equality could then be justifiable.[17] While the difference between strict overall equality and an equal minimum could in the end be significant, in the transition—which is to say, now—it is insignificant.

I take it, then, that our goal is to allow every human being an equal minimum level of greenhouse gas emissions.[18] We can even assume for the sake of this argument, in order to make things easier for ourselves, that the minimum level is quite low, roughly subsistence. What can we say about the transition period that starts now? Since this is, strictly speaking, a minimum standard, its implementation may not be compromised or delayed. We should immediately see to it that no one is ever denied her minimum share of emissions. What does this mean concretely?

Here is where the concrete problem begins to get interesting. A broad consensus of scientists tells us that the global *total* of greenhouse gas emissions must be *reduced* below the 1990 total if serious global warming is to be avoided—emissions are not, I emphasize, to be stabilized *at* the 1990 level (which was the Clinton administration's great new commitment as of Earth Day 1993) but to be reduced well *below* it.[19] The original report of the

---

'Bangladesh Environmentalist Calls for Division of Global Emissions Allowances', *International Environment Reporter: Current Reports* 16 (1993): 574.

[16] For a pioneering attempt to deal with the ethics of transition in the case of climate change by a progressively changing weighing of current emissions and population, see Michael Grubb, James Sebenius, Antonio Magalhaes, and Susan Subak, 'Sharing the Burden', in *Confronting Climate Change: Risks, Implications, and Responses*, ed. Irving M. Mintzer (Cambridge: Cambridge University Press for the Stockholm Environment Institute, 1992), pp. 305–22 (for the formula, see note 7 on p. 321).

[17] 'Avoidable necessity: global warming, international fairness, and alternative energy', this volume.

[18] This is my answer to question (4a). I have developed other implications of this answer in 'After you: may action by the rich be contingent upon action by the poor?', this volume.

[19] See J. T. Houghton, G. J. Jenkins, and J. J. Ephraums, eds, *Climate Change: The IPCC Scientific Assessment* (New York: Cambridge University Press for the Intergovernmental Panel on Climate Change (WMO/UNEP), 1990); the brief 'Policymakers' Summary' is also available in a separate and more colourful form as *Scientific Assessment of Climate Change: The Policymakers' Summary of the Report of Working Group I to the Intergovernmental Panel on Climate Change* (Geneva: Intergovernmental Panel on Climate Change (WMO/UNEP), 1990). Also see J. T. Houghton, B. A. Callander, and S. K. Varney, eds, *Climate Change 1992: The Supplementary Report to the IPCC Scientific Assessment* (New York: Cambridge University Press, for the Intergovernmental Panel on Climate Change, 1992). For the Clinton Earth Day speech, see 'Reaffirming the U.S. Commitment to Protect Global Environment', *U.S. Department of State Dispatch* 4 (26 April 1993): 277–80 (reduction to 1990 levels by 2000: 278, col. 3).

Intergovernmental Panel on Climate Change (IPCC) said that CO_2 emissions have to be reduced 60 per cent below the 1990 total! If a political commitment is made to act upon that scientific consensus, the total of emissions becomes a *shrinking* zero-sum: the room under the global ceiling for any additional emissions to sustain a newborn child, or to move a malnourished child back from the brink of irreparable physical damage, must be created by reductions in someone else's current emissions, and then significant further reductions must be made in current emissions in order to make the total shrink. Everyone understands that the waste of fossil fuels in most rich countries is so egregious that there are 'no-regrets' reductions in fossil fuel energy consumption to be made—the initial cuts in carbon emissions would be not merely painless but profitable.[20] Nevertheless, this provides only a kind of grace period before the crunch comes.

The problem remains dramatic even if we wave a magic wand again and make it much easier still. Assume, to ease the strain, that the scientific consensus, as originally reported by the IPCC, is wildly alarmist and that we do not need to go below 1990 levels but need only to maintain 1990 levels. Perhaps, to continue the fairy tale, the elimination of waste in the rich countries would make enough space under the global ceiling (constituted by the 1990 total) to accommodate subsistence levels of economic activity by everyone currently living in absolute poverty. Still, even with a few such fantastic favourable assumptions, it is clear that, given current fuels for economic activity, sooner rather than later *people in rich countries must voluntarily and indefinitely reduce their level of economic activity* in order to allow for additions to the human population, who share in the universal human right to a minimum level of economic activity. Even with all my magical weakening assumptions, we may be contemplating revolutionary change.[21]

Here are the primary options, explicit and implicit (civilized and barbaric):

1. We can simply renounce any shred of commitment to human equality, abandoning even any universal right to an equal minimum, and instead acknowledge that in our hearts we actually do think that there are two kinds of people, people worthy of life and people not necessarily worthy of life, with the 'preferences' of the latter to be excluded from our calculations if need be.

---

[20] Alliance to Save Energy et al., *America's Energy Choices: Investing in a Strong Economy and a Clean Environment* (Cambridge, MA: Union of Concerned Scientists, 1991), p. 2; Michael Shepard, 'How to Improve Energy Efficiency', *Issues in Science and Technology* 7 (Summer 1991): 85–91.

[21] And I am for now leaving aside entirely a host of possibly compounding problems about the interests of nonhumans rightly raised by Wendy Donner in ch. 3, *Earthly Goods: Environmental Change and Social Justice*.

2. We can acknowledge the scientific consensus about what is likely to happen if we do not adopt mitigation strategies now, but opt nevertheless for a policy favouring adaptation later over mitigation now and thus not requiring the capping of total emissions; in short, deny that the science justifies the policy of mitigation.

3. We can voluntarily and indefinitely reduce our level of economic activity in order to reduce emissions. Or:

4. we can create energy technology that does not produce (as much) greenhouse gas emissions.

As far as I can see, these are the most eligible choices (and I would be delighted to be shown another good one).

The first option, I hope, is out of the question. Until the spectres of fascism and apartheid become too dim any longer to haunt us, we are, I hope, emotionally incapable of abandoning all pretence to a commitment to some minimal equality across humanity, however little exertion we expend in implementing our rhetorical stand and however tortured some of our interpretations of what equality means. The equanimity with which Western Europe and the United States are tuning out the 'ethnic cleansing' of the Bosnian Muslims in the former Yugoslavia is, however, not encouraging. Still, if we can be optimistic about our commitment to equality, we effectively have three options: 2, 3, and 4.

Unlike option 1, option 2 (denying that the policy of immediate steps toward mitigation of global warming is well enough supported by the science) strikes me as a perfectly sane *type* of argument. Since scientific findings alone can never dictate what we ought to do but can only give us (the absolutely vital) information about the paths that are physically possible and the comparative risks and benefits of the alternative paths, it is always in principle perfectly reasonable to challenge any policy prescription based explicitly only on the science, since implicit values are necessarily being assumed where they are not explicitly invoked.[22] Nevertheless, in this particular instance, the arguments are quite weak for continuing to increase the rate at which we inject greenhouse gases into the atmosphere in spite of the magnitude of uncontrolled and unpredictable climate change that is clearly possible and that many scientists think is virtually inevitable.

The strongest case I have seen for later adaptation instead of present mitigation is constructed by Oxford economist Wilfrid Beckerman. He advocates continuing to accumulate wealth through productive investment of the resources that might otherwise be dedicated to mitigation and relying later

---

[22] For rich development of the fundamental issues here, see Sheila Jasanoff's and Steven Yearley's contributions to *Earthly Goods: Environmental Change and Social Justice*.

upon the additional wealth thereby accumulated to pay for whatever specific adaptation is required by the actual effects of whatever warming does in fact occur.[23] One fatal flaw in this proposal is that those who seem most likely to have their lives threatened by the effects of warming—for example, Bangladeshis displaced by rising sea levels—are precisely those least likely to have accumulated the resources to pay for adaptation if no measures have been taken in the meantime to achieve that end. That the worst risks fall upon the most vulnerable is especially clear if Cynthia Rosenzweig and Martin L. Parry are correct that although temperature changes may be smallest in the tropics, tropical agriculture is least able to adapt to even small changes fast enough to avoid disruption in food supplies.[24]

So far I have said nothing about option 4, the option of aggressively pursuing non-fossil energy sources. Economic activity is a problem, as far as global warming is concerned, only because it injects greenhouse gases—above all, CO_2—into the atmosphere. *This* problem, the carbon emissions from economic activity, would be solved by a switch to photovoltaic, solar/thermal, wind, nuclear fusion, or any other non-carbon-based energy source. I do not think it is technological romanticism to think that a carbon tax invested in R&D on non-carbon energy could simultaneously raise the price of coal, oil, and gas and lower the price of their competitors through technological improvements until some non-carbons would be competitive with the carbons. If we can have the economic activity without the greenhouse gases, there is no problem (as far as global warming goes). Why choose option 3, reductions in economic activity including reductions in employment for an enlarging population, when one could choose option 4, elimination of our carbon addiction?

The primary obstacles appear to be political, not technological. The world is currently run, it seems to me, to a considerable extent to satisfy the preferences of Saudi Arabia, Kuwait, BP, Exxon, the automobile industry, and related

---

[23] Wilfred Beckerman, 'Global Warming and International Action: An Economic Perspective', in *The International Politics of the Environment*, ed. Andrew Burrell and Benedict Kingsbury (Oxford: Oxford University Press, 1992), pp. 253–89. Also see, in the same collection, Richard N. Cooper, 'United States Policy towards the Global Environment', especially 'Choice of a Discount Rate', pp. 300–4. On the narrow issue of setting a discount rate, contrast William R. Cline, 'Time Discounting', in *Global Warming: The Economic Stakes*, Policy Analyses in International Economics, no. 36 (Washington, DC: Institute for International Economics, 1992), pp. 72–5. For a brilliant general ethical critique of discounting as a technique, see Derek Parfit, 'Energy Policy and the Further Future: The Social Discount Rate', in *Energy and the Future*, ed. Douglas MacLean and Peter G. Brown, Maryland Studies in Public Philosophy (Lanham, MD.: Rowman & Littlefield, 1983), pp. 31–7.

[24] On the all-important danger to agriculture and food, especially in the tropics, see Cynthia Rosenzweig and Martin L. Parry, 'Potential Impacts of Climate Change on World Food Supply: A Summary of a Recent International Study', in *Agricultural Dimensions of Global Climate Change*, ed. Harry Kaiser and Thomas E. Drennen (Delray Beach, FLA: St. Lucie Press, 1993), pp. 87–116, and, for fuller detail, C. Rosenzweig et al., *Climate Change and World Food Supply* (Oxford: Environmental Change Unit, 1993).

interests.[25] The Saudi and Kuwaiti 'royal' families/governments were major obstructionists in Rio, and the chair of the Intergovernmental Panel on Climate Change, Dr Bert Bolin (Sweden), eventually admitted publicly that Saudi Arabia and Kuwait continued to place political obstacles in the way of the IPCC's research;[26] US fossil fuel interests orchestrated a brilliant campaign of disinformation to defeat Clinton's modest 1993 BTU-tax proposal and nearly defeated what became the pitifully small ($0.043 per gallon) gasoline tax. Are entrenched interests so deeply rooted that serious pursuit even of option 4 will not happen (soon enough)? If so, we would be left with option 3, voluntarily choosing less—perhaps even, no—economic growth.

Yet there are powerful reasons to doubt that option 3 will in fact be chosen. Chris Reus-Smit, in Chapter 5 of *Earthly Goods: Environmental Change and Social Justice*, has marshalled one set of them, which cuts very deep. What justification the modern state has for its pretensions to sovereignty rest on its claims to achieve its 'moral purpose', which in this international system (unlike others) is economic growth. To ask states whose raison d'être is the promotion of economic growth to adopt a policy of constraining economic growth might be, in effect, to invite them to commit suicide by undercutting their own rationale, something states rarely do on purpose. This is one fundamental reason to doubt that option 3 is going to be adopted.

Because option 4 seems not to challenge the underlying ideology of the contemporary state but only (!) the preferences that support the current energy regime, the interests arrayed against it are narrower, although obviously very powerful. That it might be adopted is, however, the only glimmer of light I can see in an otherwise possibly dark future. Preferences that block option 4 should, then, I think, be challenged. This is not as radical, Reus-Smit correctly observes, as a challenge to the interests and ideology blocking option 3 would be. There is also a case for mounting that challenge as well, but I hope that aggressive action to moderate global warming does not need to wait until a challenge to the justificatory ideology of the modern state could succeed.

---

[25] Obviously this is intentionally a great oversimplification. The classic political treatment is Robert Engler, *The Politics of Oil: A Study of Private Power and Democratic Directions* (Chicago: University of Chicago Press, 1961; Phoenix edition, 1967). A more recent economic study, concluding with (generally ignored) recommendations for public policy, is Duane Chapman, *Energy Resources and Energy Corporations* (Ithaca, NY: Cornell University Press, 1983). A Pulitzer-prize-winning history of the modern world centred upon oil is Daniel Yergin, *The Prize: The Epic Quest for Oil, Money, and Power* (New York: Simon and Schuster, 1991).

[26] 'IPCC Chairman Says "Political Wrangling" Now Gives Way to Science and Technology', *International Environment Reporter: Current Reports* 16 (1993): 544. Saudi Arabia and Kuwait— and OPEC more generally—have been the exceptions to a trend of cooperativeness. At the first Conference of the Parties to the Framework Convention on Climate Change (Berlin 1995), the OPEC countries were effectively ejected from the G77 developing-country group 'as having maintained an immoral and impractical position too obdurately for the rest of the developing world to bear'. See Grubb, 'Seeking Fair Weather', pp. 481–2.

This hope rests, of course, on the possibility that option 4, major moves toward alternative energy sources, can be made viable very soon.

Whose preferences, in the end, should count? The 'preference'—more accurately, the need—of every human being to engage in enough economic activity to sustain a decent standard of living. Whose preferences should not count? The preferences of those who wish to see the current energy regime, centred around oil, continue undisturbed.

Have I, then, in making option 4 the place to start, settled for a dubious technological fix? Option 4 would do absolutely nothing about, for instance, background international injustices. Option 4 speaks not at all to the many possible reasons other than carbon emissions for fundamental economic change; it would not lead to general economic restructuring. And it is limited in many other respects.

It is crucial, however, to recall the question to which option 4 is the answer: question 4(b), what is a fair allocation of emissions of greenhouse gases during the transition to a fair long-term allocation? To this question emerged the answer: in order to avoid violating a minimum standard of fairness, rich countries must begin promptly to reduce progressively their emissions of greenhouse gas rapidly enough that a minimum level of emissions is always available for every human being under the maximum global total of emissions compatible with avoiding climate change to which the poorest could not adjust. Of the two ways to reduce rich country emissions, options 3 and 4, the less painful for everyone without a vested interest in the current energy regime is option 4, to create alternatives to the current energy technologies that spew greenhouse gases into the atmosphere. Immediately.

I have no quarrel with a number of more radical proposals that solve additional problems. But for a start I would be very pleased to see this method of mitigating global warming implemented now.

7

Eroding sovereignty: the advance of principle*

The worship of the great man, or perhaps the idea of sovereignty, paralyses the moral sense of humanity.

> Lord Wright, 'War Crimes under International Law', reprinted in History of the United Nations War Crimes Commission and the Development of the Laws of War (1948)

Much argument about the sovereignty of states is made hopelessly simplistic by its generality. Should we recognize state sovereignty or not? Do states have too much sovereignty or just about the right amount? And so forth. In order to get a firm grasp, we must examine specific matters over which states could be permitted or denied sovereignty of specific kinds one at a time. Sovereignty is not some mystical cloud that either envelops the state entirely or dissipates completely; there are bits and pieces of asserted sovereignty. These assertions can be granted or contested one by one and accepted in this era and rejected in the next, or vice versa. Sovereignty should, I would think, be treated more like a (crazy) quilt that can be left to cover some things but pulled off from others.

Purported causal connections and alleged conceptual relations, in both directions and various mixtures, between nationalism and sovereignty abound. One incessant refrain goes as follows:

1. Every nation has a right to self-determination.
2. The best, or only, means by which any nation can protect and guarantee its right to self-determination is to control a sovereign state of its own.

In order for a nation to be genuinely self-determining, this argument goes, it needs its own sovereign state. A nationalist wants a state for (what he or she takes to be) his or her nation. Why must Croatia secede from Yugoslavia and form a separate state? Because Croatians (taken to be a nation) will not be safe

* 'Eroding Sovereignty: The Advance of Principle', in *The Morality of Nationalism*, edited by Robert McKim and Jeff McMahan (New York: Oxford University Press, 1997), 340–59. This chapter has endnotes rather than footnotes.

under a Yugoslav state in which they are a minority; Croatians need to form a sovereign Croatian state. Why must Bosnian Serbs secede from Bosnia and join a state controlled by Serbs? Because Serbs (taken to be a nation) will not be safe under a Bosnian state in which they are a minority; Serbs need to join a sovereign Serbian state. A self-respecting nation must never settle, according to nationalists, for being a minority in a state controlled by some other nation. A sovereign state is an indispensable weapon for the protection of a nation's right to self-determination.

I consider the dominant contemporary conceptions of both the nation and the state to embody extremely dangerous myths; a state with a nationalistic basis is, I believe, doubly dangerous. Other chapters here discuss the problematic character of so-called nations, or peoples, whose integrity is claimed to ground the argument for the separate state. What I will try to establish are some limits on the state, whose sovereignty is supposed to guard the nation's alleged integrity. If I am correct about state sovereignty, nationalists—and non-nationalists—could be a little less desperate to control a state of their own.

A separate state is most desirable to a nationalist if it is a sovereign state. A sovereign state is thought to be the best, or only possible, guardian of its people. Why? As a first approximation, we can say that a sovereign state can be valuable for a people because the sovereignty of the state embodies a right to give the interests of the people represented by the state special attention. A sovereign state may grant special attention to the interests of its own people. The interesting question is: What exactly can justifiably be meant here by 'special'?

Hereafter I will be discussing the acceptable attitude of a state toward everyone outside its territory, which is roughly what is customarily called external sovereignty. Before moving to my own topic of external sovereignty, it is worth noting quickly in passing that part of what is frightening to people like me about the internal sovereignty of a state with a nationalistic grounding is its likely understanding of who 'its own people' are. If the Croatian state, for instance, believes that the people to whose interests it may, and ought to, give 'special' attention are exclusively the ethnic Croatians, then it is liable to be willing to drive ethnic Serbs out of their houses and off their land in order to turn the houses and land over to ethnic Croatians, precisely as the newly rearmed Croatian army did, in fact, do in 1995 in the region of Krajina. Thus self-determination for the favoured group brings denial of the self-determination of individuals of other groups. Internally, 'special' attention with a nationalistic grounding seems extremely likely to lead, at best, to inequality in the self-determination of individuals, based upon which group they belong to, and often, in fact, leads to persecution and expulsion of individuals from other groups. This is a domestic application of special attention that is unacceptable. Domestic policies, however, are explored in other chapters.

What about the foreign policy of a sovereign state? The state's sovereignty entitles it to give, I initially said vaguely, 'special' attention to the interests of its own people. A partial, more definite account of special attention for nationals (however they are specified domestically—and I will worry no further here about that specification), as contrasted with foreigners, that would seem sensible to many people is that a state is obligated to take positive measures to promote the interests only of its own nationals. A state is never bound to act positively to advance the interests of foreigners; for any given foreigner, the task of promoting his or her interests falls upon his or her state. Whatever else is included under the sovereignty of the state, a state is free to ignore the interests of foreigners in the precise sense that it is bound to do nothing at all to advance the interests of foreigners. Its positive goals all have inward-looking justifications.

A sharp and exclusive division of labour among states applies when they are seen as sovereign in this sense. For instance, in the case of jobs, the state is in the business of promoting jobs strictly for its own nationals; it should promote jobs for non-nationals only when, and only to the extent to which, the indirect means of producing additional jobs for foreigners is the most efficient way to produce additional jobs for citizens, as it might, in an economically inter-dependent world, often be. According to conceptions of sovereignty, a state may, and ought to, promote the interests of its own nationals exclusively. May it also promote the interests of its nationals without restraint?

LIMITING SOVEREIGNTY EXTERNALLY

Consider one familiar kind of restraint. The doctrine of just war is centuries older than the doctrine of state sovereignty.[1] A glance at the former will provide valuable perspective on the latter. When Bodin, Hobbes, and the others were first concocting their notions of state sovereignty, the case had already been made long ago that what political units (empires, city states, nation states, and whatnot) may rightly do is constrained by principled limits. The limits were understood to be principled in being external to the political units (city states or whatever) and existing independently of judgements made about their meaning by the individual units limited. That is, the principles were genuine limits with an internationally shared meaning, not merely whatever the heads of the individual units limited might claim they meant, which would have made them effectively no limits at all.

If whatever I judged to be the case were the case just because I judged it to be, I could not—logically could not—be wrong; there would be no intersub-jective understanding beyond my own judgement itself on the basis of which my judgement could be judged by others. The all-important feature of the criteria of just war is their analogous independence of any one state or other unit to which they applied. The criteria were part of an international— cross-unit—understanding. None of this is to deny either disagreement or disregard; of course sovereigns (and their scholars) often disagreed with each other about what the principles of just war meant, and sovereigns frequently chose to disregard what the principles plainly did mean. War crimes are not a modern invention, although we have improved the technology. The crucial point is simply that there was an international meaning—a cross-cultural understanding—to be disagreed about and to be disregarded. It is intersubjective meaning external to individual judgement that gives principles power and allows them to function as external limits on otherwise arbitrary judgement. This is true whether the 'subjects' are individual persons or single states.

No one, to my knowledge, ever thought the principles of just war meant for each state whatever that state said they meant—no one thought these principles were subject to self-certifying arbitrary interpretations, criticisms of which could only be groundless. No one, that is, until Hobbes, who did try to claim that a sovereign could modify intersubjective meanings as he (subjectively) wished. Totalitarians and Madison Avenue still try to twist socially established intersubjective understandings, but this is hard even for institutions with their power because some meanings are very deeply entrenched in culture and therefore socialized into individual understandings from very early on, as language is acquired. Here Hobbes wildly underestimates the social character of language and meaning and the difficulty of deconstructing the most fundamental social constructions.

Leaving aside bizarrely asocial Hobbesian theories of meaning, we find that while it was recognized that de facto each state would decide for itself whether it had just cause to go to war, this de facto judgement was by no means self-certifyingly de jure. On the contrary, other states could decide that a first state did not have just cause for war and precisely for that very reason—that the de facto judgement of the attacker was wrong—decide that it was right to resist the attack, which was clearly understood to be a conceptually distinct judgement from whether it was in its interest to resist, especially for some third parties deciding whether to assist the victim of the attack (in the role of defender of justice de jure). The limits constituted by the principles of just war were seen to be external to the judgements of sovereigns, such that a sovereign could get them wrong. As Michael Walzer has eloquently shown, normative as well as strategic concepts, 'massacre' as well as 'retreat', have established meanings that cannot be changed simply by, say, issuing an announcement at a televised briefing.[2]

This quick reminder about the doctrine of just war is intended as evidence that the thought that sovereignty does not involve complete arbitrariness and that even sovereigns are subject on at least some points to principled limitation is not a new, dreamy, or academic notion. Perfectly sensible and tough-minded heads of state and commanders of military forces have for centuries acknowledged that it is not the case that indiscriminate slaughter is in the eye of the beholding sovereign. The dawn of principle in international relations was centuries ago, however cloudy some matters have subsequently become.

Like all concepts, the normative concepts in the principles of just war are in principle subject to change, and conversations are always under way about, for example, whether the city busting by Allied bombers during World War II should be taken as precedent for or as violation of the meaning of 'discriminate attack'. But these very conversations presuppose that international consensus is both attainable and desirable and that a consensus once reached is authoritative until changed by another consensus. Meanings change, but they do not change until *they* have changed. A social principle such as the prohibition of indiscriminate attacks can of course be deconstructed and reconstructed but only socially—not by arbitrary individual assertion and not for alleged 'reasons' that are not compelling for other parties to the consensus and certainly not by the unreasoning behaviour of individual sovereigns, however powerful. It is the purpose of principles such as these to contest and limit power, especially putatively sovereign power.

What I am calling the externality, or independence, of principle is recognized in international law in the distinction between conventional law and customary law. A convention, like a treaty, is initially binding only upon the sovereigns who choose voluntarily to become parties to it by ratifying it. Customary law, by contrast, is binding on particular sovereigns involuntarily, whether they like it or not. A principle will have become a matter of customary law only if, very roughly speaking, a large majority of sovereigns have accepted it in both theory and practice; only well-entrenched practices come to represent customary law. This is the legal analogue of coming to be the content of an intersubjective, but not necessarily unanimous, consensus. Significant portions of just war doctrine now constitute customary international law.[3]

Whether they wish to use torture, to take a second kind of example, is simply not up to sovereign states to decide for themselves. Torture remains widespread, as everyone knows, but it is a widespread violation of customary (and conventional) law, not a widespread practice that is on its way to becoming an entrenched precedent for acceptable state conduct. In *Filártiga v. Peña-Irala*, in a domestic US court, the judge ruling that torture is a violation of customary international law commented: 'The torturer has become—like the pirate and slave trader before him—*hostis humani generis*, an enemy of all mankind.'[4] An additional reason why torture has no prospect of being accepted as a precedent is that even those who use it in practice avoid defending it in theory. (Instead, they lie about their practice.) Yet even a sovereign willing to admit and defend this practice would have to change many other minds before anything changed in customary law. R. J. Vincent noted that some principles of human rights, such as the right not to suffer racial discrimination, have attained not only the status of customary law but also, as recognized in *Barcelona Traction* by the International Court of Justice, the status of *jus cogens*, making them peremptory as well as universal.[5]

The principles of just war, then, are a demonstration that external sovereignty can be, and in practice long has been, limited by international principles; more recently, international principles of human rights have begun to limit internal sovereignty. Naturally restraints on both kinds of sovereignty are often honoured in the breach, just as much violation of domestic law occurs and goes unpunished. But we can nevertheless distinguish an unpunished breach of these international principles governing both external and internal sovereignty from a disputed interpretation because the core meanings of the fundamental principles are clear. We can even, as in the rare case of Nuremberg—and possibly the Bosnian war crimes proceedings—enforce the limits by punishing violators.[6]

LEAVING SOVEREIGNTY UNCHALLENGED

Next I would like to examine a current instance in which so far, I think, less is happening than meets the eye. One frequently hears these days expressions of the hope that international cooperation in dealing with global environmental problems, such as severe ozone depletion and rapid climate change, each of which I discuss briefly below, is already proving to be 'social learning' from which more general lessons will be drawn about the value of multilateral action and coalition formation, leading in the end to some kind of decline of sovereignty. Since the hope is that we will come to agree that genuine global environmental problems are problems that we all really must face together, we will be driven by necessity to cooperate more fully than we might otherwise like. Once we have had the experience, the hope continues, we may find that we like it, or at least can tolerate it, and may then carry the new modes of cooperation into other issue areas quite unlike the environment.[7]

I think that any hope for a gradual evolution by this route away from sovereignty rests on a misunderstanding about sovereignty, but the misunderstanding is instructive. I am not convinced that international cooperation on global environmental problems will provide even small victories over sovereignty for we must not assume that cooperation is somehow a compromise of sovereignty. Sovereigns have always formed alliances, but temporary cooperation and ad hoc coalitions for specific purposes and for limited periods are scarcely harbingers of a new world order.

This is not to deny that some of the environmental dangers that we have only belatedly understood, such as rapid climate change, are unique.[8] The speed of climate change is accelerated by increased emissions of gases that are produced by almost all productive human activities, the worst being the carbon dioxide from industrial production (as long as we stupidly cling to fossil fuels as our energy sources) but also the methane from rice paddies and from the digestive processes of ruminants (the famous cow-belch problem, which is no joke, as this planet hosts a lot of cows).[9] Because the economic activities of the poor, such as the herding of ruminants, and even more the economic activities of the now-poor attempting to enrich themselves a bit through processes such as carbon-based industrial activity, could cancel out the environmental initiatives of the rich and because enhanced global warming is the result of the atmospheric swirl of greenhouse gases (GHGs) irrespective of point of origin on the surface of the planet of any particular quantity of gas, any significant reduction in the current acceleration of warming trends would require very widespread international cooperation. China alone, simply by using its own coal for its own industrialization, can produce more carbon dioxide than all of the rest of the world's nations together can eliminate from their emissions by any methods anyone is willing to consider.[10]

One feature distinctive of the problem of rapid climate change, then, is that the rich states cannot simply buy their way out of the problem with or without the cooperation of the poor states. Every state with a large economy and every state with a large population that demands a higher standard of living will have to cooperate if global warming is to be slowed (as long as we insist on using fossil fuel for transportation and manufacturing—Saudi Arabia's oil is, along with China's coal, one of the greatest threats to human welfare). Assuming that rapid climate change is a threat to agriculture and the existing world food system, the states of the world face a common threat that can be handled only through widespread—virtually universal—cooperation. Costs will have to be borne, and sacrifices will have to be made.

Yet none of this necessarily limits state sovereignty, insofar as sovereignty consists in the exclusive promotion of the interests of a state's own nationals, which I earlier said constituted special attention. To the extent that any international cooperation is basically voluntary and each state decides for itself whether the terms on which it is asked by the others to cooperate serve the interests of its own people, sovereignty is untouched. The exclusive focus of sovereignty on promoting interests of nationals is uncontested and, so far, unqualified. If a state wishes to decide that allowing its citizens the capacity to prevent the temperature in their homes from ever rising above sixty-eight degrees even in July and August is more important than the extinction of the uncounted plant and animal species that will result from the electricity generation emissions, that state is free, according to the usual doctrines of sovereignty, to do so. Any level of comfort that a rich state's economy allows its people to pay for goes unquestioned whatever its probable effects on planetary climate, just as the Brazilian state remains free to allow 'developers' to turn the Amazon into gold mines and strawberry patches unless Brazil chooses, for reasons of the interests of Brazilians, not to do so.

As my tone already suggests, I will be trying to establish that the colossal global damage that can be done by 'domestic' economic policy is a compelling reason why state sovereignty ought to be limited in this area; I spell out some of the normative case in the next section. As things are now, however, threats to the global environment are simply (mis)handled by assembling representatives of sovereign states to discuss possible multilateral treaties and protocols.[11] The states determine whether there is anything they all consider to be in their respective interests; if there is, they agree to do it and perhaps even create enforcement mechanisms and sanctions to provide assurance that the mutually advantageous agreement will be complied with.

Sometimes, in short, there are common goods that are correctly understood to be attainable only if mutual restraint is agreed to and practised. Recognizing that common goods sometimes are attainable only through mutual restraint in *no* degree qualifies sovereignty, however; Hobbesian denizens of the state of nature recognize this much. Sometimes agreed-upon restraint is forthcoming and sometimes not. While this is practically important, it is theoretically uninteresting; sovereigns have always been considered, according to dominant conceptions of sovereignty, to be bound by their own word only as long as being bound can be made to be in their own interest.

A crucial distinction must be drawn here. It is one thing to claim, in accord with one common-sense interpretation of special attention, that a sovereign state may promote only the interests of its own nationals; call this the thesis of *exclusive promotion*.[12] It would be quite another matter to claim that a sovereign state may promote the interests of its own nationals no matter what the effects are on the interests of others; call this the thesis of *unlimited promotion*. It might well be that although a state need never make a point of advancing the interests of non-nationals, it must refrain from harming at least some interests of non-nationals however much harming the non-nationals' interests would serve to promote nationals' interests. Unlimited promotion certainly does not follow from exclusive promotion. On the contrary, exclusive promotion could turn out to be conditional upon its avoidance of certain harms—upon the prohibition of unlimited promotion.

The same distinction can be made in another way. Allowing the claim (1) that each state is free to play along or not to play along with environmental agreements entirely on the basis of its own *interests* may appear to be allowing the very different claim (2) that each state is free to play along or not to play along with environmental agreements entirely on the basis of its own *reasons*, meaning that each state is entirely its own judge of which considerations it pays attention to. But the latter would mean not simply that each state was free of any obligation to promote interests of non-nationals but that each state was entirely free to ignore the interests of non-nationals, harming them where such harm promoted the interests of its own nationals. The slide from 1 to 2 is a slide from self-centredness to arbitrariness. For instance, Saudi Arabia, whose oil was pulled out of the fire for it in 1991 in the Gulf War by an international coalition, has done its arbitrary best since (as it did before) to spread havoc within the negotiations about global warming, because the top Saudi priority is to sustain, whatever the harm to everyone else's interests, the long-term market for fossil fuel, far and away the most significant cause of global warming.[13] I do not think this kind of arbitrary behaviour, with its flagrant disregard of all interests except Saudi interests, need be accepted as a responsible exercise of sovereignty, and I will now try to explain why not.

CONTESTING SOVEREIGNTY 'AT HOME'

It may be that when states are making war outside their own territory, they must act with some restraint, but surely, it will be said, a sovereign state operating at home may look out for the economic welfare of exclusively its own people. One of the defining purposes of the modern state is the management of the domestic economy for the enrichment of its own constituents.[14] Will not other people's states in turn look out for them? The short answer is: no, because the interests of the other people affected may be vital, and their own states may be powerless to defend them against fatal harm initiated abroad. A longer answer follows.

It has been a decade and a half since Charles R. Beitz pointed out that the Hobbesian picture of international relations could have been correct only if four false propositions had all been true; one of those decisive mistaken Hobbesian assumptions is the following: 'States are independent of each other in the sense that they can order their internal (i.e., nonsecurity) affairs independently of the internal policies of other actors.'[15] That, for instance, even the richest and most powerful states, the Group of Seven, think they need regular meetings in order to coordinate their economic policies is one of innumerable symptoms of the depths to which the so-called domestic economic policies of the most nearly independent states that exist can be buffeted by the so-called domestic economic policies of the others. Much remains to be understood about the modalities of economic interdependence, but its depth seems beyond question. The Hobbesian argument was headed for a conclusion about domestic authority, but this particular erroneous Hobbesian premise was about domestic power—about the actual extent of effective control over an 'inside' that was much too sharply dichotomized from an 'outside'.

That the actual control over their economic fortunes that can be successfully exercised by modern states is, in fact, limited is, of course, not at all the same thing as their sovereignty over economic matters being limited, since sovereignty is a matter of the proper authority to try to exert power even where the exertions may turn out to be futile. Domestic criminal law is regularly broken, but until the violations approach the point of general disorder, only the power, not the authority, of the state is put in question by the lawbreaking. That domestic economic policy regularly fails is similarly no grounds for challenging the authority, as distinguished from the competence, of the state. If, as Beitz maintains, the Hobbesian conclusion about authority rests upon this premise about domestic power that is falsified by the penetration of the economic 'inside' that comes with economic interdependence, a pillar of the Hobbesian argument for sovereignty is undermined. But sovereignty itself is not undermined unless the Hobbesian case for it was the best case there was.

Unlike Beitz, I want to argue not against an argument for sovereignty but against sovereignty directly. I want to contest it, as I have already indicated, not wholesale but in specific aspects in a specific area. Namely, I shall argue that there ought to be external limits on the means by which domestic economic ends may be pursued by states, limits that ought to become binding on individual sovereigns irrespective of whether those sovereigns wish to acknowledge them, just as sovereigns are already bound by both legal rights and moral rights against the domestic use of torture whatever their own opinions on the subject of torture may be—the sovereign's own opinion about torture is of no consequence legally or morally. The same should be true of some particular means of pursuing economic ends. I take this to be a controversial thesis.

An analogy may also be drawn between my thesis and the portion of just war doctrine traditionally called *jus in bello*, the principles governing the means, as distinguished from the ends, of warfare. While the competence of the modern state to succeed in attaining its narrow economic ends is very much in doubt, I shall not challenge, for the sake of this argument, its authority to promote exclusively the welfare of its own people, or exclusive promotion, as it was called at the end of the previous section.[16] That is, I shall try to show that even if it were the case that a sovereign government is entitled, as a matter of its proper choice of ends, to pursue actively the economic welfare only of its own people, there nevertheless ought to be limits on the means acceptable in the pursuit of that end. The formal structure of the position advanced here, then, is the same as the formal structure of *jus in bello*: even once ends are accepted, many means remain unacceptable.[17] I emphasize this analogy in order to establish that while I may be advocating a new limit on sovereignty, I am not advocating a new kind of limit. Sovereignty's coverage is simply being contested in a new area. Since long before anyone thought up the doctrine of state sovereignty, civilized people have understood that even in the pursuit of fully legitimate ends, including extreme cases such as self-defence against arbitrary attack, there are some harms one may not inflict upon uninvolved parties.

A further aspect of the unoriginality of my thesis calls for emphasis. At least since Locke it has been generally acknowledged in this part of the world that the liberty to pursue one's own legitimate ends does not include or entail any liberty to inflict serious harm upon others. This is a negative requirement—no positive assistance whatsoever is mandated, and exclusive promotion is not challenged. My argument for the thesis that pursuit of the end of economic welfare for one state, like the end of defence of a state against military attack, is subject to limits upon means has the general shape of showing that the pursuit of economic goals unlimited in the ways to be specified would inflict unacceptable harms on what amount to innocent bystanders in other states.

In order to begin, then, to make this case, one needs to consider an intolerably severe harm that is inflicted across state boundaries—that is, a severe harm that is caused by economic activities within the territory controlled by one state that affects people who live on the territory of other states. Skin cancer is such a harm; no one's liberty to pursue his or her economic welfare entitles him or her to inflict skin cancer on others. So, too, is clouding of the cornea of the eye to such an extent that sight is impaired. Both these harms result from increases in ultraviolet-B (UV-B) radiation from the sun, and ultraviolet radiation is increasing sharply because of the progressive thinning of the stratospheric layer of ozone, which is the only element in the earth's atmosphere that can block UV-B.[18] The destruction of the atmospheric ozone is caused, in turn, by the long-lived molecules of chlorofluorocarbons (CFCs) that are continuing to migrate and reside there. In spite of regular denials by ill-informed members of Congress, the mechanisms are increasingly well understood.[19] Only true kooks any longer doubt either the seriousness or the reality of the problem of ozone depletion, and international agreement to deal with it was reached in 1987 under the Montreal Protocol on Substances that Destroy the Ozone.[20] The manufacturing of CFCs is being greatly reduced, although at a pace (apparently designed to guarantee that no manufacturer actually loses any money) that is rather leisurely, given the long atmospheric residence times of the various CFCs and the seriousness of the afflictions caused by the increased accumulations of CFCs in the upper atmosphere.

Apart from the slow pace of its implementation, the Montreal Protocol is in several respects an admirable precedent. People are doing basically the right thing, even if not at the right time; even under the terms of the Montreal Protocol some manufacturing of CFCs continues, and it will be decades before all of the existing air conditioners and refrigerators rust enough to release their CFCs for the journey to the stratosphere, where the CFCs will gobble up ozone for about another century. The question here about the international action taken to protect the ozone layer lies entirely with the interpretation of its significance. It seems evident that while on the whole the right thing is, in fact, slowly being done, there was morally no other choice. One is no more at liberty to rain malignant melanomas than nuclear missiles down from the heavens upon random people in other states. Consequently, any economic dislocations suffered by the manufacturers (and their employees, who may be as 'innocent' in the relevant sense as the affected foreigners and may of course themselves develop skin cancer as well as lose their current jobs—and therewith their medical coverage for cancer treatments) will simply have to be absorbed. Who exactly absorbs them is a matter of justice in distribution that bears looking into. Here what matters is only that it is plainly unconscionable to continue to manufacture and distribute a substance whose presence on the planet (CFCs are not a naturally occurring compound—they are an entirely human creation developed in 1930 by Thomas Midgley Jr at General Motors Research Laboratories)[21] will cause cancer in people who are not using the substance and, in fact, never heard of it or went near it, merely on the grounds that jobs will be lost and other economic penalties will be suffered as the substance is restricted. Other jobs will have to be created (preferably, I would think, at the expense of those, like Dupont, who profited most from the substance in question, for the sake of the incentive effects of following the principle 'the polluter pays', even apart from consideration of justice).

The fundamental fact here is the utter falsity of the Hobbesian assumption of internal control, as noted years ago by Charles R. Beitz. Beitz emphasized the fact that the success of 'internal' economic policies is dependent upon 'internal' economic policies elsewhere and upon multinational and transnational actors. The CFC case illustrates how 'internal' health policies and health itself are dependent on choices among industrial technologies made on the other side of the planet by foreign states or left by those states for their firms to make as those firms wish. The national energy policies that are driving rapid global climate change are a second, more complex but fundamentally similar illustration.[22] My question about sovereignty, then, is: Is a state any freer—or indeed as free—to decide for itself whether it will cooperate with reasonable measures (that is, do its fair share) to restrict CFCs, or to slow and reduce global warming, than it is to decide for itself whether it will slaughter innocents during a war fought for a just cause? That my suggested answer is no is, I hope, no surprise.

To insist upon whatever economic policies are best for one's own economy while ignoring the distinct possibility that an indirect but predictable effect may be that children in the tropics may develop malignancies caused by excessive UV-B radiation or starve from crop failures produced by climate shifts is an assertion of arbitrary wilfulness for which I can imagine no grounds. Of course one is entitled to pursue one's own economic well-being to some degree and in some ways but not at absolutely any cost to others. What is wrong (beyond criminality) with being a cocaine dealer? It is not wrong to want to do well, even perhaps very well, financially. But when it is possible to do well enough in alternative ways that are not destructive of other people's lives, even perhaps in ways that are mildly useful, it is wrong to choose to do that much damage. A state that enriches itself in utter disregard of indirect effects that make survival impossible for innocent strangers seems to me to be no different morally from a cocaine dealer. Some means of pursuing economic goals—namely, means that lead to the severe disruption of the environment within which others, too, must try to survive and prosper—should be removed from the range of options over which states have discretion.[23] We can specify a guiding principle by returning to the consideration of warfare and embracing a theoretical innovation from Michael Walzer's *Just and Unjust Wars*.

TAKING DUE CARE

To my knowledge, Michael Walzer first introduced into the doctrine of double effect as employed in just war doctrine an explicit requirement of 'double intention', which he grounded in what he nicely called 'a right that "due care" be taken'.[24] As Walzer, following Kenneth Dougherty, had spelled out the received doctrine of double effect, one of the four conditions that must be satisfied by an act that is likely to have evil consequences in order for the act nevertheless to be permissible is that 'the intention of the actor is good, that is, he aims only at the acceptable effect; the evil effect is not one of his ends, nor is it a means to his ends'.[25] Walzer's proposal is that this requirement be replaced by the following more demanding requirement: 'the intention of the actor is good, that is, he aims narrowly at the acceptable effect; the evil effect is not one of his ends, nor is it a means to his ends, *and, aware of the evil involved, he seeks to minimize it, accepting costs to himself*'.[26]

This is a compelling and important strengthening of the condition. It moves from requiring, in addition to the intention to do the good, the mere absence of any intention to do the evil that will, in fact, also result to requiring the presence of an active intention to keep the unintended evil as restricted as possible. Consequently, Walzer requires a 'double intention': an intention to do the good and an intention to hold the unintended evil to a minimum in spite of significant risks to oneself required by the second intention. Part of the beauty of Walzer's reformulation is that he ties a strong requirement of proportionality tightly together with the requirement of discrimination. One is permitted the unintended death and destruction—it may be credited as not indiscriminate—*only if* it is proportional in a more than usually demanding sense: as little as one can make it, significant costs to oneself notwithstanding. Otherwise, the doctrine of double effect enables one in practice to get away with simply waving one's hand over death and destruction that one could have avoided if one had taken more care, and perhaps had run more risk, and declaring, 'I wish it had not happened.' Walzer's requirement has the further virtue of naturally focusing attention on concrete measures that might be taken in order to reduce seriously unintended bad consequences and their respective risks and benefits. Then the issue becomes not an intractable one about one's intentions, hopes, and wishes but a palpable one about measures taken and measures not taken to protect the innocent: was a (very risky) ground patrol sent in to see who was actually there before the (very safe) air strikes were ordered, and so forth.

A 'right that "due care" be taken' would be a natural and helpful explication of the already mentioned Lockean right not to be harmed; seriousness about not harming involves taking some considerable care to be sure that one does not harm or, where the harm is unavoidable in doing something that must be done, taking some care to be sure that the harm is the least that is compatible with what must be done. The most frightening thing I ever learned from talking to people in the Pentagon came from the civilian who revealed that he relied on the principle that 'it is easier to get forgiveness than permission'. In itself the principle is chillingly correct—it *is* much easier since the damage is then already done. But any serious conviction that others have a right not to be harmed leads one not into preparations for saying that one is sorry but into measures taken in advance to prevent the harm.

The general application to economic policy seems clear enough.[27] If one can expect soldiers fighting in a just cause to bear extra danger in order to inflict less damage, one can expect those with other worthwhile but less urgent and less dangerous pursuits to take at least as much care to minimize 'collateral' damage. It may not be convenient for us to switch to electric cars whose batteries need frequent recharging (although battery technology is rapidly improving) before the cheap Saudi oil runs out. Yet if that is a good way to cut GHG emissions quickly and sharply enough to avoid shocks to the fragile agricultural systems of states where life is generally a lot less convenient than it would be for us even with electric cars, allegedly sovereign control over domestic economic and energy policy seems an extraordinarily self-indulgent excuse for totally discounting severe effects upon other human beings.

In the utter discounting of the profound human effects across political borders lies the arbitrariness of assertions of sovereignty over 'domestic' economic policy. Sometimes some interests of uninvolved strangers may be discounted, and sometimes, as when they will otherwise be the unintended victims of military action, they may not be discounted. The fact that vital interests of outsiders will suffer if our economic policies ignore dangers such as global warming and trivial preferences of our own for historically unprecedented levels of comfort and convenience are protected strikes me as good grounds for judging that those interests ought to take priority over these preferences. Defenders of any economic sovereignty that extends beyond exclusive promotion to unlimited promotion—that is, promotion without due care for people outside the state's borders—need to explain the grounds for their view.

Perhaps I have overlooked some good argument to show that the vital needs of all individual human beings are best served in some cases by a division of labour under which each state completely ignores the welfare of all humans who are not its own citizens. In the instances of ozone depletion and climate change, in particular, this could not possibly be true; no state is able to protect its own people against the effects on the planetary atmosphere of the choices of other states about whether to use CFCs and to use fossil fuels. In fact, these so-called domestic economic decisions have four critical features that together add up to a compelling case for demanding that states take responsibility for *everyone affected* rather than for only those who live on their own political territory:[28]

1. The policies contribute substantially to harm to people living outside the territory of the state that controls the policies.

2. The states that govern the territories in which the people harmed live are powerless to block this harm.

3. The harm is to a vital human interest such as physical integrity (a physically sound body).

4. An alternative policy is available that would not harm any vital interest of anyone inside or outside the state that controls the choice among policies.

For a state to assert a right to continue to contribute to such harm in such circumstances is by all common-sense standards wildly irresponsible and arbitrary. Any state that considers itself entitled to inflict severe harms outside its borders by some doctrine of sovereignty that slides from exclusive promotion to unlimited promotion needs to explain to those outside, in the line of fire, the grounds of its asserted entitlement to inflict severe harm.

Second only to the arbitrariness of declaring that palpable harms do not exist is the arbitrariness of declaring that in the moral calculus they do not matter. States that violate domestic human rights often make people disappear physically (through arbitrary executions). States that promote the economic interests of their own people unconstrained by due care for the fundamental interests of outsiders are attempting to make the outsiders disappear morally (through dropping out of the moral calculations).[29] Either way, making people 'disappear' is unacceptable.

Why do we ask young soldiers to walk toward the snipers in the trees to check for innocent civilians also perhaps hiding nearby when we can incinerate the whole forest, snipers and civilians alike, from a safe distance with air power?[30] Why should young soldiers risk death when old civilians will not incur discomfort in order to take care not to inflict avoidable harm upon the uninvolved? Perhaps our economic policies should be as civilized as our methods of warfare are supposed to be.[31]

NOTES

For penetrating critiques that radically reshaped my argument, I am abashedly grateful to the editors of *The Morality of Nationalism* and to Gerald Dworkin and his comment on the original paper, 'Ought? Replies Khan'. Additional helpful suggestions were made at the workshop on this book by Belden Fields, Robert Goodin, Amy Gurowitz, and Michael Walzer.

1. See Norman Kretzmann, Anthony Kenny, and Jan Pinborg, *The Cambridge History of Later Medieval Philosophy: From the Rediscovery of Aristotle to the Disintegration of Scholasticism, 1100–1600* (Cambridge: Cambridge University Press, 1982), pp. 771–84.
2. Michael Walzer, *Just and Unjust Wars: A Moral Argument with Historical Illustrations*, 2nd edn with a new preface (New York: Basic Books, 1993), p. 14.
3. For an overview, with emphasis on the use of air power, see Robert K. Goldman, 'The Legal Regime Governing the Conduct of Air Warfare', in *Needless Deaths in the Gulf War: Civilian Casualties during the Air Campaign and Violations of the Laws of War*, A Middle East Watch Report (New York: Human Rights Watch, 1991), pp. 25–64.
4. Richard Pierre Claude, 'The Case of Joelito Filártiga and the Clinic of Hope', *Human Rights Quarterly* 5 (1983): 291. Claude powerfully relates the dramatic story of how a Paraguayan policeman who tortured to death the son of a doctor (who ran the Clinic of Hope) came to be tried in a US court, establishing an important legal precedent about the universality of the right not to be tortured (pp. 275–300). Significantly, the US Department of State submitted a brief arguing that the US court did indeed have jurisdiction even though the torture-death occurred in Paraguay. On the customary international law of human rights more generally, see Paul Sieghart, *The International Law of Human Rights* (Oxford: Clarendon Press, 1983); for a more introductory treatment, see Paul Sieghart, *The Lawful Rights of Mankind: An Introduction to the International Legal Code of Human Rights* (New York: Oxford University Press, 1985).
5. R. J. Vincent, *Human Rights and International Relations* (New York: Cambridge University Press, 1986), p. 46. For *Barcelona Traction*, see Louis B. Sohn and Thomas Buergenthal, eds, *International Protection of Human Rights* (Indianapolis: Bobbs-Merrill, 1973), pp. 18–19. On *jus cogens*, see Ian Brownlie, *Principles of Public International Law*, 3rd edn (Oxford: Clarendon Press, 1979), pp. 512–15, 596; and Lauri Hannikainen, *Peremptory Norms* (*Jus Cogens*) *in International Law: Historical Development, Criteria, Present Status* (Helsinki: Finnish Lawyers' Publishing Company, 1988), pp. 1–19.
6. For a penetrating, and chilling, analysis of the US effort to undermine the Nuremberg precedent and maintain that national security decisions are not

subject to external review in order to get off the hook for the mining of Nicaraguan harbours, see Paul W. Kahn, 'From Nuremberg to the Hague: The United States Position in *Nicaragua v. United States* and the Development of International Law', *Yale Journal of International Law* 12, no. 1 (Winter 1987): 1–62.

7. For optimism about 'the introduction of novel elements of *governance* into the international system', although no thesis specifically about sovereignty, see Martin List and Volker Rittberger, 'Regime Theory and International Environmental Management', in *The International Politics of the Environment*, ed. Andrew Hurrell and Benedict Kingsbury (New York: Oxford University Press, 1992), pp. 108–9.

8. The fundamental studies are J. T. Houghton, L. G. Meira Filho, et al., eds, *Climate Change 1995: The Science of Climate Change*, Contribution of Working Group I to the Second Assessment Report of the Intergovernmental Panel on Climate Change (Cambridge: Cambridge University Press, 1996); Robert T. Watson, Marufu C. Zinyowera, and Richard H. Moss, eds, *Climate Change 1995: Impacts, Adaptations, and Mitigation of Climate Change*, Contribution of Working Group II to the Second Assessment Report of the Intergovernmental Panel on Climate Change (Cambridge: Cambridge University Press, 1996); and James P. Bruce, Hoesung Lee, and Erik F. Haites, eds, *Climate Change 1995: Economic and Social Dimensions of Climate Change*, Contribution of Working Group III to the Second Assessment Report of the Intergovernmental Panel on Climate Change (Cambridge: Cambridge University Press, 1996).

9. On the significance of methane, see Thomas E. Drennen, 'After Rio: Measuring the Effectiveness of the International Response', *Law and Policy* 15, no. 1 (January 1993): esp. pp. 28–31; see also Thomas E. Drennen and Duane Chapman, 'Negotiating a Response to Climate Change: The Role of Biological Emissions', *Contemporary Policy Issues* 10 (July 1992): 49–58.

10. See National Environmental Protection Agency of China, State Planning Commission of China, United Nations Development Programme, and World Bank, *China: Issues and Options in Greenhouse Gas Emissions Control*, Summary Report (Washington, DC: World Bank, 1994).

11. For a brief critique of the 'treaty-and-protocol' method, see Lawrence Susskind and Connie Ozawa, 'Negotiating More Effective International Agreements', in Hurrell and Kingsbury, *The International Politics of the Environment*, pp. 142–65. For a longer discussion, see Lawrence E. Susskind, *Environmental Diplomacy: Negotiating More Effective Global Agreements* (New York: Oxford University Press, 1994).

12. I am not convinced of even the thesis of exclusive promotion, which is to say that I reject conventional conceptions of state sovereignty, as I have indicated in *Basic Rights: Subsistence, Affluence, and U.S. Foreign Policy*, 2nd edn with a new afterword (Princeton: Princeton University Press, 1996), ch. 6. In the present chapter, however, I am accepting state sovereignty as exclusive promotion for the sake of argument to show that unlimited promotion would still not follow.

13. See 'IPCC Chairman Says "Political Wrangling" Now Gives Way to Science and Technology', *International Environment Reporter: Current Reports* 16, no. 15 (28 July 1993): 544. Saudi Arabia and Kuwait are the exceptions to a trend of general cooperativeness.

14. See Christian Reus-Smit, *The Moral Purpose of the State* (Princeton: Princeton University Press, 1999).
15. See Charles R. Beitz, *Political Theory and International Relations* (Princeton: Princeton University Press, 1979), p. 36, for all four of Hobbes's mistaken assumptions.
16. I very much challenge, on other grounds, the acceptability of assigning absolute priority to the satisfaction of superfluous preferences entertained by one's compatriots over the fulfilment of urgent needs suffered by strangers. See Shue, *Basic Rights*, esp. chs 5, 6. If the argument there that rights to economic subsistence are as fundamental as rights not to be tortured were accepted, it would follow straightaway that even sovereign governments are not at liberty to ignore the harm they do to non-citizens' economic welfare in their pursuit of the economic welfare of their citizens. In this chapter, however, I make no appeal to the subsistence rights I defended there.
17. Michael Walzer and Jeff McMahan have each separately pointed out to me that there is no general reason not to challenge state ends by analogy with *jus ad bellum*, which restricts acceptable wars to, for instance, those with 'just cause', in addition to challenging state means by analogy with *jus in bello*. I hope to explore that elsewhere.
18. A highly accessible account is in Jonathan Weiner, *The Next One Hundred Years: Shaping the Fate of Our Living Earth* (New York: Bantam, 1990), pp. 153–5. 'The E.P.A. estimates that every 1 percent depletion in ozone will cause something like a 2 or 3 percent increase in UV-B and a 5 percent increase in skin cancer, including a 1 percent increase in malignant melanomas. In our lifetimes, the thinning of the ozone shield may lead to a 60 per cent rise in the incidence of skin cancers of all kinds in the United States' (p. 155). Weiner cites John S. Hoffman, ed., *Assessing the Risks of Trace Gases That Can Modify the Stratosphere* (Washington, DC: Office of Air and Radiation, US Environmental Protection Agency, 1987); and 'Skin Cancer Facts and Figures', newsletter, The Skin Cancer Foundation (May 1988).
19. See R. L. Miller, A. G. Suits, et al., 'The "Ozone Deficit" Problem: $O_2(X,v \geq 26) + O(^3P)$ from 226-nm Ozone Photodissociation', *Science* 265 (23 September 1994): 1831–8; and 'Ozone Depletion: Environmentalists Say Nobel Prize Demonstrates Need for More Action', *International Environment Reporter: Current Reports* 18, no. 21 (18 October 1995): 796.
20. Useful analysis is in Jason M. Patlis, 'The Multilateral Fund of the Montreal Protocol: A Prototype for Financial Mechanisms in Protecting the Global Environment', *Cornell International Law Journal* 25 (1992): 181–230.
21. Midgley's life approaches the dimensions of a Greek tragedy. In addition to his disastrous invention of CFCs, he also conceived of adding lead to gasoline. After being paralysed in the legs by polio, he invented, in order to move himself in and out of bed, a harness with pulleys, in which he one day became entangled and strangled to death. It is difficult now not to see his brilliantly innovative chemistry as the global analogue of his clever bed harness. See Weiner, *The Next One Hundred Years*, pp. 45–7.

22. Like the firing of an intercontinental missile and like the release of CFCs into the atmosphere, the processes that are building toward changes in surface-level temperatures around the world will cause severe harm to people far away from the territory from which any particular GHGs originate. The extent of temperature change within a state's territory will in no way correlate with the amount of GHGs released within that territory. On the all-important danger to agriculture and food, especially in the tropics, see Cynthia Rosenzweig and Martin L. Parry, 'Potential Impacts of Climate Change on World Food Supply: A Summary of a Recent International Study', in *Agricultural Dimensions of Global Climate Change*, ed. Harry Kaiser and Thomas E. Drennen (Delray Beach, FLA: St. Lucie Press, 1993), pp. 87–116; and, for fuller detail, C. M. Rosenzweig et al., *Climate Change and World Food Supply* (Oxford: Environmental Change Unit, 1993). My own attempts to contribute to analysis of the ethical issues about climate change are (all this volume): 'The unavoidability of justice'; 'Subsistence emissions and luxury emissions'; 'After you: may action by the rich be contingent upon action by the poor?'; 'Avoidable necessity: global warming, international fairness, and alternative energy'; 'Equity in an international agreement on climate change'; and 'Environmental change and the varieties of justice'.

23. Thomas W. Pogge has made a similar point: 'If we follow Rawls's brief sketch . . . we repeat a failing that is common to all historical social-contract doctrines. In assessing the institutional structure of a society by looking merely at how it affects (distributes benefits and burdens among) *its members*, we fail to come to terms with how our society affects the lives of foreigners (and how our lives are affected by how other societies are organized)—we disregard the (negative) externalities a national social contract may impose upon those who are *not* parties to it' (Thomas W. Pogge, *Realizing Rawls* (Ithaca: Cornell University Press, 1989), p. 256, emphasis in original). In this chapter I concentrate on a specific subset of negative externalities, physical harms, as I did in 'Exporting Hazards', in *Boundaries: National Autonomy and Its Limits*, ed. Peter G. Brown and Henry Shue, Maryland Studies in Public Philosophy (Lanham, MD: Rowman and Littlefield, 1982), pp. 107–45. For a society to ignore the physical harms its institutions inflict, provided that the harms are suffered by foreigners, is, I shall be arguing here, a foreign policy that is barbaric.

24. Walzer, *Just and Unjust Wars*, pp. 155, 156. I very briefly discussed a more general notion that I called 'a responsibility to take due care', probably unconsciously influenced by the first edition of *Just and Unjust Wars* (1977), in 'Subsistence Rights: Shall We Secure These Rights?', in *How Does the Constitution Secure Rights?*, ed. Robert A. Goldwin and William A. Schambra (Washington, DC: American Enterprise Institute for Public Policy Research, 1985), p. 92.

25. Walzer, *Just and Unjust Wars*, p. 153.

26. Walzer, *Just and Unjust Wars*, p. 155. I have italicized Walzer's addition; he also changed 'only' to 'narrowly' in the second clause.

27. For provocative thoughts about legal change, see Alfred C. Aman, Jr, 'The Earth as Eggshell Victim: A Global Perspective on Domestic Regulation', *Yale Law Journal* 102, no. 8 (June 1993): 2107–22; and Christopher D. Stone, *The Gnat Is Older than Man: Global Environment and Human Agenda* (Princeton: Princeton University

Press, 1993). Aman's title refers to the 'thin skull' rule of legal liability: you take your victim as you find him. Stone's title is from lines in the Talmud noted by his father, I. F. Stone: 'The world was made for man, though he was the latecomer among its creatures . . . Let him beware of being proud, lest he invite the retort that the gnat is older than he.'

28. These conditions are jointly sufficient for responsibility. Whether any of them are individually necessary conditions is another issue.

29. Jeff McMahan has pointed out that if one pressed the analogy with just war restrictions hard enough, the conclusion reached would be not that no harm at all (with the four features I have listed) could be inflicted but that once one had taken due care, one could inflict only *proportionate* harm (as unintended 'collateral damage'). I think this can be adequately dealt with in either of two ways. First, there is no compulsion to press the analogy that hard into every twist and turn of just war doctrines. My only essential thesis is that even in matters of such extreme significance that a state could be justified in going to war over them, the state must nevertheless constrain how it attempts to win; it is not the case that in war everything is justified. Second—and essentially equivalent in substance—even if one were somehow forced to press the analogy to the point of granting that these kinds of harm (with the four features outlined) were not totally prohibited but restricted to being proportional to the ends being pursued, it is most unlikely that the ends being pursued by normal economic policies are sufficiently momentous to make the infliction of skin cancer (through ozone destruction) or the flooding of agricultural land and homes (through sea-level rise produced by global warming) anywhere near proportional unless one discounts the welfare of the billions of non-compatriots at some outrageously implausible rate. The single greatest cause of GHGs is superfluous driving of excessively powerful automobiles with extremely inefficient internal combustion engines (namely, sport utility vehicles and light trucks) that guzzle fossil fuel. See William K. Stevens, 'With Energy Tug of War, U.S. Is Missing Its Goals', *New York Times*, 28 November 1995, A1, B8. This is not even economically rational—it is sheer waste—much less urgently important enough to justify physically harming people (and causing species extinctions, as climate change surely will). Severe damage to the vital interests of billions would not conceivably be a proportionate side effect qualitatively or quantitatively to the pursuit of frivolous whims by hundreds of millions. See 'Subsistence emissions and luxury emissions', this volume.

30. See Walzer, *Just and Unjust Wars*, pp. 154–5, 188–96.

31. A more moderate conception of state sovereignty also has implications for military intervention, as I have tried to show in 'Let Whatever Is Smoldering Erupt?: Conditional Sovereignty, Reviewable Intervention, and Rwanda 1994', in *Between Sovereignty and Global Governance: The State, Civil Society, and the United Nations*, ed. Christian Reus-Smit, Anthony Jervis, and Albert Paolini (New York: St. Martin's Press, 1998).

8

Bequeathing hazards: security rights and property rights of future humans*

LITTLE SURPRISES

My hobby is building model landmines. Like model electric trains, my little landmines are miniaturized but functioning replicas of the originals. When I was young, I just played with matches like everyone else. Now, in the summertime, I take a few of the model landmines I have assembled during the winter along with me in my backpack and bury some along a different section of the Fingerlakes Trail each weekend. I call them my 'little surprises'. You do not need to worry, though, even if you are an avid hiker and use the Fingerlakes Trail, because I install over the detonator a special metal cover that will not deteriorate for several generations. So not even your grandchildren need to worry about my little surprises because, by the time the mines become active through the deterioration of the detonator cover, the people who will then be alive will be some totally unknown people of a distant generation—probably with very different preference schedules from ours (who knows what they will want?) and certainly with much better technology (they will probably have terrific prosthetic limbs by then). And won't the Boy Scouts (if they still have Boy Scouts) be surprised the summer that my little mines start exploding all along the Trail—it will be a riot! I wish I could actually still be around, but I get a lot of satisfaction from just imagining the chaos.

Now, the criminality and possible insanity of the project just described, had I been serious, are evident. The possible insanity derives from the sheer maliciousness of the project, and the delight in other people's injury, shock, and dismay. However, we could imagine different activities in which the devices I was leaving scattered through the world were not designed precisely so that they would injure and shock others, but were instead designed

* 'Bequeathing Hazards: Security Rights and Property Rights of Future Humans', Henry Shue, in *Limits to Markets: Equity and the Global Environment*, ed. Mohammed Dore and Timothy Mount (Malden, MA: Wiley Blackwell Pubs., 1998), pp. 38–53. This chapter has endnotes rather than footnotes.

primarily to provide some benefit now, and would cause the future damage and dismay only as a by-product—as what, in war, we call 'collateral damage'.

Of course, we do not need to imagine such activities, which combine present benefit with future damage, because we are in fact engaged in them left and right: our nuclear power plants create a previously non-existent hazard in the form of spent fuel that we do not have a clue how to dispose of safely,[1] our fossil fuel driven power plants inject long-resident CO_2 into the upper atmosphere in constantly increasing amounts that will lead to rises in surface temperature sufficient to disrupt agriculture, and so on. These real activities differ from my imaginary landmines in having good intentions, not malicious intentions. In general, however, good intentions do not carry a lot of weight in cost–benefit analyses or other economic calculations, so there is no reason to think they should save anyone here.

Before we leave behind the simple landmines of imagination for the complex landmines of reality, we might pause to see what we can learn from the clarity of the simple case. Why is it so obviously totally wrong for anyone to have a hobby of planting even landmines that would explode only in the distant future? Because the bodies of human beings will be damaged. Some people sometimes engage in the intellectual exercise of denying that there is any such thing as a right at all; certain types of utilitarians, for example, like to deny rights. It is, of course, perfectly possible in theory to deny consistently the existence and even the meaningfulness of all assertions of rights, including the right to physical security—to the integrity of one's body. One can try to imagine, say, a 'state of nature' in which assault, beating, rape, torture, and mayhem violate no rights and break no rules, because there are no such rights or underlying rules. If, however, one approaches a person engaged in general theoretical scepticism about the existence of rights with an instrument well designed to do damage to his body, e.g. a butcher knife, it quickly turns out that, however reluctant on theoretical grounds the person may be to use the word 'right', the person has no doubt whatsoever that it is unacceptable for a person's body to be damaged. It is simply not possible for a sane person to act in practice as if he or she believes that his or her body is not entitled to the kind of special protection against the depredations of others that a right constitutes, whatever intellectual doubts he or she may have about concepts of rights.

I think we would consider a person who genuinely acted as if there were no good reason why his or her body should not be used by other people for those other people's purposes to be suffering from a serious psychological problem, perhaps involving a pathological deficiency in self-esteem. A surgeon may be given permission to cause temporary damage to one's body when there is good reason to believe that this temporary damage is the best available means to longer-term benefit. Yet it is precisely because we do in practice think that we ought to be immune to assaults upon our bodies that surgeons must have permission, must be prepared to defend in court any deviations from standard practices, and so on. We have no serious—non-academic—doubt that a person has a right to physical security (and that one of the minimal police functions of government is to provide a high level of protection for our bodily integrity).

The preceding no doubt says more than enough about the obvious. I have belaboured the obvious to this extent because I think this practically undeniable basic right to physical security has more implications—even for environmental policy—than it may seem to have. For a start, it seems evident that we should operate on the assumption that all the people who will in future actually exist will have this same fundamental right to bodily integrity. Once again, one can, as always, play logical conceivability games: one can imagine a possible world in which, perhaps through genetic engineering, people have come to have, in effect, bodies of Teflon, such that however much others stab at or beat upon them, no damage can be done. In that imagined world, the right to physical security would have lost its point, and the police could cease to concern themselves with physical attacks. It would remain the case that insofar as it remained possible for any person to damage the body of another, every person would retain a right not to suffer the damage that remained possible; in that imagined case, however, I see no objection to saying that the right to bodily integrity had disappeared, along with the function it had served (protecting once highly vulnerable bodies against damage).

However, this imagined case, with its bare logical possibility that the right to physical security could someday lose its point, has zero implications for how we should actually conduct ourselves for the indefinite future, and toward the indefinite future. Absolutely nothing follows from the mere conceivability of change. What we know now is that human bodies are extremely easy to tear, fold, and spindle; consequently, it is extremely important that they be protected. When, and if, we have genuine reason to believe that the human apparatus has become less vulnerable, we can provide it less protection; if invulnerable, no protection at all. Yet, until we have real evidence that things are in fact going to change, it is reasonable to assume only that people in the future—all foreseeable ones—will have the same right to physical security that we have. If, in defence of planting mines along the Fingerlakes Trail, someone offered for no good reason the bare hypothesis that future people might have feet of steel, we would lock him up just as fast.

SURPRISE ANNUITY

All right, then, human beings have basic rights to physical security, which I will have to take into account if I want to pursue my model landmine hobby. Still, I cannot help laughing out loud when I think how amazed those little Boy Scouts would be in around 2098 suddenly to hear those explosions going off all up and down the Fingerlakes Trail, when they thought they were all alone communing with nature. This is too good a scheme to abandon. So here is what I will do. I will endow an annuity—let us call it the 'Surprise annuity'—with enough principal now so that the compounded interest will, by 2098 or so when the mines are likely to become active, have added up to a large enough sum to pay for all the medical expenses and prosthetic limbs and what-not needed by those who suffer from my little joke. No one will end up out of pocket by any amount, and whoever needs one will get a nice new high-tech foot. We could discuss the criteria for adequate compensation; perhaps 'making them whole', as the lawyers call it, by paying all their expenses in restoring normal functioning would not be sufficient compensation, and I should throw in, say, a college scholarship. I am willing to pay what it is worth.

This scheme is becoming more and more insane, but it is important to understand exactly the reason why. The reason, in old-fashioned terminology, is: inalienability. That means: some things are not for sale—especially not some rights. This is not a feature of many rights, of course. Property rights, in particular, are mostly transferable for compensation; marketability is their main function. Not all rights are property rights, however, and the right to physical security is a prime example of a right that is not a property right. This is the main reason why, in the USA, it is illegal to sell bodily organs for use in transplants.[2] Kidneys, hearts, corneas, etc. are non-marketable: there is, by federal statute, no legal way to buy or sell a part of a—or, since the abolition of slavery in 1865, a whole—human being.[3] I cannot transfer the ownership, or use,[4] of a part of my body and—with regard to the imaginary landmine example—I cannot sell anyone the right to inflict damage on my body. By 'cannot', I naturally mean legally and morally cannot; there is usually someone who will in fact trade in just about anything.

If no one can buy a right to inflict damage upon someone else's body, even in the case of people who are currently alive and therefore able to agree to the sale, no one can buy the same right in the case of people of future generations. Someone who thought, for example, that surrogate motherhood is an acceptable practice now could go through the usual exercises and come up with some calculations about, for example, how much in current funds would have to be set aside to pay for a certain amount of surrogacy at some future date. Any given woman at that future date would of course still have to agree—assuming for the sake of argument that surrogacy is the performance of a service, not the sale of a child—to perform the service (at the rate made possible by the amount set aside today). Even if the transaction is permissible, it is permissible only on condition that the service is performed voluntarily. In the imaginary landmine case, the Boy Scouts of 2098 would presumably not voluntarily walk a trail containing active landmines. However, that is not the point.

The point is that it does not matter whether some Scouts could be found in 2098 who would agree to walk the Trail for a large enough package of compensation. A society that still acknowledged a right to physical security would not permit such a transaction then, just as we would not permit it now. Submitting your body to the risk of this kind of injury is a 'service' that we do not allow to be performed; that someone would be willing to perform it has no effect legally or morally.[5] To repeat: the problem for the landmine scheme is not that one cannot get the voluntary agreement (because the people whose agreement would be needed do not yet exist); the problem is that voluntary agreement is immaterial because such agreements are ruled out by our fundamental attitude to the human body.

Do we not only permit but also require financial compensation for bodily injury? What about all the lawsuits following automobile accidents? Why did Dow Chemical agree to set aside billions (without technically acknowledging having done anything wrong, of course) to compensate women who received breast implants made of material created by Dow and who are now suffering various kinds of physical distress? Of course, we do require compensation for physical injury. However, there is, in both legal and moral terms, an absolutely fundamental distinction between compensation for injuring something, e.g. an eye, and the purchase of a right to injure it. If I negligently knock your eye out, I will have to compensate you for the loss of sight, but I cannot make a contract with you to pay you that same amount in advance to purchase the right to knock your eye out. In reality, either way you lose your eye and I pay, but one should not underestimate the magnitude of the legal and moral difference between ex post compensation and ex ante purchase. Compensation is a distinct second best after the best has become impossible; purchase would be the legitimated rejection of the fully possible best.

The difference may seem to lie in voluntariness: in the accident, you lose your eye involuntarily while, in the sale, you would lose it, if you were to agree to the transaction, voluntarily. However, as I indicated already, the deeper element of our legal and moral structure is that voluntariness does not come into it. Voluntariness is ruled out, because the sale of such rights in one's body is disallowed, even for—especially for—those who would agree to such transactions. The traditional way to describe the conceptual structure here was to say that the basic right to physical security is inalienable. Now we would probably say simply that the right in question is non-marketable.

WHY DISCOUNTING IS NOT THE PRIMARY ISSUE

The moral position adopted by economist Richard N. Cooper is to recommend the following principle: 'An appropriate minimum standard is to leave a world no worse than ours in terms of income, and with more options for action, and more knowledge about the implications of these options' (Cooper, 1992, p. 301). I will call this Cooper's minimum standard. Although in the chapter I am citing he simply announces it as if it needed no justifying arguments, Cooper's minimum standard is in itself intuitively appealing. Cooper goes on immediately to say: 'That standard requires comparing mitigation actions with alternative forms of investment, and choosing investments with the highest returns, appropriately measured. What discount rate should be applied?' However, although we certainly have to compare, in the case of climate change that he is discussing, mitigation actions with alternative forms of investment, we certainly do not have simply to choose the investments with the highest returns. To assume that we either must, or even may, choose the investments with the highest returns is to assume that we must or may maximize returns *without* constraint and that is to assume, without argument, that no one has any non-marketable rights that stand in the way of maximization. The assumption that there are no non-marketable rights that might constrain our maximization of returns is, to say the least, extremely controversial—it is in fact flatly contradicted by the widespread assumption that all persons have non-marketable rights to physical security.

Let us back up a little. First, it is essential to be absolutely clear that Cooper's minimum standard is a moral principle, whatever else it may be in addition. To say 'an appropriate minimum standard is to' means 'we must do at least'— a requirement for our (economic) behaviour is being laid down. I am not complaining because a moral principle has been ventured. Life is complicated and we need well-grounded guiding principles; since guiding principles have economic assumptions and economic implications, we need the help of economists (certainly not exclusively of philosophers) in formulating reasonable principles. While principles such as Cooper's minimum standard have economic grounds and economic implications, they nevertheless also tell us how we *may* or *must* conduct ourselves. Consequently, they need to mesh with our other considered judgements about morality as well as our other considered judgements about economics; and one of our most unshakable moral judgements is that people have a (non-marketable) fundamental right not to have their bodies damaged by the actions of others, when the damage is preventable. When bodily damage is unavoidably done, compensation must be paid ex post; this does not mean that one may plan to do the damage ex ante as long as one also plans to compensate, because that would be equivalent to buying what may not be sold. It is not acceptable for me to plant landmines provided only that I also establish an annuity for the medical expenses of my victims.

Cooper's proposed standard is, therefore, too weak a requirement to be a *minimum* standard. It may be that we ought not to do anything that would violate Cooper's standard, but there are activities that would not violate

Cooper's standard that we nevertheless ought not to engage in. One kind of such activities that ought to be ruled out is activities that will predictably inflict physical damage upon other human beings, including human beings who happen not yet to be alive.

Cooper's minimum standard runs into difficulty in part because it is stated with too much generality and consequently covers too many kinds of cases that are significantly different from each other. Brian Barry has argued for a superficially similar-seeming principle, but Barry's principle is in fact quite different because he restricts it specifically to resource depletion. Here is Barry's minimum standard: 'We can now venture a statement of what is required by justice toward future generations. As far as natural resources are concerned, depletion should be compensated for in the sense that later generations should be left no worse off (in terms of productive capacity) than they would have been without the depletion' (Barry, 1983).[6]

The fundamental idea that, I suspect, both Cooper and Barry are attempting to capture is what is often called the 'no-harm principle': one ought not to harm others; and, therefore, one ought not to leave others any worse off than they would have been but for one's actions, with regard to any of their rights.[7] In one respect, Cooper's minimum standard is stronger than Barry's minimum standard, but in another respect Cooper's is weaker than Barry's. On the one hand, Cooper's standard is stronger in that while Barry's principle is strictly negative—'be left no worse off'—Cooper's requires improvements in both options for action and knowledge and settles for the purely negative requirement of no harm only in the case of income. On the other hand, Barry's principle is tightly drawn to apply only to depletion of natural resources—and he makes clear that he is not willing to have everything homogenized into 'utility'—while Cooper's, through being formulated with maximum generality in terms of 'returns', allows any trade-offs among more specific categories that would yield the highest overall 'returns'. Implicitly Cooper's standard allows— because it does not prohibit, in what is explicitly put forward as a 'minimum standard'—trade-offs between:

1. possible current expenditures to prevent, or mitigate, climate change, and

2. investments to provide future generations with greater resources, which they could, if they choose at the time, use for adapting to the effects of unmitigated climate change.

My thesis is that any of those predictable effects that would constitute physical harm to human beings must be prevented and may not be merely compensated for, no matter how great the proposed compensation, if prevention is possible. Before returning to climate change, let us glance at a relatively simpler case, ozone depletion, which, I am assuming, Cooper might approach in a similar fashion.

Emissions of chlorofluorocarbons (CFCs) damage human bodies by causing skin cancer, for one thing—cornea damage is also done. The causal process has only three basic steps, making it extremely simple and clear as disease etiologies go:

1. Having risen into the earth's atmosphere, molecules of the various CFCs spend decades destroying molecules of ozone.

2. The progressively thinner layer of atmospheric ozone allows progressively increased levels of ultraviolet-B (UV-B) radiation from the sun to penetrate to the earth's surface.

3. The increased levels of UV-B produce human skin cancer (as well as impaired sight, plus damage to other animals and to plants) that would otherwise not be expected to occur, including highly dangerous malignant melanomas.

This is about as straightforward as it gets, and the mechanisms are relatively well understood:

> The E.P.A. [US Environmental Protection Agency] estimates that every 1 percent depletion in ozone will cause something like a 2 or 3 percent increase in UV-B and a 5 percent increase in skin cancer, including a 1 percent increase in malignant melanomas. In our lifetimes, the thinning of the ozone shield may lead to a 60 percent rise in the incidence of skin cancers of all kinds in the United States.[8]

A report by the Japan Meteorological Agency (Bureau of National Affairs, 1994, p. 302) found continuing thinning in the Northern Hemisphere, over Japan. Creating CFCs, which are entirely man-made and are a compound not found in nature, creates human cancers, through a long-distance but well-grasped set of reactions.

The production of CFCs is, as everyone knows, gradually being phased out under the Montreal Protocol on Substances that Destroy the Ozone, although it will be a long time before all the 'freon' (one kind of CFC) from all the air conditioners and refrigerators already filled with it has drifted into the sky to begin its long period of ozone-eating. Thus ozone destruction will continue for decades after the end of CFC production. Cooper's minimum standard, if applied to ozone destruction (he was writing about climate change), would require us to choose, using discounting, between:

1. expenditures on the prevention, or mitigation, of CFC-caused future skin cancers, and

2. investments that would provide future generations with greater resources, which 'they' could, if 'they' choose, use for treating unprevented skin cancers (or, if 'they' choose, use some other way).

Our choice now is to be made by 'comparing mitigation actions with alternative forms of investment, and choosing investments with the highest returns, appropriately measured'. In contrast, our common-sense conviction, that no one is free to inflict cancer upon other people, requires that we not trade off prevention measures for the sake of some nebulous overall highest returns to be controlled by an unspecified 'they'. This conviction actually contains two points.

First, a person's right to physical security against, among other things, anthropogenic malignancies is non-marketable and can, therefore, not be purchased for any amount of returns from alternative investments. Physical security is not for sale; one cannot buy rights to inflict disease on humans any more than one can buy rights to blow off the left feet of Boy Scouts.[9] It is, for example, the realization that 'second-hand' tobacco smoke causes cancer that is beginning to put tobacco companies on the defensive. The right to physical security is slowly triumphing in this case over an alleged right to liberty, the 'right to smoke', i.e. the right to force others to inhale your carcinogenic second-hand smoke; the difficulty for the 'right to smoke' is that liberties are conditional upon fulfilment of the no-harm principle, but smoking clearly harms other people in addition to the smoker.

Second, Cooper's proposal is severely weakened by the indeterminateness of the future 'they' who would control the returns from our choice not to mitigate but to invest. Some people in the future will get skin cancer, and some people will control the returns from our current investments on 'their' behalf, but will the people who control the returns know the people who get the cancer? Will they even be in the same nations? Will they care? Here is one place not to ignore property rights. Will the specific 'they' who control the returns from investment, who have the relevant property rights, be willing to spend those returns on the specific 'them' who have the skin cancer? Is there any reason to think they would? To ignore this problem is to ignore property and politics, that is, to ignore both the distribution of wealth and the distribution of power. That is hopelessly apolitical.

Because the institutions created under the Montreal Protocol have already been in operation for a while, ozone destruction is a less interesting case than climate change, to which we will presently return. First, one implication is worth drawing so as to bring out more fully the meaning of the right to physical security in this case: the reduction of ozone production ought to be proceeding as rapidly as possible. If I discover that, somehow unknown to me, I am standing on someone's neck, my duty is not to get off his neck sometime soon—my duty is to get off his neck now.[10] Roughly the same is true of ending the production of CFCs. Such a move cannot, of course, literally be taken immediately, but it could have been taken much faster than it was and could still be greatly speeded up. As is well-known, the negotiations on the Montreal Protocol stalled until after some firm (as it happened, Dupont) came up with a (relatively) safer substitute; this is like my keeping my foot on your neck until I find somewhere else equally soft and comfortable to rest it. To my knowledge no one has been forced, against his wishes, to spend a single summer night uncomfortably warm due to lack of air conditioning for the sake of reducing the ultimate total accumulation of CFCs in the atmosphere, although there is every reason to think that additional skin cancers have been caused by our continuing purchases of CFC-filled air conditioners for cooling as usual. Similarly, the phase-out of production of CFCs appears to be moving at a leisurely pace designed to avoid causing any manufacturers any problems. If we genuinely believe we are not entitled to inflict cancer on others, the phase-out—and the crackdown on the mushrooming black market in freon—ought to be proceeding with more urgency.[11]

Does a respect for rights, then, mean throwing efficiency to the winds? No, but it does challenge the conventional way, typified by Cooper's treatment of climate change, of setting up our decisions. Any pursuit of highest returns that is to be acceptable, I have argued above, must be conducted within the constraints set by established non-marketable fundamental rights.[12] Since the non-marketable rights have no price, they cannot without distortion be incorporated into a single comprehensive efficiency calculation. Their having no price does not mean, as is often suggested, that they have an infinite price— that suggestion is another attempt to insist on incorporating into a single set of calculations what cannot be incorporated.[13] Non-marketable rights are con-straints that must be honoured, even during the pursuit of highest returns.

Therefore, one may not calculate the path with highest returns by throwing potential expenditures for ceasing, preventing, and correcting violations of non-marketable rights into a single soup with potential expenditures for the satisfaction of ordinary preferences. Decisions need instead to be partitioned into at least two sets:

1. alternative ways of ceasing, preventing, and correcting violations of non-marketable rights, and

2. alternative paths to the highest returns using resources not allocated to the best choice from set 1.[14]

Discounting then becomes a secondary, although still important, issue within each set, but not between the two sets, because trade-offs are not routinely made between sets. What is primary is the partition between:

1. resources for the prevention of new violations of non-marketable rights and for the restoration of non-marketable rights already violated, and

2. resources not required by respect for non-marketable rights.

Do rights belong, then, to some ethereal realm in which costs, including opportunity costs, somehow do not matter? No; if an attempt to honour a particular putative right can be known to be excessively costly, we do not acknowledge it as a right. Ascertaining whether a proposed right is 'excessively costly' does indeed involve considering its opportunity costs, in a general sense—that is, comparing the importance of securing the right in question to the importance of the next best use of the same resources.[15] The judgement about whether we can afford to treat something as the content of a right, however, is a prior, stable judgement that is not reopened every time that we must choose between consuming resources in the enforcement of the right and consuming the same resources in some other way. Usually, it is a judgement that cannot be quantitatively calculated. In principle, the question whether we can afford to treat any given matter as the content of a right can be reopened at any time. However, if the question of the affordability of something simply stays open all the time, that matter is being treated merely as the content of one more preference, not treated as the content of a right.[16] We do have to allocate resources between fundamental rights and ordinary preferences, as well as to make allocations within the sphere of rights and within the sphere of preferences. Yet the allocation between spheres takes precedence. First, we provide for basic rights; then, preference satisfaction uses whatever resources are left.[17]

Efficiency of allocation among alternative uses of resources remains important within each of the two segregated sets of decisions. Inside the first set, the rights-fulfilling set, allocations need to be made as efficiently as anywhere else: it is no more rational to waste resources partitioned off to provide for rights than to waste any other resources. What kind of discounting, if any, is appropriate then becomes the next issue. Within the second, or return-maximizing, set, appropriate kinds of efficiency are also called for. What kind of discounting, if any, is appropriate becomes the next issue there too.

DISCOUNTING AND PROPERTY RIGHTS

Daniel W. Bromley (1991, pp. 84–103) in a chapter called 'Property rights, missing markets, and environmental uncertainty', has reflected on one truly bizarre feature of conventional discounting: it allocates *all* property rights to the current generation.[18] Observing that between generations what one has is not market failure but an entirely missing market, Bromley says: 'The economic analysis has implicitly assumed that the present generation has a right to impose costs on the future and can only be denied that right if it is more efficient to do otherwise' (Bromley, 1991, p. 88). The present 'has a right' to impose the costs in the sense that it has no property rights-based duty not to impose them; and the present has no property rights-based duty not to impose them only if future generations have no property rights at all. Bromley (1991, p. 93) goes on to note that this sharp dichotomy of all-rights and no-rights defines a very specific decision environment and specifies a completely asymmetrical benchmark:

> Notice that the structure of property rights not only determines how decisions are framed and choices made, but also determines how we shall assess the impacts of those choices. If the future is regarded as having no legitimate case to have its losses covered by the present[,] then one might be tempted to view the problem as one of the present generation having to sacrifice present and future income in order to make the future better off. On the other hand, if the future is regarded as having a *right* not to incur losses at the hands of the present, then those now living will be viewed as doing what is correct—giving up ill-gotten income—in order not to impair the well-being of the future.

Now, I have attempted in the bulk of this chapter to show that future humans, just like current humans, have non-property rights, most notably non-marketable rights to physical security. Bromley, by contrast, is discussing property rights and suggesting that future humans have property rights too, or at the very least that it is passing strange that absolutely all property rights should happen to belong to the humans who happen to be alive at any given time. This assignment of property rights, Bromley notes, predetermines the outcome of any discounting.

It is important to distinguish between control and benefit.[19] Insofar as property rights are construed as a matter of control, they must be assigned entirely to the current generation. For better or for worse, any given current generation does in fact exercise control. Control by the living is inevitable. Benefit, however, is a different matter. The living need not exercise their exclusive control for their own exclusive, or even primary, benefit, which is what is done when one maximizes net present value. The technique of discounting values everything from one's own point of view exclusively, which seems to be roughly the opposite of morality, which mandates somehow including the points of view of others, presumably not excluding future generations, along with one's own. Maximizing net present value utterly ignores all rights, property rights as well as non-marketable rights, of every person who will ever live after the present. Whatever may be the correct approach, simply maximizing net present value is certainly not it.

Suppose we grant that all persons, including persons yet to be born, have at least basic security rights—and some (unspecified) property rights, although nothing in my positive argument turns on the property rights. What are the implications of this proposition about minimum universal rights for our current policies regarding climate change? Obviously space here permits merely a couple of hints toward a full answer, but it turns out that what we can establish fairly easily does a great deal to set the direction for policy. First, if all persons universally share some rights, then we in the current generation have those rights too. Those of us now alive are living the only life we will ever get, so we ought to have now everything that everyone is entitled to. This immediately establishes an ultimate ceiling on how much we can be expected to sacrifice for future generations no matter how many more of them than of us there will be.[20] The most that can be demanded of us is a level of sacrifice that does not compromise our secure enjoyment of the same rights that, we are acknowledging, belong as well to persons in the future. To deprive ourselves of basic rights in order to guarantee those same basic rights to people in the future would be in effect to treat our generation as inferior—as somehow entitled to less than equal minimum rights.[21]

Ironically, this undercuts one of the main reasons often given to justify the alleged need to discount future generations: we ought not to have to make excessive sacrifices for the sake of future generations; therefore, we ought to discount their welfare.[22] The premise is true, but the inference is invalid. It is the case that the presumably indefinitely large numbers of persons in all future generations will completely swamp the numbers in any one current generation if one engages in simple aggregation without discounting, and discounting is indeed one method for (radically) reducing the weight given to people in the future even where several generations are aggregated. It is, however, certainly not the only one. Another is to respect universal basic rights, including the rights of people in the present. That is the immediate good news for the current generation.

The bad news is that we are prohibited from engaging in any practices that will make it impossible for people in the future to enjoy their basic rights when, but for our choosing to engage in those practices rather than other practices open to us, they could have enjoyed their basic rights too. The doubtlessly obvious reason for the qualification, 'when, but for our engaging in those practices rather than other practices open to us', is to rule out the following: it was what we did that closed the option for them, but we had no alternative. For example, we consumed the last energy available to the planet, but we ourselves could not have survived without consuming that much energy: we have the same basic rights as people in the future, and no alternative way to enjoy our rights was open to us. If there were such an absolute, irremediable scarcity of something vital such as energy, it would not be wrong for some people—such as us—to consume minimum shares of it even though this would cause other people to be forced to do without it. If the supply of something vital actually were zero-sum and the total were grossly inadequate for everyone to have their minimum share, it would not be wrong to go ahead and consume your own minimum share. Any alternative view would entail a general duty to commit suicide.[23]

However, while this may be a moderately interesting theoretical possibility, it does not even remotely resemble what is in fact happening. Far from only reluctantly and sadly (because we were so sensitive to the fact that we were tragically dooming others with the same rights as we) consuming precisely our minimum share of the planet's resources and pollution sinks, many of us in the rich countries are, I would think, more nearly engaged in an orgy of self-indulgent consumption and unbridled pollution with little or no thought about the fate of anyone more than approximately two generations after us. The bad news, then, is that we ought to be avoiding consumption and pollution that are superfluous for us and are threatening to basic rights of people in the future.

And exactly which consumption and pollution is that? Obviously, one cannot establish concretely what this limitation means entirely through the kinds of purely theoretical argument with which economists and philosophers feel most at home, and which constitute this chapter. Factual information is, unfortunately for us theorists, needed in addition to theoretical argument. However, uncertainties about anthropogenic climate change are being reduced (Houghton et al., 1996). The not-so-unlikely threats to the physical security of people in the not-so-distant future from fossil fuel consumption by people in the present include:

- the migration away from the equator toward both poles of semi-tropical habitats suitable for mosquitoes and consequently of the mosquito-borne diseases that currently wreak havoc only in the tropics;
- the infiltration by saltwater of the fresh groundwater supplies of gigantic population centres such as Shanghai;
- the modification in some impoverished regions of the crucial parameters for agriculture such as rainfall patterns and length of growing season, in addition to temperatures themselves, more rapidly than seed types, irrigation, and fertilizer can in fact be adjusted with the resources likely to be available to many of the people affected;
- and much else.

Uncertainties remain and research needs to continue. But where present practices that are superfluous, trivial, or frivolous—and could consequently be changed at the sacrifice only of preferences of no intrinsic value—are likely to contribute to the infliction of physical harms on people who will live later, we are bound to change our practices to protect their rights. This is an essential element in any minimum standard to guide our behaviour where it will affect the physical security of people who will succeed us in this environment.

NOTES

1. The best discussion of the follies in current US management of nuclear waste is given by Kristin Shrader-Frechette (1993). We always have new schemes, of course. One is to pay ethnic minorities, who so badly need the money because they have become impoverished as a result of decades of discrimination and

violations of treaty agreements, to let us store it near them instead of near us; see Erickson et al. (1994).

2. If we did allow the purchase in the Third World, and then the import, of human organs for use in transplants, the implication would be clear: people here are human beings in a sense that is incompatible with the sale and purchase of their physical parts, but people in the 'source' [!] countries are something else deserving less respect. In 1994, a US tourist was beaten to death in a small town in Guatemala (as reported on national public radio). When the local people were asked why, they said they had heard that people in the USA were adopting Guatemalan children and, once getting them to the USA, cutting them up to use their vital organs for transplants into ill US children. They suspected the woman they beat to death was there to adopt a child for this purpose. This woman was in fact a completely innocent tourist, not even inquiring about adoption, and the general rumour was not only false but was being calculatedly spread by an exceptionally virulently anti-US political party. It is still noteworthy how believable the rumour turned out to be.

3. In the states in which contracts for surrogate motherhood are legally enforceable, their enforceability rests on a legal fiction that what is occurring is not the sale of a baby but the performance of a service (which happens to result in a baby) by the surrogate mother. Whether surrogate motherhood is a violation of our basic views about rights remains to be settled. The payments involved in old-fashioned adoptions are also treated as compensations for adoption services, not payments for the purchase of a baby. Lots of illegal payments change hands, of course.

4. Prostitution, where legal, is treated—correctly or incorrectly—as payment for services, not as rental of body parts.

5. Professional boxing could be argued to come very close.

6. See Barry (1983, p. 23). Josh Farley (1998) has persuasively suggested that no compensation could be adequate for the avoidable complete destruction of a renewable resource.

7. John Locke is customarily credited with the no-harm principle, and he certainly embraced it, but what about Hippocrates? The Hippocratic Oath contains a no-harm principle for medical people. I would have thought most morality was at bottom about taking other people seriously and at least, therefore, not inflicting harm on them.

8. Jonathan Weiner (1990, p. 155) cites John S. Hoffman (1987).

9. Here is a possible exception: although medical experimentation may normally be conducted only upon diseases that people already have, not by purposely infecting them with disease for the sake of experimentation, people do sometimes volunteer to have a disease inflicted upon them. This appears to make this case a matter of permission (which we do not have from future generations for the infliction of skin cancer), not exactly traditional inalienability. On the other hand, such people would normally volunteer and be honoured, but not be paid; so it is not marketability either.

10. I have discussed such 'extrication ethics', as Tony Coady has called it, in Shue 'Environmental change and the varieties of justice', this volume, 129–30.

11. I am inclined to think that no nation has a right to refuse to do its fair share in a task of this kind, namely the prevention of the severe physical harm to humans that will be caused by a nation's failure to do its share. The existence of a black market, of course, demonstrates anew the difficulty of state prohibitions on purchases of commodities that people with the money strongly desire to buy, but if the nation state system fails to protect our physical health wherever it is profitable for someone to inflict disease, we need new political structures.

12. Are rights simply side constraints? No, they are side constraints upon the pursuit of preferences, but they are also goals to be attained. The provision of adequate security for basic rights is a goal that takes priority over the satisfaction of preferences, which is why rights function as side constraints on the satisfaction of preferences.

13. In a somewhat similar instance, that kidneys for transplant are non-marketable does not mean that they are very expensive; it means one cannot legally or morally buy them. Period. Insofar as the legal and moral prohibition fails, there will be a 'black' market, in which organs for transplant will indeed be expensive. This black-market price is a violation of their status as non-marketable, however, not an illustration of what their status means.

14. For the distinction among ceasing, preventing, and correcting, see Shue, *Basic Rights: Subsistence, Affluence, and US Foreign Policy*, 2nd edn. (Princeton: Princeton University Press, 1996), ch. 2.

15. The clearest account of how the consideration of costs should come into accounts of human rights that I am aware of is given by James W. Nickel (1987, pp. 120–30).

16. I have argued against the 'homogenization' by economists of all considerations as so-called preferences—in a misguided effort to create a single, comprehensive calculation about climate change—in Shue, 'Subsistence emissions and luxury emissions', this volume, 62–4.

17. This does not mean that no further normative issues arise once provision has been made for basic rights. The provision for basic rights will include provision for urgent needs, but questions of distributive justice still arise as we choose the mechanisms that determine whose preferences will in fact be satisfied to what extent.

18. I am grateful to Josh Farley for this reference and for provocative discussions of discounting.

19. The value of carefully separating control and benefit is emphasized by Thomas W. Pogge (1989, pp. 251–2), in a critique of Beitz's resource redistribution principle. I think, however, Beitz observes the distinction to a greater extent than Pogge allows—see Charles R. Beitz (1979, p. 138).

20. The final ceiling, all things considered, on the extent of our sacrifice may be still lower than this initial ceiling, but it will not be any higher.

21. This is not to say that it is not the case that an earlier generation could ever be expected to make sacrifices in its overall quality of life, where the quality was well above the minimum, so that later generations could have a still higher quality of life. My point is about the minimum guaranteed by basic rights: an earlier generation cannot be expected to accept less than the minimum so that later generations can have more than the minimum. The minimum is the minimum for everyone. Where the earlier generation and the later generations are all above the minimum, the theoretical issues become much more complex, especially if

moderate sacrifices by a few early generations would produce vast improvements for indefinitely many later generations. At present, given the extent of the threats to the environment, the health of which is necessary to sustaining minimum quality of life, one can be forgiven for thinking that the practically relevant questions are, not about a 'just saving rate' to make future generations better off, but about a just pollution rate to prevent future generations from being worse off or indeed falling below the minimum.

22. Derek Parfit (1983, pp. 31–7) calls this 'the argument from excessive sacrifice'; Parfit also refutes five other principal justifications for discounting. The same arguments and examples are repeated, with additional material, in Cowen and Parfit (1991, pp. 144–61).

23. It is not even clear that there is a coherent alternative consisting of a general duty for a whole generation to commit suicide. Duty to whom? Future generations? What future generations? Anyway, as I next indicate in the text, this is hardly our immediate problem, however tantalizing theoretically.

REFERENCES

Barry, B. (1983) 'Intergenerational justice in energy policy', in MacLean and Brown (1983); reprinted as: 'The ethics of resource depletion', in Barry, B. (1991) *Liberty and Justice, Essays in Political Theory*, vol. 2, Oxford: Oxford University Press at the Clarendon Press.

Beitz, C. R. (1979) *Political Theory and International Relations*. Princeton: Princeton University Press.

Bromley, D. W. (1991) *Environment and Economy: Property Rights and Public Policy*. Cambridge, MA: Blackwell Publishers.

Bureau of National Affairs, Washington (1994) 'Report cites largest ozone hole ever as thinning occurs at "significant pace",' *International Environment Reporter: Current Reports*, 17 (7) 6 April.

Cooper, R. N. (1992) 'United States policy towards the global environment', in A. Hurrell and B. Kingsbury (eds) *The International Politics of the Environment: Actors, Interests, and Institutions*. Oxford: Oxford University Press at the Clarendon Press.

Cowen, T. and Parfit, D. (1991) 'Against the social discount rate', in P. Laslett and J. S. Fishkin (eds) *Justice between Age Groups and Generations*. New Haven: Yale University Press.

Erickson, J. D., Chapman, D. and Johnny, R. E. (1994) 'Monitored retrievable storage of spent nuclear fuel in Indian country: Liability, Sovereignty and Socioeconomics', *American Indian Law Review*, 19 (1), 73–103.

Farley, J. (1998) '"Optimal" deforestation in the Brazilian Amazon—theory and policy: The local, national, international and international viewpoints'. PhD dissertation Cornell University.

Hoffman, J. S. (ed.) (1987) *Assessing the Risks of Trace Gases that can Modify the Stratosphere*. Washington, DC: Office of Air and Radiation, US Environmental Protection Agency, December.

Houghton, J. T., Meira Filho, L. G., et al. (1996) *Climate change 1995: The science of climate change*. Contribution of Working Group 1 to the Second Assessment Report of the Intergovernmental Panel on Climate Change. Cambridge and New York: IPCC.

MacLean, D. and Brown, P. G. (eds) (1983) *Energy and the Future*. Lanham, MD: Rowman and Littlefield.

Nickel, J. W. (1987) *Making Sense of Human Rights: Philosophical Reflections on the Universal Declaration of Human Rights*. Berkeley: University of California Press.

Parfit, D. (1983) 'Energy policy and the further future: The social discount rate', in MacLean and Brown (1983).

Pogge, T. W. (1989) *Realizing Rawls*. Ithaca, NY: Cornell University Press.

Shrader-Frechette, K. (1993) *Burying Uncertainty: Risk and the Case Against Geological Disposal of Nuclear Waste*. Berkeley, CA: University of California Press.

Weiner, J. (1990) *The Next One Hundred Years: Shaping the Fate of our Living Earth*. New York: Bantam Books.

9

Global environment and international inequality*

My aim is to establish that three common-sense principles of fairness, none of them dependent upon controversial philosophical theories of justice, give rise to the same conclusion about the allocation of the costs of protecting the environment.

Poor states and rich states have long dealt with each other primarily upon unequal terms. The imposition of unequal terms has been relatively easy for the rich states because they have rarely needed to ask for the voluntary cooperation of the less powerful poor states. Now the rich countries have realized that their own industrial activity has been destroying the ozone in the earth's atmosphere and has been making far and away the greatest contribution to global warming. They would like the poor states to avoid adopting the same form of industrialization by which they themselves became rich. It is increasingly clear that if poor states pursue their own economic development with the same disregard for the natural environment and the economic welfare of other states that rich states displayed in the past during their development, everyone will continue to suffer the effects of environmental destruction. Consequently, it is at least conceivable that rich states might now be willing to consider dealing cooperatively on equitable terms with poor states in a manner that gives due weight to both the economic development of poor states and the preservation of the natural environment.

If we are to have any hope of pursuing equitable cooperation, we must try to arrive at a consensus about what equity means. And we need to define equity not as a vague abstraction but concretely and specifically in the context of both development of the economy in poor states and preservation of the environment everywhere.

* 'Global Environment and International Inequality', Henry Shue, *International Affairs*, vol. 75, no. 3 (1999), Wiley, pp. 531–45.

FUNDAMENTAL FAIRNESS AND ACCEPTABLE INEQUALITY

What diplomats and lawyers call equity incorporates important aspects of what ordinary people everywhere call fairness. The concept of fairness is neither Eastern nor Western, Northern nor Southern, but universal.[1] People everywhere understand what it means to ask whether an arrangement is fair or biased towards some parties over other parties. If you own the land but I supply the labour, or you own the seed but I own the ox, or you are old but I am young, or you are female but I am male, or you have an education and I do not, or you worked long and hard but I was lazy—in situation after situation it makes perfectly good sense to ask whether a particular division of something among two or more parties is fair to all the parties, in light of this or that difference between them. All people understand the question, even where they have been taught not to ask it. What would be fair? Or, as the lawyers and diplomats would put it, which arrangement would be equitable?

Naturally, it is also possible to ask other kinds of questions about the same arrangements. One can always ask economic questions, for instance, in addition to ethical questions concerning equity: would it increase total output if, say, women were paid less and men were paid more? Would it be more efficient? Sometimes the most efficient arrangement happens also to be fair to all parties, but often it is unfair. Then a choice has to be made between efficiency and fairness. Before it is possible to discuss such choices, however, we need to know the meaning of equity: what are the standards of equity and how do they matter?

Complete egalitarianism—the belief that all good things ought to be shared equally among all people—can be a powerfully attractive view, and it is much more difficult to argue against than many of its opponents seem to think. I shall, nevertheless, assume here that complete egalitarianism is unacceptable. If it were the appropriate view to adopt, our inquiry into equity could end now. The answer to the question, 'What is an equitable arrangement?' would always be the same: an equal distribution. Only equality would ever provide equity.

While I do assume that it may be equitable for some good things to be distributed unequally, I also assume that other things must be kept equal—most importantly, dignity and respect. It is part of the current international consensus that every person is entitled to equal dignity and equal respect. In traditional societies in both hemispheres, even the equality of dignity and respect was denied in theory as well as practice. Now, although principles of equality are still widely violated in practice, inequality of dignity and of respect

---

[1] Or so I believe. I would be intensely interested in any evidence of a culture that seems to lack a concept of fairness, as distinguished from evidence about two cultures whose specific conceptions of fairness differ in some respects.

have relatively few public advocates even among those who practise them. If it is equitable for some other human goods to be distributed unequally, but it is not equitable for dignity or respect to be unequal, the central questions become: 'Which inequalities in which other human goods are compatible with equal human dignity and equal human respect?' and 'Which inequalities in other goods ought to be eliminated, reduced, or prevented from being increased?'

When one is beginning from an existing inequality, such as the current inequality in wealth between North and South, three critical kinds of justification are: justifications of unequal burdens intended to reduce or eliminate the existing inequality by removing an unfair advantage of those at the top; justifications of unequal burdens intended to prevent the existing inequality from becoming worse through any infliction of an unfair additional disadvantage upon those at the bottom; and justifications of a guaranteed minimum intended to prevent the existing inequality from becoming worse through any infliction of an unfair additional disadvantage upon those at the bottom. The second justification for unequal burdens and the justification for a guaranteed minimum are the same: two different mechanisms are being used to achieve fundamentally the same purpose. I shall look at these two forms of justification for unequal burdens and then at the justification for a guaranteed minimum.

UNEQUAL BURDENS

Greater contribution to the problem

All over the world parents teach their children to clean up their own mess. This simple rule makes good sense from the point of view of incentive: if one learns that one will not be allowed to get away with simply walking away from whatever messes one creates, one is given a strong negative incentive against making messes in the first place. Whoever makes the mess presumably does so in the process of pursuing some benefit—for a child, the benefit may simply be the pleasure of playing with the objects that constitute the mess. If one learns that whoever reaps the benefit of making the mess must also be the one who pays the cost of cleaning up the mess, one learns at the very least not to make messes with costs that are greater than their benefits.

Economists have glorified this simple rule as the 'internalization of externalities'. If the basis for the price of a product does not incorporate the costs of cleaning up the mess made in the process of producing the product, the costs are being externalized, that is, dumped upon other parties. Incorporating into the basis of the price of the product the costs that had been coercively socialized is called internalizing an externality.

At least as important as the consideration of incentives, however, is the consideration of fairness or equity. If whoever makes a mess receives the benefits and does not pay the costs, not only does he have no incentive to avoid making as many messes as he likes, but he is also unfair to whoever does pay the costs. He is inflicting costs upon other people, contrary to their interests and, presumably, without their consent. By making himself better off in ways that make others worse off, he is creating an expanding inequality.

Once such an inequality has been created unilaterally by someone's imposing costs upon other people, we are justified in reversing the inequality by imposing extra burdens upon the producer of the inequality. There are two separate points here. First, we are justified in assigning additional burdens to the party who has been inflicting costs upon us. Second, the minimum extent of the compensatory burden we are justified in assigning is enough to correct the inequality previously unilaterally imposed. The purpose of the extra burden is to restore an equality that was disrupted unilaterally and arbitrarily (or to reduce an inequality that was enlarged unilaterally and arbitrarily). In order to accomplish that purpose, the extra burden assigned must be at least equal to the unfair advantage previously taken. This yields us our first principle of equity:

> When a party has in the past taken an unfair advantage of others by imposing costs upon them without their consent, those who have been unilaterally put at a disadvantage are entitled to demand that in the future the offending party shoulder burdens that are unequal at least to the extent of the unfair advantage previously taken, in order to restore equality.[2]

In the area of development and the environment, the clearest cases that fall under this first principle of equity are the partial destruction of the ozone layer and the initiation of global warming by the process of industrialization that has enriched the North but not the South. Unilateral initiatives by the so-called developed countries (DCs) have made them rich, while leaving the less developed countries (LDCs) poor. In the process the industrial activities and accompanying lifestyles of the DCs have inflicted major global damage upon the earth's atmosphere. Both kinds of damage are harmful to those who did not benefit from Northern industrialization as well as to those who did. Those societies whose activities have damaged the atmosphere ought, according to the first principle of equity, to bear sufficiently unequal burdens henceforth to correct the inequality that they have imposed. In this case, everyone is bearing costs—because the damage was universal—but the benefits have been overwhelmingly skewed towards those who have become rich in the process.

---

[2] A preliminary presentation of these principles at New York University Law School has been helpfully commented upon in Thomas M. Franck, *Fairness in international law and institutions* (Oxford: Clarendon, 1997), pp. 390–1.

This principle of equity should be distinguished from the considerably weaker—because entirely forward-looking—'polluter pays' principle (PPP), which requires only that all future costs of pollution (in production or consumption) be henceforth internalized into prices. Even the OECD formally adopted the PPP in 1974, to govern relations among rich states.[3]

Spokespeople for the rich countries make at least three kinds of counterarguments to this first principle of equity. These are:

1. The LDCs have also benefited, it is said, from the enrichment of the DCs. Usually it is conceded that the industrial countries have benefited more than the non-industrialized. Yet it is maintained that, for example, medicines and technologies made possible by the lifestyles of the rich countries have also reached the poor countries, bringing benefits that the poor countries could not have produced as soon for themselves.

Quite a bit of breath and ink has been spent in arguments over how much LDCs have benefited from the technologies and other advances made by the DCs, compared to the benefits enjoyed by the DCs themselves. Yet this dispute does not need to be settled in order to decide questions of equity. Whatever benefits LDCs have received, they have mostly been charged for. No doubt some improvements have been widespread. Yet, except for a relative trickle of aid, all transfers have been charged to the recipients, who have in fact been left with an enormous burden of debt, much of it incurred precisely in the effort to purchase the good things produced by industrialization.

Overall, poor countries have been charged for any benefits that they have received by someone in the rich countries, evening that account. Much greater additional benefits have gone to the rich countries themselves, including a major contribution to the very process of their becoming so much richer than the poor countries. Meanwhile, the environmental damage caused by the process has been incurred by everyone. The rich countries have profited to the extent of the excess of the benefits gained by them over the costs incurred by everyone through environmental damage done by them, and ought in future to bear extra burdens in dealing with the damage they have done.

2. Whatever environmental damage has been done, it is said, was unintentional. Now we know all sorts of things about CFCs and the ozone layer, and about carbon dioxide and the greenhouse effect, that no one dreamed of when CFCs were created or when industrialization fed with fossil fuels began. People cannot be held responsible, it is maintained, for harmful effects that they could not have foreseen. The philosopher Immanuel Kant is often quoted in the West for having said, 'Ought presupposes can'—it can be true that one ought to have done something only if one actually could have done it. Therefore, it is

---

[3] OECD Council, 14 November 1974c (1974), 223 (Paris: OECD, 1974).

allegedly not fair to hold people responsible for effects they could not have avoided because the effects could not have been predicted.

This objection rests upon a confusion between punishment and responsibility. It is not fair to punish someone for producing effects that could not have been avoided, but it is common to hold people responsible for effects that were unforeseen and unavoidable.

We noted earlier that, in order to be justifiable, an inequality in something between two or more parties must be compatible with an equality of dignity and respect between the parties. If there were an inequality between two groups of people such that members of the first group could create problems and then expect members of the second group to deal with the problems, that inequality would be incompatible with equal respect and equal dignity. For the members of the second group would in fact be functioning as servants for the first group. If I said to you, 'I broke it, but I want you to clean it up,' then I would be your master and you would be my servant. If I thought that you should do my bidding, I could hardly respect you as my equal.

It is true, then, that the owners of many coal-burning factories could not possibly have known the bad effects of the carbon dioxide they were releasing into the atmosphere, and therefore could not possibly have intended to contribute to harming it. It would, therefore, be unfair to punish them—by, for example, demanding that they pay double or triple damages. It is not in the least unfair, however, simply to hold them responsible for the damage that they have in fact done. This naturally leads to the third objection.

3. Even if it is fair to hold a person responsible for damage done unintentionally, it will be said, it is not fair to hold the person responsible for damage he did not do himself. It would not be fair, for example, to hold a grandson responsible for damage done by his grandfather. Yet it is claimed this is exactly what is being done when the current generation is held responsible for carbon dioxide emissions produced in the nineteenth century. Perhaps Europeans living today are responsible for atmosphere-damaging gases emitted today, but it is not fair to hold people responsible for deeds done long before they were born.

This objection appeals to a reasonable principle, namely that one person ought not to be held responsible for what is done by another person who is completely unrelated. 'Completely unrelated' is, however, a critical portion of the principle. To assume that the facts about the industrial North's contribution to global warming straightforwardly fall under this principle is to assume that they are considerably simpler than they actually are.

First, and undeniably, the industrial states' contributions to global warming have continued unabated long since it became impossible to plead ignorance. It would have been conceivable that as soon as evidence began to accumulate that industrial activity was having a dangerous environmental effect, the industrial states would have adopted a conservative or even cautious policy of cutting back greenhouse gas emissions or at least slowing their rate of increase. For the most part this has not happened.

Second, today's generation in the industrial states is far from completely unrelated to the earlier generations going back all the way to the beginning of the Industrial Revolution. What is the difference between being born in 1975 in Belgium and being born in 1975 in Bangladesh? Clearly one of the most fundamental differences is that the Belgian infant is born into an industrial society and the Bangladeshi infant is not. Even the medical setting for the birth itself, not to mention the level of prenatal care available to the expectant mother, is almost certainly vastly more favourable for the Belgian than the Bangladeshi. Childhood nutrition, educational opportunities and lifelong standards of living are likely to differ enormously because of the difference between an industrialized and a non-industrialized economy. In such respects current generations are, and future generations probably will be, continuing beneficiaries of earlier industrial activity.

Nothing is wrong with the principle invoked in the third objection. It is indeed not fair to hold someone responsible for what has been done by someone else. Yet that principle is largely irrelevant to the case at hand, because one generation of a rich industrial society is not unrelated to other generations past and future. All are participants in enduring economic structures. Benefits and costs, and rights and responsibilities, carry across generations.

We turn now to a second, quite different kind of justification of the same mechanism of assigning unequal burdens. This first justification has rested in part upon the unfairness of the existing inequality. The second justification neither assumes nor argues that the initial inequality is unfair.

Greater ability to pay

The second principle of equity is widely accepted as a requirement of simple fairness. It states:

> Among a number of parties, all of whom are bound to contribute to some common endeavour, the parties who have the most resources normally should contribute the most to the endeavour.

This principle of paying in accordance with ability to pay, if stated strictly, would specify what is often called a progressive rate of payment: insofar as a party's assets are greater, the rate at which the party should contribute to the enterprise in question also becomes greater. The progressivity can be strictly proportional—those with double the base amount of assets contribute at twice the rate at which those with the base amount contribute, those with triple the base amount of assets contribute at three times the rate at which those with the base amount contribute, and so on. More typically, the progressivity is not strictly proportional—the more a party has, the higher the rate at which it is expected to contribute, but the rate does not increase in strict proportion to increases in assets.

The general principle itself is sufficiently fundamental that it is not necessary, and perhaps not possible, to justify it by deriving it from considerations that are more fundamental still. Nevertheless, it is possible to explain its appeal to some extent more fully. The basic appeal of payment in accordance with ability to pay as a principle of fairness is easiest to see by contrast with a flat rate of contribution, that is, the same rate of contribution by every party irrespective of different parties' differing assets. At first thought, the same rate for everyone seems obviously the fairest imaginable arrangement. What could possibly be fairer, one is initially inclined to think, than absolutely equal treatment for everyone? Surely, it seems, if everyone pays an equal rate, everyone is treated the same and therefore fairly? This, however, is an exceedingly abstract approach, which pays no attention at all to the actual concrete circumstances of the contributing parties. In addition, it focuses exclusively upon the contribution process and ignores the position in which, as a result of the process, the parties end up. Contribution according to ability to pay is much more sensitive both to concrete circumstance and to final outcome.

Suppose that Party A has 90 units of something, Party B has 30 units, and Party C has 9 units. In order to accomplish their missions, it is proposed that everyone should contribute at a flat rate of one-third. This may seem fair in that everyone is treated equally: the same rate is applied to everyone, regardless of circumstances. When it is considered that A's contribution will be 30 and B's will be 10, while C's will be only 3, the flat rate may appear more than fair to C who contributes only one-tenth as much as A does. However, suppose that these units represent $100 per year in income and that where C lives it is possible to survive on $750 per year but on no less. If C must contribute 3 units—$300—he will fall below the minimum for survival. While the flat rate of one-third would require A to contribute far more ($3,000) than C, and B to contribute considerably more ($1,000) than C, both A (with $6,000 left) and B (with $2,000 left) would remain safely above subsistence level. A and B can afford to contribute at the rate of one-third because they are left with more than enough while C is unable to contribute at that rate and survive.

While flat rates appear misleadingly fair in the abstract, they do so largely because they look at only the first part of the story and ignore how things turn out in the end. The great strength of progressive rates, by contrast, is that they tend to accommodate final outcomes and take account of whether the contributors can in fact afford their respective contributions.

A single objection is usually raised against progressive rates of contribution: disincentive effects. If those who have more are going to lose what they have at a greater rate than those who have less, the incentive to come to have more in the first place will, it is said, be much less than it would have been with a flat rate of contribution. Why should I take more risks, display more imagination, or expend more effort in order to gain more resources if the result will only be that, whenever something must be paid for, I will have to contribute not merely a larger absolute amount (which would happen even with a flat rate) but a larger percentage? I might as well not be productive if much of anything extra I produce will be taken away from me, leaving me little better off than those who produced far less.

Three points need to be noticed regarding this objection. First, of course, being fair and providing incentives are two different matters, and there is certainly no guarantee in the abstract that whatever arrangement would provide the greatest incentives would also be fair.

Second, concerns about incentives often arise when it is assumed that maximum production and limitless growth are the best goal. It is increasingly clear that many current forms of production and growth are unsustainable and that the last thing we should do is to give people self-interested reasons to consume as many resources as they can, even where the resources are consumed productively. These issues cannot be settled in the abstract either, but it is certainly an open question—and one that should be asked very seriously—whether in a particular situation it is desirable to stimulate people by means of incentives to maximum production. Sometimes it is desirable, and sometimes it is not. This is an issue about ends.

Third, there is a question about means. Assuming that it had been demonstrated that the best goal to have in a specific set of circumstances involved stimulating more production of something, one would then have to ask: how much incentive is needed to stimulate that much production? Those who are preoccupied with incentives often speculate groundlessly that unlimited incentives are virtually always required. Certainly it is true that it is generally necessary to provide some additional incentive in order to stimulate additional production. Some people are altruistic and are therefore sometimes willing to contribute more to the welfare of others even if they do not thereby improve their own welfare. It would be completely unrealistic, however, to try to operate an economy on the assumption that people generally would produce more irrespective of whether doing so was in their own interest—they need instead to be provided with some incentive. However, some incentive does not mean unlimited incentive.

It is certainly not necessary to offer unlimited incentives in order to stimulate (limited) additional production by some people (and not others). Whether people respond or not depends upon individual personalities and individual circumstances. It is a factual matter, not something to be decreed in the abstract, how much incentive is enough: for these people in these circumstances to produce this much more, how much incentive is enough? What is clearly mistaken is the frequent assumption that nothing less than the maximum incentive is ever enough.

In conclusion, insofar as the objection based on disincentive effects is intended to be a decisive refutation of the second principle of equity, the objection fails. It is not always a mistake to offer less than the maximum possible incentive, even when the goal of thereby increasing production has itself been justified. There is no evidence that anything less than the maximum is even generally a mistake. Psychological effects must be determined case by case.

On the other hand, the objection based on disincentive effects may be intended—much more modestly—simply as a warning that one of the possible costs of restraining inequalities by means of progressive rates of contribution, in the effort of being fair, may (or may not) be a reduction in incentive effects. As a caution rather than a (failed) refutation, the objection points to one sensible consideration that needs to be taken into account when specifying which variation upon the general second principle of equity is the best version to adopt in a specific case. One would have to consider how much greater the incentive effect would be if the rate of contribution were less progressive, in light of how unfair the results of a less progressive rate would be.

This conclusion that disincentive effects deserve to be considered, although they are not always decisive, partly explains why the second principle of equity is stated not as an absolute but as a general principle. It says: 'the parties who have the most resources *normally* should contribute the most'—not always, but normally. One reason why the rate of contribution might not be progressive, or might not be as progressive as possible, is the potential disincentive effects of more progressive rates. It would need to be shown case by case that an important goal was served by having some incentive and that the goal in question would not be served by the weaker incentive compatible with a more progressive rate of contribution.

We have so far examined two quite different kinds of justifications of unequal burdens: to reduce or eliminate an existing inequality by removing an unfair advantage of those at the top and to prevent the existing inequality from becoming worse through any infliction of an unfair additional disadvantage upon those at the bottom. The first justification rests in part upon explaining why the initial inequality is unfair and ought to be removed or reduced. The second justification applies irrespective of whether the initial inequality is fair. Now we turn to a different mechanism that—much more directly—serves the second purpose of avoiding making those who are already the worst off yet worse off.

GUARANTEED MINIMUM

We noted earlier that issues of equity or fairness can arise only if there is something that must be divided among different parties. The existence of the following circumstances can be taken as grounds for thinking that certain parties have a legitimate claim to some of the available resources: (a) the aggregate total of resources is sufficient for all parties to have more than enough; (b) some parties do in fact have more than enough, some of them much more than enough; and (c) other parties have less than enough. American philosopher Thomas Nagel has called such circumstances radical inequality.[4] Such an inequality is radical in part because the total of available resources is so great that there is no need to reduce the best-off people to anywhere near the minimum level in order to bring the worst-off people up to the minimum: the existing degree of inequality is utterly unnecessary and easily reduced, in light of the total resources already at hand. In other words, one could preserve considerable inequality—in order, for instance, to provide incentives, if incentives were needed for some important purpose—while arranging for those with less than enough to have at least enough.

Enough for what? The answer could of course be given in considerable detail, and some of the details would be controversial (and some, although not all, would vary across societies). The basic idea, however, is of enough for a decent chance for a reasonably healthy and active life of more or less normal length, barring tragic accidents and interventions. 'Enough' means the essentials for at least a bit more than mere physical survival—for at least a distinctively human, if modest, life. For example, having enough means owning not merely clothing adequate for substantial protection against the elements but clothing adequate in appearance to avoid embarrassment, by local standards, when being seen in public, as Adam Smith noted.

In a situation of radical inequality—a situation with the three features outlined above—fairness demands that those people with less than enough for a decent human life be provided with enough. This yields the third principle of equity, which states:

> When some people have less than enough for a decent human life, other people have far more than enough, and the total resources available are so great that everyone could have at least enough without preventing some people from still retaining considerably more than others have, it is unfair not to guarantee everyone at least an adequate minimum.[5]

---

[4] See Thomas Nagel, 'Poverty and food: why charity is not enough', in Peter G. Brown and Henry Shue, eds, *Food policy: the responsibility of the United States in the life and death choices* (New York: Free Press, 1977), pp. 54–62. In an important recent and synthetic discussion Thomas W. Pogge has suggested adding two further features to the characterization of a radical inequality, as well as a different view about its moral status—see Thomas W. Pogge, 'A global resources dividend', in David A. Crocker and Toby Linden, eds, *Ethics of consumption: the good life, justice and global stewardship*, in the series 'Philosophy and the Global Context' (Lanham, MD, Oxford: Rowman & Littlefield, 1998), pp. 501–36. On radical inequality, see pp. 502–3.

[5] This third principle of equity is closely related to what I called the argument from vital interests in 'The unavoidability of justice', in this volume. It is the satisfaction of vital interests that constitutes the minimum everyone needs to have guaranteed. In the formulation here the connection with limits on inequality is made explicit.

Clearly, provisions to guarantee an adequate minimum can be of many different kinds, and, concerning many of the choices, equity has little or nothing to say. The arrangements to provide the minimum can be local, regional, national, international, or, more likely, some complex mixture of all, with secondary arrangements at one level providing a backstop for primary arrangements at another level.[6] Similarly, particular arrangements might assign initial responsibility for maintaining the minimum to families or other intimate groups, to larger voluntary associations such as religious groups, or to a state bureau. Consideration of equity might have no implications for many of the choices about arrangements, and some of the choices might vary among societies, provided the minimum was in fact guaranteed.

Children, it is worth emphasizing, are the main beneficiaries of this principle of equity. When a family drops below the minimum required to maintain all its members, the children are the most vulnerable. Even if the adults choose to allocate their own share of an insufficient supply to the children, it is still quite likely that the children will have less resistance to disease and less resilience in general. And of course not all adults will sacrifice their own share to their children. Or, in quite a few cultures, adults will sacrifice on behalf of male children but not on behalf of female children. All in all, when essentials are scarce, the proportion of children dying is far greater than their proportion in the population, which in poorer countries is already high—in quite a few poor countries, more than half the population is under the age of fifteen.

One of the most common objections to this third principle of equity flows precisely from this point about the survival of children. It is what might be called the overpopulation objection. I consider this objection to be ethically outrageous and factually groundless, as explained elsewhere.[7]

The other most common objection is that while it may be only fair for each society to have a guaranteed minimum for its own members, it is not fair to expect members of one society to help to maintain a guarantee of a minimum for members of another society.[8] This objection sometimes rests on the assumption that state borders—national political boundaries—have so much moral significance that citizens of one state cannot be morally required, even by considerations of elemental fairness, to concern themselves with the welfare of citizens of a different political jurisdiction. A variation on this theme is the contention that across state political boundaries moral mandates can only be

---

[6] On the importance of backstop arrangements, or the allocation of default duties, see 'Afterword' in Henry Shue, *Basic rights: subsistence, affluence, and US foreign policy*, 2nd edn (Princeton, NJ: Princeton University Press, 1996).

[7] *Basic rights*, ch. 4.

[8] This objection has recently been provided with a powerful and sophisticated Kantian formulation that deserves much more attention than space here allows—see Richard W. Miller, 'Cosmopolitan respect and patriotic concern', *Philosophy & Public Affairs* 27: 3, Summer 1998, pp. 202–24.

negative requirements not to harm and cannot be positive requirements to help. I am unconvinced that, in general, state political borders and national citizenship are markers of such extraordinary and overriding moral significance. Whatever may be the case in general, this second objection is especially unpersuasive if raised on behalf of citizens of the industrialized wealthy states in the context of international cooperation to deal with environmental problems primarily caused by their own states and of greatest concern in the medium term to those states.

To help to maintain a guarantee of a minimum could mean either of two things: a weaker requirement (a) not to interfere with others' ability to maintain a minimum for themselves; or a stronger requirement (b) to provide assistance to others in maintaining a minimum for themselves. If everyone has a general obligation, even towards strangers in other states and societies, not to inflict harm on other persons, the weaker requirement would follow, provided only that interfering with people's ability to maintain a minimum for themselves counted as a serious harm, as it certainly would seem to. Accordingly, persons with no other bonds to each other would still be obliged not to hinder the others' efforts to provide a minimum for themselves.

One could not, for example, demand as one of the terms of an agreement that someone make sacrifices that would leave the person without necessities. This means that any agreement to cooperate made between people having more than enough and people not having enough cannot justifiably require those who start out without enough to make any sacrifices. Those who lack essentials will still have to agree to act cooperatively, if there is in fact to be cooperation, but they should not bear the costs of even their own cooperation. Because a demand that those lacking essentials should make a sacrifice would harm them, making such a demand is unfair.

That (a), the weaker requirement, holds, seems perfectly clear. When, if ever, would (b), the stronger requirement to provide assistance to others in maintaining a minimum for themselves, hold? Consider the case at hand. Wealthy states, which are wealthy in large part because they are operating industrial processes, ask the poor states, which are poor in large part because they have not industrialized, to cooperate in controlling the bad effects of these same industrial processes, such as the destruction of atmospheric ozone and the creation of global warming. Assume that the citizens of the wealthy states have no general obligation, which holds prior to and independently of any agreement to work together on environmental problems, to contribute to the provision of a guaranteed minimum for the citizens of the poor states. The citizens of the poor states certainly have no general obligation, which holds prior to and independently of any agreement, to assist the wealthy states in dealing with the environmental problems that the wealthy states' own industrial processes are producing. It may ultimately be in the interest of the poor states to see ozone depletion and global warming stopped, but in the medium term the citizens of the poor states have far more urgent and serious problems—such as lack of food, lack of clean drinking water, and lack of jobs to provide minimal support for themselves and their families. If the wealthy states say to the poor states, in effect, 'Our most urgent request of you is that you act in ways that will avoid worsening the ozone depletion and global warming that we have started,' the poor states could reasonably respond, 'Our most urgent request of you is assistance in guaranteeing the fulfilment of the essential needs of our citizens.'

In other words, if the wealthy have no general obligation to help the poor, the poor certainly have no general obligation to help the wealthy. If this assumed absence of general obligations means that matters are to be determined by national interest rather than international obligation, then surely the poor states are as fully at liberty to specify their own top priority as the wealthy states are. The poor states are under no general prior obligation to be helpful to the wealthy states in dealing with whatever happens to be the top priority of the wealthy states. This is all the more so as long as the wealthy states remain content to watch hundreds of thousands of children die each year in the poor states for lack of material necessities, which the total resources in the world could remedy many times over. If the wealthy states are content to allow radical inequalities to persist and worsen, it is difficult to see why the poor states should divert their attention from their own worst problems in order to help out with problems that for them are far less immediate and deadly. It is as if I am starving to death, and you want me to agree to stop searching for food and instead to help repair a leak in the roof of your house without your promising me any food. Why should I turn my attention away from my own more severe problem to your less severe one, when I have no guarantee that if I help you with your problem you will help me with mine? If any arrangement would ever be unfair, that one would.

Radical human inequalities cannot be tolerated and ought to be eliminated, irrespective of whether their elimination involves the movement of resources across national political boundaries: resources move across national boundaries all the time for all sorts of reasons. I have not argued here for this judgement about radical inequality, however.[9] The conclusion for which I have provided a rationale is even more compelling: when radical inequalities exist, it is unfair for people in states with far more than enough to expect people in states with less than enough to turn their attention away from their own problems in order to cooperate with the much better off in solving their problems (and all the more unfair—in light of the first principle of equity—when the problems that concern the much better off were created by the much better off themselves in the very process of becoming as well off as they are). The least that those below the

---

[9] And for the argument to the contrary see Miller, 'Cosmopolitan respect and patriotic concern'.

minimum can reasonably demand in reciprocity for their attention to the problems that concern the best off is that their own most vital problems be attended to: that they be guaranteed means of fulfilling their minimum needs. Any lesser guarantee is too little to be fair, which is to say that any international agreement that attempts to leave radical inequality across national states un-touched while asking effort from the worst off to assist the best off is grossly unfair.

OVERVIEW

I have emphasized that the reasons for the second and third principles of equity are fundamentally the same, namely, avoiding making those who are already the worst off yet worse off. The second principle serves this end by requiring that when contributions must be made, they should be made more heavily by the better off, irrespective of whether the existing inequality is justifiable. The third principle serves this end by requiring that no contribu-tions be made by those below the minimum unless they are guaranteed ways to bring themselves up at least to the minimum, which assumes that radical inequalities are unjustified. Together, the second and third principles require that if any contributions to a common effort are to be expected of people whose minimum needs have not been guaranteed so far, guarantees must be provided; and the guarantees must be provided most heavily by the best off.

The reason for the first principle was different from the reason for the second principle, in that the reason for the first rests on the assumption that an existing inequality is already unjustified. The reason for the third principle rests on the same assumption. The first and third principles apply, however, to inequalities that are, respectively, unjustified for different kinds of reasons. Inequalities to which the first principle applies are unjustified because of how they arose, namely some people have been benefiting unfairly by dumping the costs of their own advances upon other people. Inequalities to which the third principle applies are unjustified independently of how they arose and simply because they are radical, that is, so extreme in circumstances in which it would be very easy to make them less extreme.

What stands out is that in spite of the different content of these three principles of equity, and in spite of the different kinds of grounds upon which they rest, they all converge upon the same practical conclusion: what-ever needs to be done by wealthy industrialized states or by poor non-industrialized states about global environmental problems such as ozone destruction and global warming, the costs should initially be borne by the wealthy industrialized states.

10

Climate*

Environmental challenges do not wear philosophical name tags, and the judgement about what kinds of question they raise is more fateful, because more difficult to shake loose from, than any particular answers offered to the questions once assumed. Much philosophical writing about climate change so far has assumed that the fundamental issues are distributive: issues of justice between rich and poor and between present and future. Issues of distributive justice, trans-spatial and trans-temporal, are certainly important, but other kinds of question arise as well.

INFLICTING HARM

A change in the climate is a profound change, a modification in some of the interacting forces that, in combination with its distance from the sun, give this planet its distinctive environment. What may be harmed by human choices to bring about avoidable changes in the climate depends upon what is morally considerable, the range of things whose interests deserve to be taken into account. However, since the climate is a planetary-level phenomenon, virtually anything that is indeed morally considerable may well be affected by the extent or speed of climate change. For example, the species that flourish on earth now are species that found a hospitable spot in the climate we have now. Species adapt—some migrate—and speciations are constantly occurring, but natural adaptation and natural evolution occur on a timescale far slower than the timescale that appears likely if increases in the atmospheric concentration of anthropogenic greenhouse gases (GHGs) continue. Few species can adapt nearly as rapidly as the climate may change. The primary cause of species extinction is loss of habitat; rapid climate change could be the ultimate destroyer of habitats. Insofar as existing species, or the ecosystems in which

* 'Climate', in *A Companion to Environmental Philosophy*, ed. by Dale Jamieson (Malden, Mass.: Blackwell Pubs., 2001), pp. 449–59.

they live, or the natural processes through which they would otherwise change, have some value and are worthy of protection from avoidable human destruction or distortion—insofar as anthropogenic extinction is an ethical issue—anthropogenic rapid climate change is morally suspect for the harm it will inflict beyond the human species.

Here the focus will be on harm to humans, which comes in at least two kinds, the second somewhat less evident than the first. First is straightforward, if indirectly caused, physical harm of many kinds. Climate change will centrally involve relatively simple changes in weather, some of which will be dangerous. Consider water and then insects. Some locations that now have abundant rainfall will have inadequate rainfall; drinking water may become inadequate. Other locations that now suffer drought may receive heavy rains, producing flooding and causing people to drown. Changes such as deficits and surpluses in rainfall and drinking water are uncomplicated and utterly ordinary, but people—and crops and livestock, and thus indirectly more people—perish from droughts and floods. And wars, in which of course more people (and crops and livestock) die, are fought over access to water for drinking, irrigation, and manufacture.

Some insects, such as certain mosquitos, cannot range farther than a certain distance from the equator because the winters are too cold or long for them. Those of us who live in the cooler temperate zones rather complacently think of the infections carried by these species as 'tropical' diseases and invest far less in their prevention and cure than in diseases that now threaten us, in spite of the fact that some 'tropical' diseases—such as malaria—are among humanity's greatest scourges. One of the most likely components of climate change will be a lessening of the severity of the winters in some temperate areas, with the incursion of unfamiliar 'tropical' diseases by way of an extension of the range of their vectors. Human illness and death are very likely to result. Changes in rainfall patterns and disease patterns resulting from weather changes which in turn result from changes in atmospheric concentrations of GHGs raise just as many—but no more—issues about causation, and about the relation between causal responsibility and moral responsibility, as more familiar cases such as second-hand cigarette smoke and cancer. If the smoker can be morally responsible for the cancer of his non-smoking children who inhale the smoke, it is not apparent why the sources of the avoidable GHGs cannot at least sometimes be responsible for the inundations or infections to which the increased atmospheric concentrations contribute. This appears to be the very long-distance but fully real infliction of physical harm on innocent strangers.

The second kind of harm is not so straightforward, but is, if anything, more serious: it is the harm of preventing people from obtaining the necessary minimum of a resource vital for their survival. Before explaining how in this instance some people make it impossible for others to secure a minimum needed, we should notice the nature of the resource in question here, which may initially seem odd. Every human being needs, in order to survive, some minimum quantity of certain obvious resources such as drinkable water and breathable air. The quantity of each resource needed obviously varies from individual to individual depending upon features of each individual, such as age and level of activity, and features of the individual's situation, including climate. Yet any given individual at any given place and time needs some minimum amount of essential resources such as safe water and safe air. Less obviously, most people now also need what amounts to a minimum amount of the planet's capacity to deal safely with GHG emissions. The vast majority of people alive today must, in order to survive, engage in economic activities that generate GHG emissions: they, for example, raise livestock or rice, both of which create methane, or they consume fossil fuel, which releases CO_2. It may seem slightly odd to treat absorptive capacity as a vital resource, but absorptive capacity for GHG emissions is not only a necessity but an increasingly scarce one, as the research on climate change has demonstrated.

The shortage of this absorptive capacity is of course not an unchangeable feature of the universe; on the contrary, it is an artefact produced by the energy regime on which we have based our economic system. At one technological extreme, denizens of a pre-industrial, tropical society might survive simply by gathering uncultivated fruits and vegetables, without heating or cooling their habitations and without transportation. They might produce no GHG emissions of any consequence, making their minimum need for GHG absorptive capacity effectively zero. At the other technological extreme, we might construct an advanced industrial (or post-industrial) economy powered by solar, wind, and/or geothermal energy that also produced no GHG emissions because, although it consumed large quantities of energy, none of the energy was carbon-based. There, too, GHG absorptive capacity would not be needed by everyone engaged in economic activity. Our current need for GHG absorptive capacity is, therefore, an avoidable necessity: it is perfectly possible for matters to become otherwise than they in fact are. It is avoidable, however, only on the social level, not the individual level. If, for example, one needs surgery, the surgeon will probably drive to the hospital in a vehicle powered by a combustion engine and the operating theatre will be lit by electricity more than likely generated by burning fossil fuel. An individual could forgo the surgery in order not to be responsible for the GHG emissions involved, but that would, given the kind of economy we have, constitute an insanely large sacrifice for a relatively small saving in emissions. Yet the society as a whole, with as advanced a medical technology as one likes, could be operated on non-carbon-based energy. The reasonable avoidability is only at the social level, and not a matter for individual martyrdom.

Consequently, whether an individual needs to generate GHG emissions in order to live—and thereby needs some minimum absorptive capacity for those emissions—is a function of the time and place, and specifically the energy regime, into which she is born. For practically everyone at present, and for the immediate future, survival requires the use of GHG emissions absorptive capacity. No reasonable, immediate alternative exists. Strange as it may initially sound, emission absorptive capacity is as vital as food and water and, virtually everywhere, shelter and clothing. The central finding of research about climate change, however, is that there is a global maximum on safe GHG emissions because there is a physical limit on the quantity of GHGs that can be absorbed without changes that affect surface climate. Emissions in excess of this safe maximum produce climate change; and, of course, current total emissions are already far in excess of the safe global maximum and rapidly growing every year (worse, at an accelerating rate of increase). The underlying logic of the situation is, then, abundantly clear: (a) the global total of emissions must (if we are to avoid climate change—in fact, increasingly rapid climate change) be reduced, while (b) every human being depends for survival on the production (direct or indirect) of a minimum amount of GHG emissions. So the global minimum total is the product of the per capita minimum (given the existing energy regime) times the existing population, while the global maximum total is the amount the planet can recycle safely, i.e. recycle without climate change.

Actually, three global emissions totals are in play: (1) the socially determined minimum total (for survival of current population with current energy regime); (2) the naturally determined maximum total (for a sustainable economy that does not produce climate change); and (3) the actual total (current emissions). The relations among the three totals are matters of fact. The research of the Intergovernmental Panel on Climate Change (IPCC) has found that (3), the current total, is far in excess of (2), the sustainable total that would avoid climate change. The size of (1) depends critically on the energy sources on which the world economy relies. In a world economy driven by solar, wind, and geothermal (or nuclear) energy, (1) could be extremely low; agriculture, as we know it, will still produce methane, but the methane would not be a problem without CO_2 produced by electricity generation and transportation. In a world economy such as ours, driven by fossil fuels, (1) may be larger than (2).

Indeed, (1) may even be larger than (3); this depends on the ratio between the amount of energy now frivolously consumed, or carelessly wasted, by the affluent and the additional amount now needed by the desperately poor for decent lives. It is logically possible that sheer redistribution of current energy use could provide for everyone's basic needs, while keeping the existing fossil fuel regime, i.e. that (1) is no larger than (3), while (3) is now simply too unequally distributed. The obvious difficulties with taking consolation in the possibility that (1) is no larger than (3), provided that the distribution of energy consumption were sufficiently less unequal internationally than the current wildly unequal distribution, include: (a) the affluent and powerful would have to be willing to accept a redistribution of energy consumption that would severely reduce their current standard of living (given the existing carbon-based regime)—a willingness rarely in sight, whatever its ethical merits; (b) the redistribution would have to be feasible without throwing the world economy into a tailspin that made matters even worse for the worst off (by, for example, large net elimination of jobs); and (c) none of this matters here in any case because (3) is much larger than (2)—redistributing emissions that are already in excess of what the planet can handle without changes in the climate would not prevent climate change (nor, over the long run, help the poorest, who will be least able to afford to adapt to climate change).

Where is there any infliction of harm? The billion or so poorest human beings on the planet need sound and sustainable economic development. They need to engage in economic activity that will, given that the current world economy runs on fossil fuel, add GHG emissions to the earth's atmosphere by consuming fossil fuel. They need 'space' for their increased emissions, and this space needs to be within a global total less than (2), the maximum sustainable total compatible with the avoidance of climate change. They need to use emission absorptive capacity, but no absorptive capacity is left because those of us in the affluent economies have taken it all (and much more). We are parked in their spaces, and no empty spaces exist. Our unnecessary emissions are blocking their vital ones, except at the price of speeded-up climate change, which will be most unmanageable for them. We are depriving them of a necessity for their survival, given the fossil fuel regime, which the poor, unlike the affluent, cannot change. The economic/energy system in which we thrive and indulge in excess prevents their enjoying a decent life. The energy regime that makes life opulent for Belgians and Saudis makes it impossible for Rwandans and Haitians, who are helpless victims of a complex global social institution in which they have absolutely no voice.

Suppose a picnic table is set with enough food for two dozen people, but only one dozen come. Any food not eaten will spoil. I 'pig out', eating enough for two or three people. I have demonstrated that I am a glutton, lacking in self-discipline and perhaps civility. But I have harmed no one—no one else is worse off as a result of my excess (although I myself may be).

Now suppose that a picnic table is set with enough food for one dozen people, and one dozen eventually come. I arrive early and, being a glutton, eat enough for two or three people, as usual. The last one or two people to come have nothing to eat because I have eaten their share. I have harmed them because I have deprived them of what they were entitled to, unless we make some heroically implausible assumption such as that they have no right to be there or that I am so special that I have a right to double the share of ordinary folks. If we add the assumption that this meal was part of the minimum those deprived needed for their survival, which would make the picnic more analogous to GHG emissions, then I have injured them severely—arguably, I have killed them. I have certainly done them harm.

INCREASING INJUSTICE

Physical necessity requires, if humanity is to break free from the spiral into rapid climate change predictable with business as usual, that practically every state in the world cooperate—most vitally, the states with the highest emissions (the USA has by far the highest) and the states with the largest populations (China's emission total, although not its per capita rate, will soon surpass that of the USA). Fairness requires that each state do its fair share. What are the criteria of a fair share in this case? This turns out to be a philosophically challenging question. For a start, fairness and unfairness arise at three theoretically separable but practically intertwined points: procedures, emissions, and costs. Further, costs will be incurred for different, though deeply interrelated, purposes.

Fair procedures. What, if anything, will be done to ward off rapid climate change is in fact now being decided under the Framework Convention on Climate Change (FCCC) (opened for signature at the Earth Summit, the United Nations Conference on Environment and Development, held in Rio de Janeiro on 4 June 1992; entered into force, 21 March 1994; and potentially modified by the Kyoto Protocol, approved by the Conference of the Parties, 11 December 1997, and awaiting a sufficient number of national approvals, with the US Executive declining to submit it for US Senate ratification). The FCCC itself requires no action. The ratification process for the Kyoto Protocol, which would for the first time require some action by affluent states (results to be averaged across 2008–12, with 'demonstrable progress' by 2005 required—Article 3), is heavily weighted, in a manner roughly similar to the weighted voting within the World Bank and the International Monetary Fund, in favour of the states with the highest current emissions of GHGs (entrance into force depends upon approval by states responsible for 55 per cent of the CO_2 emissions in 1990—Article 24). Any argument for the fairness of this ratification process seems to depend upon some claim that those who pay the piper should call the tune. While this practice is customary in international affairs, it tacitly assumes the absence of any prior grounds for responsibility for bearing costs. In other words, it prejudges the substantive questions discussed below.

The fundamental features of the FCCC procedures are that they are (a) negotiations among (b) radically unequal parties, namely national states, precisely the kind of procedure most likely to produce outcomes that merely reflect the enormous power differentials among the bargainers. The parties range from a nuclear superpower and the European Union with vast wealth and power, at one extreme, to the states most immediately vulnerable to the most indisputable results of climate change such as sea-level rise (e.g. Bangladesh and the small island states), which have no leverage, at the other extreme. Insofar as the leaders of the rich and powerful states construe their own 'national interest' within a horizon that is narrow, the poor in the weak states are largely at their mercy; insofar as they construe 'national interest' within a horizon that is short term, the poor (at least) of future generations are completely at their mercy. The actual procedure will permit as unfairly skewed an outcome as the USA, EU, and other most powerful players choose to give it; only stands on principle on behalf of the weak—current and future—taken by the bargainers with the leverage could enable this biased process to produce substantively fair outcomes.

The substantive principles need to cover both fair sharing of emissions and fair sharing of costs. One crucial complication is whether the allocation of emissions of GHGs and the allocation of costs of dealing with climate change ought to be settled separately, each on its own merits. The specific alternative to independent settlement of the two issues worth considering is allocating the costs of dealing with climate change precisely in order to bring about the allocation of emissions that is fair. This would mean treating the allocation of costs as an economic incentive structure designed to generate the allocation of emissions sought—using the allocation of costs as the means to the goal of attaining the fair allocation of emissions. First, note quickly why the allocation of emissions is an issue of fairness at all rather than a matter to be left entirely to markets.

Fair distribution of emissions. Morally permissible emissions of GHGs are extremely limited because of the harms being caused by emissions in excess of the quantity that can be recycled without climate change. Legally permissible emissions would also be extremely limited if negotiations under the FCCC ever led to serious attempts to achieve the original purpose of the treaty: gaining control over anthropogenic changes in the atmospheric concentrations. So, at least morally, and potentially legally, if we are to avoid the harms sketched above, total global emissions of GHGs are scarce. Given that they are also valuable, as long as economic welfare is kept dependent upon fossil fuel, their distribution poses a classic question of justice: what ought to be the distribution of a resource (absorptive capacity for GHGs, in this case) that is both scarce and vital for life?

As long as those in control choose not to leave behind the fossil fuel energy regime that makes GHG emissions essential to a decent life, powerful considerations support the following principle: the only morally permissible allocations of emissions are allocations that guarantee the availability of the minimum necessary emissions to every person, which entails reserving adequate unused absorptive capacity for those emissions. As already noted, the global total constituted by the minimum essential for each person might in itself equal or exceed the global total compatible with the avoidance of rapid climate change. This is a matter for scientific determination. It seems entirely possible that once every person was guaranteed minimum emissions from burning carbon-based fuels, the maximum safe global total of CO_2 would have been reached, so that no emissions beyond the minimum could be allowed to anyone. Presumably such an outcome would be completely unacceptable to the now wealthy and powerful, but then no one has a God-given guarantee that they can (1) burn fossil fuel, (2) live the kind of lives they most enjoy, and (3) avoid climate change. It may turn out that burning fossil fuels was a primitive solution, feasible for only a brief, technologically unsophisticated period of human history on this planet, and ought now to be left behind as quickly as entrenched energy interests will permit.

The most obvious reason why every person born into a fossil fuel-based world economy would be entitled to a guaranteed minimum of the emissions essential to life is quite simply that to make the political choice to impose a ceiling on total emissions, while not guaranteeing a minimum to each person, would condemn to death the poorest people on the planet. To make the social choice to continue to rely on burning fossil fuel, while capping the total emissions of the GHGs that result, would be to impose upon the world a package of international institutions—an energy regime combined with a climate-change regime—that made it impossible for considerable numbers of the poorest people on the planet to survive, when there are alternative packages of institutions that would allow them to survive, and perhaps even flourish—human society need not choose to continue burning oil until the last barrel is gone. Setting up institutions, knowing in advance that they unnecessarily condemn large numbers of people to die, is as obviously morally wrong as human conduct can be. One would like to think that this option is as politically infeasible as it is morally outrageous and is accordingly not worthy of much discussion.

Inalienable minimum emission rights are perfectly compatible with the kinds of scheme for trading in emission rights that are currently popular among intellectuals who study environmental policy. Economic considerations support leaving as much of the distribution of emission permits to the market as is compatible with fundamental moral requirements. The minimum guarantee simply requires that, however many permits for emissions beyond the minimum necessary for a decent life could safely be made available for trading (which is of course a scientific/political question), no tradeable permits should be issued for vital emissions. Minimum vital emissions could be viewed as an inalienable private property right, or simply a human subsistence right. The vital minimum ought not to be marketable.

Food is vital for life and scarce, but—someone may say—we market all the food. Why not market all the emission permits as well? For a start, strong arguments are also available for a right to food—a right to the minimum sustenance essential to normal vitality. Decent societies do not in fact market all their food but, instead, reserve significant amounts of it to be distributed by way of food stamps, or their functional equivalent, which are not themselves supposed to be marketed but distributed according to need, and reserve other significant amounts of food for relief to victims of famines, refugees from war, and other desperately helpless victims. A society in which food is available only for payment is a brutal and uncivilized place. What is suggested here is merely the equivalent of food stamps on the global level for vital emissions.

Further, even if one thought, on the contrary, that food stamps and famine relief were somehow misguided and that people without money for food (including children presumably) ought for some reason simply to be left to starve, the case of CO_2 emissions in a carbon-based energy regime with a global ceiling on the emissions is a still more compelling case even than food. The imposition of the ceiling on emissions would create an extraordinary—to my knowledge, utterly unique—case. There is no global ceiling on the quantity of food that may be produced. Consequently, when confronted by a healthy adult with neither food nor money, one might persuade oneself that if only he rose earlier, worked harder, or both, he would be able to catch, find, or grow some food or earn some money with which to buy food. This is naive in all kinds of ways (not least, that desperately hungry people are not usually very healthy), but it is at least a conceptually coherent fantasy. The food supply legally can be, and physically can be, enlarged indefinitely; indeed, the global total is increasing constantly. It is possible to obtain more for oneself without taking it away from someone else. If there were a global ceiling on CO_2 emissions, by contrast, the supply of permissible CO_2 emissions would be strictly zero-sum: more for anyone would mean less for someone else, because the supply would be fixed. (If the worst fears of the scientists studying climate change are confirmed, the total sum would, even worse, need to be progressively diminishing over the years.) Prohibited emissions by those below their minimum would have to come either at the expense of someone else or at the expense of the climate—by exceeding the safe total. It would be as if the only way for the hungry to obtain more food were to steal it. No person in her right mind would cooperate with a set of institutions that imposed a ceiling on the total by refusing to grant her a minimum she needed and could obtain in no other non-harmful way, while others were allowed to exceed the minimum. Such institutions treat her with contempt. If she has any self-respect, she will evade or resist them. It is evident that the distribution of emissions must guarantee a minimum to every person unless the climate-change regime is simply to be forced upon the poor.

Fair distribution of costs: abatement. In the short term it will be expensive to take the measures necessary to reduce total global GHG emissions to a rate that does not constantly expand their atmospheric concentrations. Large quantities of emissions now result from sheer waste, so significant reductions can initially be made at a profit. The reductions in emissions possible from reductions in waste of energy are, however, not sufficient to bring the current rate down to a sustainable rate, even on optimistic scientific assumptions, so after the initial savings from reducing energy waste there will be significant costs for abatement of climate change through, for example, improved technology.

In addition to abatement costs, two other sets of costs will result from attempts either to avoid damage from components of climate change (for example, sea walls to try to hold back rising sea levels) or to deal with damage (for example, build new inland cities to replace flooded or eroded coastal cities not saved by sea walls). These two kinds of cost from attempting to deal with climate change that in fact occurs, as distinguished from the abatement costs of attempting to mitigate the climate change itself, are usually called adaptation costs. As mentioned earlier, adaptation costs and abatement costs must be considered both separately and in relation to each other. First, however, is the other question already raised about the relation between the allocation of emissions and the allocation of—specifically—abatement costs.

If the abatement costs—the costs of measures taken to prevent or reduce climate change itself, rather than its economic and social effects—are to be structured in order to create incentives designed to bring about a fair distribution of emissions, then the greatest abatement costs will have to be borne by the states whose current emissions most exceed their fair share of emissions. The absolute quantity of a fair share depends in part, of course, on the research to determine the sustainable annual global total of emissions—one cannot calculate the absolute size of a share of a total without actually knowing the total. Nevertheless, if one accepts both (1) the finding of the research so far that whatever precisely is the sustainable rate of emissions, it is far lower than the current rate and (2) the suggestion above that every human is entitled to at least a minimum share of the total emissions, it is perfectly clear from the vast current inequalities in per capita rates of emissions that (3) in order for those below the economic minimum to reach it, those above it must reduce their emissions. Other things being equal, it would appear equitable that those with the highest per capita rates of emissions should make the largest reductions in emissions.

Whoever ought to make the largest reductions in GHG emissions—assume these are the people with the highest per capita rates—ought to bear the largest shares of the abatement costs, if the allocation of the abatement costs is to be used to create incentives that will motivate movement in the direction of a fair distribution of emissions. At least two questions immediately arise about this allocation of abatement costs: would it in turn be fair and is such a scheme likely to work, if it is fair to attempt it?

Allocating the heaviest financial burdens for abatement to those whose own emissions ought to be reduced the most would be unfair to them only if there are other grounds for the distribution of abatement costs that would yield a different distribution less burdensome to them. The most familiar alternative grounds are (1) the backward-looking considerations of (a) contribution to the problem—causal responsibility for climate change—and (b) benefit from the processes that have caused climate change, and (2) the forward-looking consideration of ability to pay. In theory it would be perfectly possible for these various grounds—contribution, benefit, and ability—each to point toward a different distribution of responsibility for contributing to the mitigation of the problem. Climate change, fortunately, is a case in which ethical reality is much less messy than ethical theory.

Consider contribution. The threat of climate change has been produced predominantly by the Industrial Revolution, fed first by coal and then by oil as well, and the consumerist societies which trade and use its products. Naturally, there are variations that matter, such as Japan's industrial production being less energy intensive than the USA's and the USA's being less in turn than the ageing Stalinist installations in Eastern Europe. The chief contributors are the wealthy industrialized states, roughly the members of the Organisation for Economic Co-operation and Development (OECD). Consider benefit. The primary beneficiaries of industrialization have been the states that have industrialized. A small exception exists to the extent that genuine benefits of industrial society have trickled down to non-industrialized states, but the lion's share of the benefits has gone to those which industrialized in pursuit of them: the OECD. Consider ability. The states that are the wealthiest and best able to pay for abatement measures mostly became so through the process of industrialization that created the excess concentrations of GHGs: the OECD. Contribution, resultant benefit, and consequent ability are largely coextensive in the states they select in this instance. The one exception worth noting consists of the major oil producers, which have become wealthy from selling oil mostly without industrializing; by not having industrialized, they have not contributed directly to the excess emissions they have fuelled, but they have benefited greatly and are well able to contribute.

And the states that have contributed and benefited most from the last century and a half of massive CO_2 emissions, and are as a result most able to bear the costs of the abatement of the threat to the climate for which they are most responsible, are also in fact the states with the largest per capita emissions. No industrial state has yet managed to leave behind the fossil fuel-based economy that made it rich, so those which emitted the most in the past are for the most part those which are emitting the most now. As noted above, the oil producers have not in the past, and are not now, directly contributing high emissions. And insofar as the relatively less industrialized states, which of course include the two most populous, industrialize by adopting the fossil fuel-based technologies of the OECD, they can soon become the worst emitters ever. It is evident that while these poorer states must improve the quality of life of their people, they must not gain their wealth in the same way that the now wealthy gained theirs. Yet only the now wealthy can afford to create and spread the alternative energy technology needed by everyone.

FURTHER READING

Banuri, T., Göran Mäler, K., et al. (1996) 'Equity and social considerations', *Climate Change 1995: Economic and Social Dimensions of Climate Change,* Contribution of Working Group III to the Second Assessment Report of the Intergovernmental Panel on Climate Change, ed. James P. Bruce. Hoesung Lee, and Erik F. Haites (New York: Cambridge University Press), pp. 79–124. (Ethical issues as understood within UN process)

Bolin, Bert (1998) 'The Kyoto negotiations on climate change: a science perspective', *Science* 279 (whole no. 5349), pp. 330–1. (Inadequacy of Kyoto Protocol briefly explained by original head of IPCC)

Broome, John (1992) *Counting the Cost of Global Warming* (Cambridge: White Horse Press). (Analysis of inter-temporal distribution construed as issue of discounting)

Coward, Harold, and Hurka, Thomas, eds (1993) *Ethics and Climate Change: The Greenhouse Effect* (Waterloo, ON: Wilfrid Laurier University Press). (Pioneering philosophical discussions)

Grubb, Michael (1995) 'Seeking fair weather: ethics and the international debate on climate change', *International Affairs* 71, no. 3, pp. 463–96. (The best single article integrating ethics, economics, and politics)

Hayes, Peter, and Smith, Kirk, eds (1993) *The Global Greenhouse Regime: Who Pays?* (London: Earthscan). (A book-length integrated interdisciplinary analysis)

Houghton, J. T., Meira Filho, L. G., et al. (1996) *Climate Change 1995: The Science of Climate Change.* Contribution of Working Group I to the Second Assessment Report of the Intergovernmental Panel on Climate Change (New York: Cambridge University Press, 1996). (The most authoritative account of scientific findings)

Kaiser, Jocelyn, and Schmidt, Karen (1998) 'Coming to grips with the world's green-house gases', *Science* 281, (whole no. 5376), pp. 504–7. (Current situation)

Kempton, Willett, Boster, James S. and Hartley, Jennifer A. (1995) *Environmental Values in American Culture* (Cambridge, MA: MIT Press). (Attitudes of citizens of most powerful state)

Kyoto Protocol to U.N. Framework Convention on Climate Change (1998) (10 December 1997), 37 ILM, pp. 22–43. (The most recently proposed international treaty)

McKibben, Bill (1989) *The End of Nature* (New York: Anchor Books), pp. 3–91. (An eloquent lament for the destruction of the autonomy from humans of climate)

Paterson, Matthew (1996) 'International justice and global warming', in *The Ethical Dimensions of Global Change,* ed. Barry Holden (London: Macmillan), pp. 181–201. (A survey of philosophical positions)

Pogge, Thomas W. (1998) 'A global resources dividend', *Ethics of Consumption: The Good Life, Justice, and Global Stewardship,* ed. David A. Crocker and Toby Linden

(Lanham, MD: Rowman and Littlefield), pp. 501–36. (A concrete constructive suggestion about how to lessen injustice)

Reus-Smit, Christian (1996) 'The normative structure of international society', in *Earthly Goods: Environmental Change and Social Justice*, ed. Fen Osler Hampson and Judith Reppy (Ithaca, NY: Cornell University Press), pp. 96–121. (International politics in brief)

Schelling. Thomas C. (1997) 'The cost of global warming', *Foreign Affairs* 76. no. 6 (November/December), pp. 8–14. (International economics in brief)

Shue, Henry, this volume 'Avoidable necessity: global warming, international fairness, and alternative energy'.

11

A legacy of danger: the Kyoto Protocol and future generations*

INTRODUCTION

Rapid climate change is capable, if not controlled soon, of causing massive harm not only to humans but to many other species of animals and plants (see J. T. Houghton et al., 2001; Mahlman, 2001). Even slowing the now accelerating rate of climate change will be expensive (McCarthy et al., 2001). This urgent effort must be paid for by a world already containing extreme inequalities in wealth and power (Shue, 1999). But no inequality in power is greater than the inequality between those of us alive today and those who will come after us. This chapter explores the dangerous implications for inequalities among generations of an institutional innovation, the Clean Development Mechanism (CDM) under the Kyoto Protocol, that is supposed to deal with climate change in a manner that takes account of present inequalities among nations.

LEAST COST FIRST AND LATER GENERATIONS

The negotiations about whether to take effective action concerning climate change are conducted by national governments; and measurement and reporting of sources, sinks, and net emissions are all categorized in national terms. Consequently it is difficult to overlook the fact that distributive justice has dimensions that are spatial: the international distributions of costs and of emissions. Many debates about the international division of the benefits and

* 'A Legacy of Danger: The Kyoto Protocol and Future Generations', in *Globalisation and Equality*, ed. by Keith Horton and Haig Patapan (London and New York: Routledge, 2004), pp. 164–78.
Earlier versions of this chapter benefited from comments offered at Albion College, the Philosophy Department at the University of Chicago, Cornell Law School, the Department of International Relations at the Australian National University, and an annual convention of the American Political Science Association. This chapter has endnotes rather than footnotes.

burdens have reflected pure conflicts of national interest, even when sometimes clothed in the rhetoric of fairness. International justice will be attained
only if powerful national states aim explicitly at it; nothing supports any hope
that the pure clash of national interest will somehow lead to international
justice. There is no Invisible Normative Hand at the international level
transmuting national interest into international justice. Nevertheless, even
the nations that are too weak to defend themselves successfully against unfair
treatment at least have official representatives at the negotiations, however
little their voices are in fact heard.

The temporal dimension of justice is more easily forgotten, but nothing
will affect more human beings than the intergenerational distributions of costs
and of emissions. Indeed, future generations are usually the unrepresented
parties, and therefore the most vulnerable parties of all, in spite of the fact that
the often unnoticed other side of the coin in most of the decisions made about
how much burden the current generation will handle is how much is consequently left for succeeding generations to bear. They are liable to the most
unequal treatment of all: being completely ignored. For example, every decision about the relative emphasis to be placed on present mitigation efforts as
opposed to future adaptation efforts is also a decision about the intergenerational allocation of effort and resources. Two connected tendencies deeply
engrained in the current approaches to climate change by the national governments who do the negotiating especially strongly shift burdens forward
toward our grandchildren, and their grandchildren.

The first tendency is the largely unquestioned assumption that it is only
rational for governments always to pursue least-cost-first solutions to mitigation. While this may often be rational if no interests are given full weight
except the interests of the current generation, it is far from evident that such
solutions are rational in light of the interests of future generations or fair
to them. This first danger to the future has two aspects. First, if the level of
expenditures were somehow fixed independently, then the choice of a least-
cost-first approach would mean that, for the level of expenditure set, the most
would be accomplished. In reality, however, nothing keeps expenditures fixed,
and consequently many parties will choose to accomplish the same mitigation
for a lower expenditure rather than accomplishing more mitigation for the
same expenditure. The result of this employment of the tactic of least cost first
is not that more is accomplished to slow climate change but only that less
is spent now. The appeal of this to electorates, and thereby to governments, is
obvious, but it effectively transfers costs forward in time to those who succeed
us. This forward transfer would fail to occur only if some costs that we chose
not to pay now would 'evaporate' if they were deferred long enough.

Now deferred costs can disappear in one sense, namely one discovers later
that some task that one earlier believed needed to be performed turns out, on
the basis of better knowledge, not to have been necessary after all. If one had

acted sooner, one would have done something that did not need to be done, as would later have been realized. Thus, one's own generation would have borne some costs itself, but would not have saved future generations any costs because the future generation would have known better than to have done what one did and so would not have borne these costs in any case.

This, however, is not a reason generally to follow the principle of least cost first. It is, most simply, a reason to try to get things right and not implement mistaken measures. Of course, if one is going to follow a mistaken policy, it is better to make a cheap mistake than to make an expensive mistake: if your only options are two dumb policies, choose the cheaper one. But obviously the main thing is not to make mistakes, any more often than necessary. The point is not to be cheap, it is to be smart. If the science is actually inadequate to the point that we do not know which measures are likely effectively to mitigate climate change, then we should invest in more research, not in dubious mitigation measures. The principle of choice is smart first—if possible, smart always—and it is not least cost first.

The second aspect of this first dangerous tendency is that least cost first can mean higher cost later, unless technological advances that reduce the costs of the currently more expensive tasks at least keep pace with the passage of time. That is, it is not merely that future generations will have to pay to perform the tasks that this generation does not pay to perform, but that the tasks that they are left to pay for are likely to be the more expensive, in real terms, among the total set of tasks that need to be done. A policy of least cost first is likely to be a policy of 'cherry picking', that is, grabbing the big juicy options—the ones, to switch metaphors, with the biggest bang for a buck—and leaving the scrawnier options, in which a larger investment brings a smaller pay-off, for later generations. This is not necessarily the case, however, and we should remember that mitigation tasks are, after all, not as static as cherries.

If one defers options that are more expensive now but will become less expensive later because of improvements in the technology required for the task, both the present generation and the future generation may be better off. The future generation is, then, better off because we will have meanwhile invested our resources on other tasks for which our technology was already good, and so accomplished more than we could have with an inferior, soon-to-be-supplanted technology, and they can then employ their superior replacement technology that has, by hypothesis, emerged in the interim and accomplish the deferred tasks more cheaply, in real terms, than we could have. We avoided wasting money by using inferior technology that would shortly be replaced.

As on the previous point, however, the sensible principle of choice does not actually turn out to be least cost first. This time it is something more like most mature—most productive—technology first. And while spontaneous technological advances might occur in time to contain costs for future generations, simply to assume net technological progress would be a largely groundless gamble for which future generations will be the ones to pay if it turns out badly. Therefore, if one judges that no mitigation technologies are mature, one ought not simply to do nothing and leave all problems to future generations, but instead to invest whatever is one's fair share of contribution (which I have here said nothing to specify) to solving the overall problem in the creation of the better technologies needed. When our generation cannot effectively mitigate because our technology is inferior, it can invest in better mitigating technology for future generations. Principles of justice among generations do not tell us whether to mitigate using the technology we have or to invest in research on better technology—that is a technological and scientific judgement, not an ethical judgement. Good research may well be preferable to ineffective action. Principles of justice only tell us not merely to do whatever costs us ourselves least irrespective of the consequences of such a policy for others who are affected, including future generations who may be affected profoundly.

So far I have merely sketched one obvious point, with two aspects. The approach to the mitigation of climate change favoured by governments tends to assume uncritically that the principle of choice should always be the least-cost option first. The least-cost-first approach does not necessitate, but encourages, two readings that are dangerous for future generations: first, spending less now rather than doing more (and better) now for the same cost, and, second, 'cherry picking'. Each of these dangers is a distortion of a single sensible caution: don't just 'do something' if that means merely thrashing around in expensive and ineffectual ways—be smart and try to determine whether it is better for everyone affected, including future people, if you invest in creating a better technology that they can employ rather than using the technology at hand.

INTERGENERATIONAL FAIRNESS AND INNOVATIVE TECHNOLOGY

A deeper point about the politics of technology lurks beneath the surface here. The basic decisions concerning climate change are choices about energy technology. We face climate change because, like the cavemen, we still generate a great deal of our energy by setting fire to lumps of coal and—slightly more sophisticated but less of an improvement than one might think—igniting gasoline in the combustion engines of our SUVs. Like the cavemen, we burn fossil fuel, spewing carbon sequestered beneath the earth for millennia into the atmosphere as carbon dioxide. Choices about which nations are to bear the current costs of technological transitions obviously implicate international fairness; but choices about the rate of technological innovation deeply affect intergenerational fairness in a less obvious way, which we can now explore.

A second tendency ingrained in current approaches to climate change, then, is not to abandon fossil fuel technologies until their continued retention becomes too expensive for the then current generation and meanwhile to avoid investing very much in research or development of alternative energy technologies, investments that would hasten the day when alternative technologies become competitive with fossil fuel. Most obviously, this favours, within the current generation, those who own, transport, process, and distribute fossil fuels, and those whose existing capital stock burns these fuels, such as the electricity generators that burn coal. Yet this reliance on the existing atmosphere-distorting technology also strongly favours the current generation more generally, compared to policies on which we would bear the costs of developing alternative energy. The issue of fairness is whether this favouritism toward ourselves is unreasonably strong. This political commitment to consume fossil fuels until they become so scarce that their prices rise significantly—a day that is at least decades, if not at least a century away—and the first tendency to select lowest-cost-first solutions are related; the first supports the second—it is correctly believed to be cheaper to continue to spread into the less developed states the current form of industrialization that runs on fossil fuel than to create and develop alternative energy sources. But many other factors, including immensely powerful, entrenched political and economic interests and lack of imagination, also underwrite retention of the familiar fossil fuel technology that is overwhelmingly the most important contributor to climate change. The primary issue of intergenerational fairness here is not, however, merely the first one already sketched of deferring costs so that they fall upon generations later than one's own, although that remains an issue too. I want now to try to uncover a deeper issue.

THE DATE OF TECHNOLOGICAL TRANSITION

Sooner or later human societies will make a transition from a reliance dominantly upon fossil fuel driven technologies to reliance on one or more alternative energy technologies. The fundamental questions are: at what point in history should this transition be made (substantive intergenerational fairness)? and how should this date be determined (procedural intergenerational fairness)?

From this point on I shall refer to 'the date of the technological transition'. By 'the date of the technological transition' I mean the year in which the burning of fossil fuels ceases to add greenhouse gas to the earth's total atmospheric concentration of greenhouse gases. In theory, this could be the year in which effective technology for carbon sequestration became so widely used that, although vast quantities of fossil fuel continued to be burned just as it is now, the carbon dioxide produced did not escape into the atmosphere. One cannot rule out the possibility of effective technology for carbon sequestration, although to my knowledge there are no particularly promising developments in this direction. Vegetative sinks are certainly not the answer. So, leaving aside the bare possibility of sequestration technology, which would enable continued fossil fuel burning to avoid its current damaging expansion of the atmospheric concentration of gases, the date of technological transition refers to the year in which the quantity of fossil fuel burned drops below the level at which the planet can recycle the resulting carbon dioxide without any further increase in the concentration of carbon dioxide in the atmosphere. The date of technological transition is the year in which we stop enhancing the greenhouse effect by adding greenhouse gas to the atmospheric total from the burning of fossil fuel.[1]

Next we need to remind ourselves about two scientific phenomena involved in climate change that have, believe it or not, features that are directly relevant to the issue of intergenerational justice. First, once molecules of carbon dioxide reach the upper atmosphere they remain for an average of approximately a century—carbon dioxide is a very long-lived gas. This means that the date at which humans stop adding to the atmospheric concentration is by no means the date at which the atmospheric concentration significantly declines, if indeed it declines at all. If we simply reduce our annual production of carbon dioxide to a quantity that does not add to the atmospheric concentration, but no less than that—what we might call an equilibrium production—the atmospheric concentration might of course stay the same indefinitely. If annual total human emissions dropped significantly below an equilibrium level, then the atmospheric concentration would eventually decline. But the point is that no matter how low the annual human production of carbon dioxide, the atmospheric concentration can at best decline only very slowly, because the carbon dioxide already in the upper atmosphere degenerates only very slowly. In this sense the atmospheric concentration has great inertia against downward movement—it can decline only over decades because individual molecules disappear, on average, only over a century. Upward, of course, is a different story: the accumulation can rise as fast as we spew more molecules into it. We are adding carbon dioxide far, far faster than it degenerates. The atmospheric concentration is now rising rapidly—and more rapidly every year than the year before: the increase is accelerating.

Second, obviously the atmospheric concentration can stabilize at any of quite a few different absolute amounts. At the beginning of the Industrial Revolution the absolute amount of greenhouse gas was roughly the same as it had been for centuries. With the currently accelerating annual increases in emissions we are now only a couple of decades away from a doubling of the pre-Industrial Revolution atmospheric concentration. And until political action is taken to see to it that annual emissions are held to an equilibrium amount, that is, an annual amount such that the total accumulated in the atmosphere ceases to grow, as it now grows every year, the atmospheric concentration will increase indefinitely—with nothing to stop it from redoubling. The pre-industrial concentration of carbon in the atmosphere, which had been the same for millennia, was 280 ppm; if we burn all currently known reserves of fossil fuel—all the coal, oil, and gas—thereby injecting into our atmosphere all the carbon now harmlessly buried in our earth, the atmospheric concentration is expected to be between 1,100 and 1,200 ppm—slightly more than two doublings of the pre-industrial level. It matters much less how rapidly we burn it than how much of it we burn all together (Kasting 1998). This is because, once the carbon dioxide is sent into the atmosphere, it stays such a long time.

The arithmetic here is exactly like the arithmetic of national population size. If a nation's population has been expanding, it will of course not become a stable size until the fertility rate drops to the replacement rate. The longer the fertility rate remains higher than the replacement rate, the larger the total population size at which the population will stabilize. In exactly the same way, the longer the annual total emissions exceed the equilibrium amount, the larger the total accumulation at which the atmospheric concentration will stabilize, if it ever does. To return to the notion of the date of the technological transition, then, the later the date of the technological transition, the larger the atmospheric total of greenhouse gas at which stabilization will occur. This is not advanced atmospheric chemistry, but simple arithmetic: for every year that the total continues to grow, the larger the total will be when it stops growing. This is exactly like world population: the longer it continues to increase, the larger it is when it stops growing and stabilizes. This much is simple.

What is far from simple, of course, is the relation between the total atmospheric concentration of greenhouse gases and the extent, and types, of climate change. Climate involves many poorly understood feedbacks, positive and negative; and we know very little about the range within which changes tend to be incremental and where abrupt switches and reversals occur, although we do know from studies of Greenland ice cores reflecting the Younger Dryas, 10,000 years ago, that extremely abrupt switches have in fact occurred (Alley, 2000). Further, climate is determined in part by factors totally independent of atmospheric chemistry, including the basic geometry of the solar system, with its cycles within cycles within cycles. Nevertheless, while there are many factors to be equal, we do have fairly good grounds for believing that, other things being equal, the higher the atmospheric concentration of greenhouse gases, the more severe climate change is likely to be. At the very least, the higher the atmospheric concentration of greenhouse gases, the farther we will have rushed, without a map, into planetary terra incognita. Consequently, fully acknowledging how little we really know, it still appears that the safest generalization, until we know much more, is this: the later the date of technological transition, the more dangerous the world we will have made and bequeathed to future generations. The later the date, the greater the danger.

I believe that a description of our actions regarding future generations only this rich in scientific detail, even though it is still quite superficial, transforms our understanding of the nature of the normative choices we face. It is still true, as we noticed in the beginning, that we face an issue recognizable as a fairly standard question of intergenerational distributive justice, namely 'Does the amount of effort and money that we are investing in either the direct mitigation of climate change, or in research on better technology for future more effective mitigation, represent our generation's fair share?' I have of course said nothing here to answer that question, but it is at least a relatively familiar type of question. I will not devote space to that question, because I think the deeper issues about the date of technological transition are more basic and more significant.

A LEGACY OF RISK OF HARM

If we adopt policies that delay the date of technological transition, then, as far as we know, we create a more dangerous world for our grandchildren and subsequent generations. We subject them to risks of unknowable probability but of enormous possible magnitude, including radical change in the very conditions of life, human and non-human, on this planet. It is vital not to make the mistaken assumption that if the size of a risk is unknown, the risk must be small—as if it could be unknown only if it were too small to see. If someone told you that there was an animal behind a closed door but that the species of the animal was not known, would you be wise to assume it was a small animal?

The imposition of such risks—of unknown (not necessarily small) probability and large magnitude—seems to me to be an inexcusable wrong. I assume that it is wrong to harm other persons without adequate reason, such as the harm's being unavoidable in self-defence and, even at that, not radically disproportional.[2] This 'no-harm' principle is of course not undeniable without contradiction or certain in any other sense. I simply take it to be an extremely compelling principle and one found in anything clearly recognizable as an ethical code. And I take it to be less uncontroversial, although equally correct, to say that it is wrong to impose a risk of harm on other persons without adequate reason. It is not as wrong to impose a risk of a harm as it is to inflict the same harm, but it is still wrong. And if the harm would be serious, the imposition of the risk is seriously wrong, even if not as seriously wrong as the infliction of the harm itself would be.

To choose policies under which climate change will probably be more severe than it will be under other feasible policies is to choose to risk additional human deaths and other disasters, not to mention additional extinctions of species, damage to coral reefs, and many other kinds of destruction of the natural world. Perhaps this would turn out in the end to be, so to speak, all risk and no harm— that is, perhaps not a single one of the harms risked will actually occur. Nevertheless, if the best information we have suggests that simply continuing to burn more and more fossil fuel will increase the maximum accumulation of greenhouse gas in the atmosphere and that increasing the accumulation of greenhouse gas is likely to produce more severe climate change that will endanger human beings, then choosing to burn more fossil fuel than we must, by relying upon it longer than we must, is, as far as we know now, choosing to endanger future persons—choosing to impose risks upon them that could be avoided by other policies. It could, for all we know, somehow turn out well, but we would still have acted wrongly, just as I would have acted wrongly by playing Russian roulette with your head even if luckily the revolver did not fire. And, of course, we would be responsible for the occurrence of any actual harms the risk of which we avoidably chose to impose.

Future generations are completely vulnerable to whatever risks of harm we impose by our choices. This is an extreme inequality in power. We are the only representatives or trustees they have—or can have. One perspective from which to see the stringency of our responsibility not to impose risks of serious harm is to notice how dramatically different the content of decisions about intergenerational responsibility has become in the case of climate change from its content as traditionally understood. For centuries the focus of intergenerational responsibility has been intergenerational justice in the literal sense of distributive justice, and specifically the issue of the just rate of savings. At what rate ought we to save on behalf of future generations, in light, especially, of the fact that they would probably be generally better off than we are? Economic progress was built into the assumptions underlying the problem in a way that now strikes those aware of environmental limits as quite naive. One of the motivations for discounting the present value of future commodities was, in effect, to protect ourselves from saving too much for the people of the future who would enjoy all these commodities and thereby to avoid unfairly depriving ourselves of consumption that we are entitled to enjoy. So thirty years ago in *The Theory of Justice* John Rawls took the problem of intergenerational justice simply to be the problem of the correct savings rate (Rawls, 1999a: 251–8). Twenty years ago, with the 'oil crisis' of the 1970s between him and Rawls, Brian Barry could plausibly suggest that the appropriate norm for conduct affecting future generations was an intergenerational analogue of the Lockean proviso: leave the next generation no worse off than you were—in particular, substitute for non-renewable resources that you unavoidably consume improved technology and greater knowledge that will enable future generations to live at least as well on a different, and smaller, resource base (Barry, 1991). What the science of climate change now suggests is that it may actually be naive to think even that we can leave the next generation as well off as we are. My generation may have had it as good as it gets. The appropriate question appears to have become not how we can leave them *no* worse off but how we can avoid leaving them *much* worse off. How can we avoid undermining the stability of the very climate to which they, and all the other living things that they value and on many of which they depend, are adapted? How can we—to return to the earlier language—avoid imposing upon them risks of serious harm?

Thus part of the basis of the stringency of the responsibility to control the extent of the harms imposed is, in effect, how low the bar has now fallen. Not long ago it was taken for granted that we ought to leave future generations better off—and of course this is still the dream most parents have for their own children. The only question then seemed to be: *how much* better off ought we to leave future generations: what is the just rate of savings, the rate at which we ought to forgo consumption of our own, for their sake? Some then lowered their sights to: how can we avoid leaving the future worse off? If one accepts that past emissions of greenhouse gases have already created an atmospheric commitment to some climate change, then, at least with regard to the climate, the question has now become: how can we avoid leaving the future *too very much* worse off? Or, in other words, how can we avoid imposing on them the risk of serious harms? We have thus lowered our goal so much that it is difficult to believe that we should not go to considerable lengths to attain it.

THE KYOTO PROTOCOL

Much more could be—and probably needs to be—said entirely at the abstract normative level, but I want now to connect these abstract normative issues to fundamental political and institutional issues concerning one central element of the Kyoto Protocol. For better or for worse, the attempts to slow climate change have copied the largely successful effort to protect the ozone layer against CFCs, copying the form of the solution even where the problems are different. In the earlier case of ozone destruction, an initial vague, largely hortatory, and now mostly forgotten 1985 Vienna Convention for the Protection of the Ozone Layer was put into business by the specific requirements spelled out two years later in the 1987 Montreal Protocol on Substances that Deplete the Ozone Layer, and the protocol's amendments in 1990 and 1992. Like the Vienna Convention, the 1992 Framework Convention on Climate Change (FCCC) simply identifies the problem of anthropogenic climate change and enunciates some important principles to guide action concerning the problem. But the FCCC does not require any nation to do anything any more than the Vienna Convention did. The 1997 Kyoto Protocol is intended to do for the FCCC what the Montreal Protocol had done for the Vienna Convention, namely, spell out who should do what. The Third Conference of the Parties to the FCCC, meeting in Kyoto in 1997, had spent a long Friday night with the diplomatic clock stopped, teetering on the brink of collapse, but had tilted away from the cliff in a compromise constructed partly on the spot by sleep-deprived negotiators around a proposal that accordingly came to be known as 'the Kyoto Surprise'—more formally, the CDM (Grubb et al., 1999: 101–3).

Here are the first three sections of Article 12 of the Kyoto Protocol:

1. A clean development mechanism is hereby defined.

2. The purpose of the clean development mechanism shall be [A] to assist Parties not included in Annex I [poor states] in achieving sustainable development and in contributing to the ultimate objective of the Convention [FCCC], and [B] to assist Parties included in Annex I [rich states] in achieving compliance with their quantified emission limitation and reduction commitments [emission caps] under Article 3.

3. Under the clean development mechanism:
   (a) Parties not included in Annex I will benefit from project activities resulting in certified emission reductions [CERs]; and
   (b) Parties included in Annex I may use the Certified emission reductions [CERs] accruing from such project activities to contribute to compliance with part of ['supplementarity'] their quantified emission limitation and reduction commitments [emission caps] under Article 3, as determined by the Conference of the Parties serving as the meeting of the Parties to this Protocol [COP/MOP].

[bracketed material added]

We need to take a couple of steps back in order to appreciate what is going on in Kyoto's Article 12. While the treaty terminology is impenetrable, the situation is fairly simple. The achievement of the Kyoto Protocol was to have been the first imposition upon the wealthy industrial states of required reductions in greenhouse gases (GHGs): the quantified emission limitation and reduction commitments mentioned in sections 2 and 3 of Article 12 are actually imposed by Article 3. The FCCC of 1992 had an Annex I, which is essentially a list of the wealthiest states. Wealthy states (Annex I) have obligatory emissions reductions (or, in exceptional cases such as Australia, which negotiated a cushy deal for itself, limits on increases); and poor states do not yet have any reductions or limits.[3] Poor states are for the time being allowed to increase GHG emissions without restriction. Emission caps were defined as annual emissions averaged over the five-year period 2008–12, which is known as the first commitment period.

The CDM is intended to be the first functioning bridge between poor states and rich states. The CDM was an imaginative solution to a problem that will never go away, namely, on what terms can rich and poor cooperate on slowing climate change? While I have some other profound disagreements with David Victor, I agree with him that the CDM is a crucial issue (Victor, 2001). Victor asserts that 'the CDM is essential to the Kyoto treaty – and to any successor to Kyoto' (103). I would put the point more sceptically: that the CDM is one possible answer to a question that must have an answer. Either there will be some mechanism to differentiate between but coordinate rich and poor states, or there will be no effective international initiative to slow climate change.

In general, far more was flexible than fixed in the Kyoto Protocol, thanks to the insistence of the OECD states, led by the United States, on introducing market, and quasi-market, mechanisms wherever possible in accord with the general post-Cold War Zeitgeist of globalization. The most important elements of the Kyoto Protocol are its multiple 'flexibility mechanisms'—the 'flex mechs'—the best known of which may be the proposed future international trading in assigned amounts of emissions (AAs) (popularly referred to as 'emissions permits'), but the most important of which may be CDM, the 'Kyoto Surprise', which David Victor, who is relentlessly critical of international emissions trading, rightly believes must survive in some form (Grubb et al., 1999). Flexibility is often a great virtue in complex institutions of the kind that any global GHG regime will certainly be, especially if it proceeds along the trajectory of complexity launched by Kyoto. Even if it be the case that, other things equal, flexibility is a good thing, other things are, of course, rarely in practice equal—and certainly not in the case of the CDM, as we shall now see.

The CDM may be the nearest possible analogue in a bilateral form to a market in tradable emission permits (strictly, AAs). But, of course, nothing bilateral can be a very close approximation to a market, because one of the greatest virtues of a true market is that the availability of multiple sellers and multiple buyers avoids one of the worst potential failings of a bilaterally negotiated transaction, namely that a great inequality of power between (single) buyer and (single) seller will allow the superior power to dictate the price to the inferior power. Thus, all too common references to CDM as 'bilateral trading' are seriously misleading, if not literally incoherent, since any bilateral exchange is a negotiation, not a market; and genuine trading, by contrast, needs to be reasonably multilateral on both sides. However, CDM is the only one of the 'flexibility mechanisms' for which the Protocol text of December 1997 provides an Executive Board (Article 12.4) that could in principle formulate and try to enforce rules of fairness, and minimum standards for the acceptability of CDM projects, that might prevent the exploitation of inequality in negotiations over the terms of projects, so the concern about CDM's being bilateral rather than multilateral need not be pressed until the procedures of the Executive Board have been set. It is possible that the CDM will follow a more multilateral, 'portfolio approach', if the Executive Board can become more than the mere clearing house for negotiated exchanges that the rich states mostly favour (Yamin, 1998: 54–6). What Farhana Yamin called the 'gaps and ambiguities contained in the Protocol text, much of which is compromise language crafted to paper over what appeared in Kyoto to be unresolvable differences of views' still partly remain to be crystallized (54).

Now, let us tie together these concrete strands about CDM and the earlier abstract strands. Recall our conclusion about the date of the technological transition: the later the date, the greater the danger. The further into the future the date of technological transition, the more dangerous the world we will have made and bequeathed to future generations, because the larger the total sum of carbon we will have moved from beneath the surface of the planet, where it does no harm, into its atmosphere, where it does the great harm of fuelling climate change. Many other things equal, the more carbon injected into the upper atmosphere, the worse climate change will be. Rather than maxi–min, this is mini–min: making the worst as bad as possible. What does this have to do with CDM?

CDM is a flexibility mechanism; what this really means is that it is a way for rich states to save money. They save money by avoiding any need to reduce their own emissions beyond the money-saving no-regrets reductions that are economically rational in any case. They avoid needing to reduce their own emissions by instead paying for CDM projects in the poorer states, which will be much cheaper per marginal unit of emission reduction, and subtracting the credits they receive for their CDM projects from any reductions they would otherwise be required to make at home. Why are emissions in the poorer states generally cheaper? Naturally, there are multiple factors, including generally lower standards of living, lower wages, lower health and safety standards, and so on. One of the biggest factors, however, is that the current technology in the richer states that would need to be replaced in order to reduce emissions—at high costs in replacement of capital stock before the end of its useful life—may well constitute a technological advance over current technology in the poor states, certainly in the poorest parts of the poor states. In purely economic terms, it is much more efficient to replicate current rich-state technology in the poor states than it is to replace current rich-state technology in the rich states with technology that would be superior measured in terms of its emissions. CDM has great short-term efficiency—it is, precisely, a least-cost solution.

Is this as clever as it may appear? We (this generation) are saving money. When one turns from efficiency narrowly construed to technology, the kicker appears. What is this familiar, easily transferable technology? It is the very fossil fuel technology that is bringing us climate change: technology fuelled by coal, oil, and gas. CDM may also be a way to guarantee that fossil fuel technology will continue to be used until all the fossil fuel is gone—gone from beneath the earth, that is, with the carbon injected into the atmosphere, providing one precondition for the worst possible degree of climate change.

Yet CDM is far from all bad. Actually, it is difficult to assess on balance even only in terms of international distribution. For it is not only achieving something good but bringing it about in a superior way. CDM will contribute to desperately needed economic development in poor states, although probably not the very poorest. It will achieve that development with relatively fewer emissions than would be required, if it could be achieved at all, without the technology transfers motivated by the CERs to be awarded to the rich states—strictly, firms located in the rich states—who finance the projects. In other words, at least some of the technology transfers demanded in the 1960s and 1970s as a matter of justice as part of the NIEO (New International Economic Order) but never granted out of any sense of justice on the part of rich states might now take place out of national, and individual firm, self-interest.

Some of this economic development in poor states might well be a good thing. And assuming only that each CDM project would enable something to be done with fewer emissions than it would otherwise have generated, if it could have been done at all, relative reductions in emissions would be generated. Indeed, it is precisely these relative reductions—reductions relative to a hypothetical development path minus the technology transfers brought about by CDM—for which the CERs are to be issued. This is the good news: (1) economic development—at least, more energy, which does not of course necessarily mean development, which, in turn, as we all know, does not necessarily mean a better life for poor people—and (2) fewer emissions per unit of development, compared to any development that hypothetically might have occurred without CDM. A worthwhile goal achieved by a better means.

But the bad news is also double. First, CDM will actually increase greenhouse gas emissions. Every development project it makes possible will produce emissions that would not have occurred if there had been no project. Now of course a CDM project may produce relatively fewer emissions than the hypothetical alternative project without CDM, which would hypothetically use inferior technology with higher emissions. But in fact in most cases the hypothetical project is simply a fantasy project, because without CDM there would be no project at all. If the poor state had been going to carry out the project on its own, in many cases it would by now already have carried it out. Development with CDM will be cleaner than it might have been, in some counterfactual world, but in this world the only real alternatives are the CDM project and no project. Many CDM projects will therefore lead to an absolute increase in emissions.

Few seriously pretend otherwise, although I think that calling the payments that rich states receive Certified 'Emission Reductions' is a slippery bit of public relations terminology—it would have been far more straightforward to have named them 'Certified Smaller Emission Increases', which unfortunately does not have quite the same ring. But the relative reductions, compared to what might otherwise have happened in future, are absolute increases, compared to what is in fact happening now. One political justification is that this is the development half of the environment/development bargain. Stockholm 1972 was the UN Conference on the Environment; Rio 1992 was the UN

Conference on the Environment and Development. There was no way that the poor states were going to be a part of any climate change initiative that did not provide for their further development. From a political bargaining standpoint, CDM is at least theoretically part of the pay-off for the poor states, although in fact the satisfaction of the rich states' determination to pursue least-cost-first strategies was, I think, even more important in the wee hours of the critical Friday night at Kyoto. CDM had—perhaps still has—some of the makings of a grand bargain of the kind that can keep diverse coalitions functioning. In any case, insofar as CDM projects actually contribute to a genuinely human development that improves the quality of the lives of the planet's poor, no sensible person is going to be against them. One might say that the absolute increase in emissions is in nearly the best possible cause, improving the lives of at least some of the poorest.

What is troubling is not the CDM projects themselves, at least not what could be the best of them. The worries arise at the other end, in the rich states whose firms would be financing the CDM projects. The reward for carrying out projects in the poor states—for doing more abroad—is doing less at home. This obviously is why the projects would lead to absolute increases in emissions. One could avoid the absolute global increases by requiring domestic reductions—reductions in the firm's home country (or elsewhere)—to balance out the additions in the poor state. But that, of course, would destroy the narrow economic rationale for CDM projects. If a firm would still have to achieve reductions at home equal to the increases produced abroad by the CDM project, it could just skip the CDM project entirely and simply bring about the required domestic reductions. Adding in the option of CDM (a) saves money, and satisfies least cost first (providing one is willing to play the game of counting relative reductions as if they were absolute reductions) and (b) promotes development assuming the CDM projects are well chosen, perhaps under the watchful eye of a multilateral Executive Board with significant poor state participation—all of which remains to be seen.

Nevertheless, the second element of bad news, and the central point of this chapter, is this: the CDM seems likely to contribute to significant delay in the date of technological transition. The ultimate motive behind the CDM is least cost first: spending as little as possible ourselves for any given reduction in emissions. Which will be cheaper for the current generation: spreading fossil fuel driven technologies to the ends of the earth or replacing them with a superior alternative technology not based on fossil fuel? There is every reason to think that the status quo is cheaper for us, at least until existing capital stock has operated for its normal expected life. The complete dynamics of the technological transition need fuller exploration, but, as far as I can see, the best that one could hope for with CDM would be the very gradual replacement of fossil fuel driven technology at the end of the expected life of the capital stock using it, but only after the installation of much more fossil fuel based capital stock across the poor states. If this is not the surest way to prolong the consumption of fossil fuel for the longest possible time, it comes close.

Whether CDM would contribute to international justice is, I think, not easy to judge. Much clearly depends on who exactly benefits in which respects from the projects. But insofar as they might constitute the diversion of investments from projects that would enhance the lives of the affluent, to projects that would improve the lives of the poor, there is something to be said for them from the standpoint of international justice. At present, however, there are no explicit requirements that the CDM project benefit the poor people in the poor states. By contrast, insofar as CDM delays the date of technological transition—delays the date when human activity stops making climate change worse—it imposes an unknown but quite possibly rather great risk of harm of significant and potentially even unmanageable kinds upon future generations. By my lights the risk of harm to future generations seems considerably more clear and serious than do the possible benefits to our impoverished contemporaries, whom we are in any case quite free to assist in ways that do not entail worsening climate change. We certainly need not reduce international inequality by increasing intergenerational inequality. If we faced an unavoidable choice between the 'present poor' and the 'future poor', that might be a difficult or even tragic choice. But only the CDM is forcing that choice upon us, and the CDM is fundamentally about letting the rich states adopt policies toward climate change that are, for them now, least cost. The real choice is between the present rich and the future poor, the most unequal pair of all.

ALTERNATIVES

Two intellectually simple but politically difficult—furiously rejected by the US government—stratagems are available to reduce the risks imposed on future generations: the 'positive list' and 'supplementarity'. The positive list would specify acceptable types of technology for CDM projects and could restrict projects to those employing alternative energy technologies (and could refuse to count sinks). A firm and definite supplementarity requirement, mentioned but not defined in Article 12.3, could restrict the use of CDM by requiring a specific percentage of an Annex I nation's efforts to be conducted domestically. Both these good ideas have been fought tooth and nail by the 'Umbrella Group', especially the governments of the United States, Australia, and Canada, and they appear for now to be politically dead, to the misfortune of future generations.

Meanwhile, poor state elites and rich state firms are rushing to negotiate CDM deals, and the World Bank has a 'CDM-Assist' Program. CDM is by far the most popular flexibility mechanism. I have tried to show here why it may also be the most dangerous, for the most vulnerable.

NOTES

1. While this particular terminology and formulation are my own, I take the fundamental idea from Grubb, 1998: 2. Grubb refers to a concern with technological transitions as 'dynamic efficiency' as distinguished from 'static efficiency' (economists' usual focus). He there draws upon Grubb et al., 1995.

2. I do not believe that one is permitted to do an unlimited amount of harm to innocent others even to save oneself, but I do not want to pursue this more basic issue here.

3. Basically, Annex I of the FCCC and Annex B of Kyoto are the same thing: a list of rich countries. While one might have expected Kyoto to employ its own terminology internally, that is, to refer to 'Annex B parties', its text usually—but, *of course*, not always—designates the rich as 'Annex I parties', as in the sections quoted earlier from Article 12.

REFERENCES

Alley, Richard B. (2000) *The Two-Mile Time Machine: Ice Cores, Abrupt Climate Change, and Our Future,* Princeton: Princeton University Press.

Barry, Brian (1991) 'The Ethics of Resource Depletion', in *Liberty and Justice: Essays in Political Theory,* Vol. 2, Oxford: Oxford University Press, 259–73.

Grubb, Michael, with Christiaan Vrolijk and Duncan Brack (1999) *The Kyoto Protocol: A Guide and Assessment,* London: Royal Institute of International Affairs.

Houghton, J.T., Y. Ding, et al. (eds.) (2001) *Climate Change 2001: The Scientific Basis,* Contribution of Working Group I to the Third Assessment Report of the Intergovernmental Panel on Climate Change. Cambridge: Cambridge University Press.

Kasting, James F. (1998) 'The Carbon Cycle, Climate, and the Long-Term Effects of Fossil Fuel Burning', *Consequences: The Nature & Implications of Environmental Change,* 4: 15–27, http://www.gcrio.org/CONSEQUENCES/vol4no1/carboncycle.html

Mahlman, Jerry D., (2001) *The Timing of Climate Change Policies: The Long Time Scales of Human-Caused Climate Warming - Further Challenges for the Global Policy Process,* Arlington, Virginia: Pew Center on Global Climate Change, http://www.pewclimate.org.

McCarthy, James J., Osvaldo F. Canziani, et al. (eds.) (2001) *Climate Change 2001: Impacts, Adaptation, and Vulnerability,* Contribution of Working Group II to the Third Assessment Report of the Intergovernmental Panel on Climate Change. Cambridge: Cambridge University Press.

Rawls, John (1999) *A Theory of Justice,* Cambridge, Mass., Harvard University Press [1971].

Shue, H. (1999) 'Global Environment and International Inequality', this volume.

Yamin, Farhana (1998) 'Operational and Institutional Challenges', in *Issues & Options: The Clean Development Mechanism,* New York: United Nations Development Program, 53–79.

12

Responsibility to future generations and the technological transition*

The nature and extent of moral responsibility toward climate change depend on the nature and extent of climate change. One cannot specify degrees of responsibility for dealing with a problem without first specifying, or implicitly assuming, the salient features of the problem itself. The likely causes and consequences of climate change are not for normative debate but, rather, for scientific investigation. Nonetheless, much discussion of ethical issues concerning climate change assumes that climate change takes one, not another, of the many forms it might be imagined to take. We normative theorists must do our best not to engage in the equivalent of assuming that the primary moral dilemma facing the passengers on the *Titanic* was the selection of the principles for the allocation of the deckchairs.

Much of what has been written about the ethics of climate change, including several of my own articles, has assumed that a central issue is the allocation of emissions of greenhouse gases (GHGs). However, it is increasingly evident that there is no allocation of GHG emissions specifically in the form of carbon dioxide that is both morally tolerable and, at present, politically feasible as long as most economies are dependent for energy upon carbon-based fuels, that is, fossil fuels.[1] In order to be morally tolerable, total GHG emissions need to be reduced to a level that will not cause climate change so rapid that societies and species cannot adjust; wherever exactly that level is, it is significantly below current levels, which are meanwhile rising.[2] To reduce GHG emissions significantly below present levels would involve sharply reducing the source of those emissions, the combustion of fossil fuel. Climate policy is energy policy. But as long as the combustion of fossil fuel is the predominant source of energy, sharply reducing the combustion of fossil fuel means sharply reducing the total use of energy. Most rich and powerful states, with the USA

* 'Responsibility to Future Generations and the Technological Transition', in *Perspectives on Climate Change: Science, Economics, Politics, Ethics*, ed. Walter Sinnott-Armstrong and Richard B. Howarth (Amsterdam and San Diego: Elsevier, 2005), pp. 265–83. This chapter has endnotes.

at the far extreme, are completely unwilling today to reduce their energy consumption to anywhere remotely near the level at which the GHGs emitted from carbon-based sources of energy would no longer cause rapid climate change.

Questions about the political and sociological bases for this reluctance—for example, the extent to which the reluctance reflects middle-class American consumers' infatuation with sport utility vehicles (SUVs) and romance with the car generally, and the extent to which it is the corrupt political influence of oil- and coal-company executives—are vital, but we will not explore them here.[3] Even so, we can work unrelentingly both to change popular attitudes and to fight elite corruption. Concurrently, it appears at least as economically and politically viable also to move away from the fossil fuel energy regime that is overwhelmingly the primary source of GHGs, to some non-carbon-based regime.

The thesis of this chapter is that, besides practical political reasons, we have, perhaps surprisingly, strong moral reasons, involving responsibilities to future generations, for an aggressive search for alternative sources of energy—sources other than coal, oil, and gas. Choices among alternative allocations of GHG emissions in the short term may turn out to matter principally because of the incentives they do or do not create for moving decisively beyond fossil fuels, the consumption of which injects such monumental, and consistently growing, emissions of GHGs into our planet's atmosphere.[4]

DATES AND THRESHOLDS

I have applied the phrase 'the date of the technological transition' to the year in human history in which the accumulated atmospheric total of all GHGs ceases to grow.[5] Carbon dioxide from the combustion of fossil fuel is only one GHG, of course, but increases in carbon dioxide have made by far the greatest contribution to the swelling of the total. Perhaps quantities of some other GHGs would even need to continue to grow, perhaps not—this is a murkier realm. But if emissions of carbon dioxide were reduced sufficiently, quantities of other GHGs could, if necessary, increase while total annual emissions of all GHGs declined because carbon dioxide is such a large part of current annual emissions of all GHGs and of annual increases in emissions of all GHGs. Reductions in emissions of carbon dioxide could 'make room' for any necessary increases in other GHG emissions.

This stabilization in the quantity of atmospheric GHGs could only begin after a cessation in the now annual increases in total global emissions.[6] The mere elimination of net increases in annual emissions would by no means produce a stabilization in the atmospheric accumulation, any more than the mere elimination of increases in annual national deficits can produce an end to growth in the national debt. As long as any annual deficit occurs (even a deficit smaller than previous deficits), the amount of that deficit will constitute an addition to the accumulated national debt. In order to stop the debt from growing, annual deficits must be eliminated entirely, not merely made smaller. Similarly, the atmospheric concentration of GHGs will stop mushrooming only when annual emissions are reduced to an amount that can be recycled in the short term so that no net additions to the accumulated total are made.[7] The rate of annual emissions must become, to use the fashionable term, 'sustainable', if the atmospheric accumulation is not to continue to expand indefinitely. The current annual rate is far above any sustainable rate, so the rate must not only stop growing but must decrease to a sustainable level, a level that makes no addition to the accumulated atmospheric total. On this much, everyone agrees.

It is also evident, although somewhat less commented on, that the stabilization of the size of the atmospheric accumulation of GHGs has a critical similarity to the more familiar case of the stabilization of the size of a national population. Suppose the populations of countries A and B are the same size and have been growing at the same rate. Country A then reduces, over a period of twenty-five years, its overall fertility rate to a replacement rate, at which the population can remain stable indefinitely; that is, it takes twenty-five years for the fertility rate to fall from the current rate to the replacement rate. Although beginning at the same place, country B moves more slowly and spends seventy-five years reaching a zero-growth fertility rate. Obviously, because country B's population expanded for fifty years longer than country A's population did, the population at which B stabilizes will be considerably larger than the population at which A stabilizes, in spite of their having begun with populations of equal size. It could easily happen, depending on the curve of the decline in fertility rates in each case, that while the population in country A doubled once during its twenty-five-year transition to stability, the population in country B doubled at least twice during its seventy-five-year transition. From the date of stabilization, the population of B will no longer grow, but its population will always be much larger than the population of country A if they both remain stable thereafter, and specifically larger by its net growth over the additional fifty years during which its population continued to expand.

Similarly, the absolute size of the atmospheric accumulation of GHGs at which stabilization occurs, if it ever does, will be dramatically different depending on how soon it occurs, depending, that is, on 'the date of the technological transition'. Many years of compound growth at even relatively small rates can yield far higher ultimate totals. This is plain arithmetic, not rocket science or even atmospheric science: the longer a total grows, the larger it ends up being. Consistent growth compounds totals. At any very significant rate of growth, a total can double several times within one century, just as a savings account doubles fairly often at any significant interest rate (doubling every twelve years at 6 per cent, for example).

The fundamental principle of the science of climate change as far as humans are concerned, since we live on the surface, is that increases in the atmospheric concentration of GHGs around a planet tend to produce increases in the surface temperature of that planet—global warming—other things being equal. Needless to say, the ratio between the two increases (in atmospheric concentration and in surface temperature) cannot be specified precisely, and there are many other factors to be, or not to be, equal. Generally speaking, however, we have solid reason to fear that the larger the atmospheric concentration of GHGs, the more severe the changes at the planet's surface to which living things will have to adapt if they are not to die. Therefore, probably—not certainly, but considerably more likely than not—the later the date of the technological transition is, the more threatening to forms of life the conditions at the surface of the planet will become. The longer the atmospheric concentration expands, the more severe the stresses upon living things will be.

Abrupt reversals in temperature trends, such as the Younger Dryas (rapid warming succeeded within a few years by rapid cooling), have occurred several times.[8] The authoritative 2002 report from the National Academy of Sciences, which the Bush administration has completely ignored after having requested it, begins: 'Large, abrupt climate changes have repeatedly affected much or all of the earth, locally reaching as much as 10 degrees C change in 10 years. Available evidence suggests that abrupt climate changes are not only possible but likely in the future, potentially with large impacts on ecosystems and societies.'[9] A rapid change in climate is never good for living things because they need time to adapt; and the only thing worse than simple rapid change is rapid change in one direction abruptly followed by rapid change back in the other direction. Zig-zag adaptation by animals or plants to abrupt reversals is especially unlikely to occur because natural selection is on a much slower timescale.[10]

Disturbing evidence is accumulating that another abrupt reversal may in fact be under way already.[11] For the date of the technological transition, then, sooner is definitely better. Just how much better, we do not know, but if there are critical thresholds within total accumulation, such as a threshold for a Younger Dryas-like abrupt reversal, sooner might be far, far better. A delay in the technological transition makes abrupt reversal more likely, other things being equal, because it means a greater accumulation of GHGs, thus likely a warmer surface temperature, more melting of Arctic ice into fresh water, and thus a greater dilution of the salinity of the North Atlantic, tending to undermine the driver of the deep ocean circulation, which is relatively heavier salty water sinking toward the bottom of the North Atlantic.[12]

In general, we do not know—and, as far as I can tell, may never know until it is too late to do anything about it—how often smooth, gradual increases in total atmospheric accumulation of GHGs will lead to smooth, gradual increases in the difficulties for forms of life on the surface, and how often various types of life will instead reach a limit that individuals cannot tolerate and to which species cannot adapt. Any plant or animal can handle gradual increases or decreases across a certain range at up to a certain rate in critical parameters such as temperature, moisture, nutrition, sunlight, top wind speed, length of growing season, and so on. But at some point for each parameter for each species, a limit is reached. And of course many species are interdependent: if a plant becomes extinct, an animal for which it was food may become extinct, and a predator for which that animal was prey may, as well. The extent of adaptation is astounding; and the resilience of individuals and species can be remarkable. Nevertheless, adaptation has severe—in fact, fatal—limits for each species on the rate at which adjustments can occur. I would suggest, therefore, that a fundamental moral responsibility regarding climate change is to take all necessary actions to make the date of the technological transition, at a minimum, soon enough to avoid the crossing of thresholds critical for very many species, including of course the human species. In the words of the Framework Convention on Climate Change, this must be soon enough to 'prevent dangerous anthropogenic interference with the climate system'.

The date of technological transition will certainly come, one way or another, but it may well come too late for individuals, for communities, or for species. At the extreme, the date of technological transition will come when fossil fuels have become sufficiently scarce for their prices to rise, purely from market forces, above the prices of alternative sources of energy. Unfortunately, the foreseeable supplies of fossil fuels that will be economically profitable for their owners and marketers to exploit are gigantic. For gas, oil, and especially coal to become sufficiently scarce for their relative prices to rise to a point at which they are no longer competitive, vast further quantities would have to be consumed. Put differently, vast amounts of carbon that have been safely sequestered out of circulation underground for millennia in coal, oil, and gas will have to be injected into the atmosphere as carbon dioxide through the combustion of these fossil fuels in order to reduce the supply enough to cause a significant purely market-induced price rise. If large percentages of the remaining stocks of fossil fuel are indeed burned at current or higher rates of consumption, the atmospheric concentration of carbon dioxide will have soared by the date of the technological transition. 'Consuming what remains of fossil fuels could well lead to a four- to eight-fold increase in CO_2.'[13]

We have identified, then, one ultimate extreme policy regarding climate change: do nothing—simply wait for combustion of fossil fuels to reduce the supply to the point at which price increases reduce consumption to the level at which GHG emissions are sustainable. The consequences of this policy of political passivity and myopia are clear: the date of technological transition drifts far into the future, and the total accumulation at which GHGs are stabilized becomes very high, making the consequences for climate change severe. For all we know, these consequences could be catastrophically beyond one or more thresholds critical for humans, certainly beyond extinction thresholds for more slowly adapting species and beyond survival thresholds for some human communities (for example, South Pacific societies whose island homes will be submerged by the rising sea level).[14]

DANGERS AND RESPONSIBILITIES

I have already suggested a moral responsibility to advance the date of the technological transition nearer in time in order to avoid the risk that later increases in atmospheric concentrations of GHGs will cause critical limits in aspects of the climate on the planet's surface to be exceeded. In order to be acceptable, this suggested responsibility needs fuller explanation: to whom would such responsibility be owed and what kind of responsibility would it be? The clearest case is future generations, that is, individual persons who will live in the future. A great deal of interesting philosophical analysis has been carried out in recent decades regarding issues arising from the fact that which policies are followed in the present on many matters will determine not only the size of future generations of human beings but the identities of the specific individuals who will constitute these generations.[15] While these analyses are important, they do not much affect the general shape and character of our responsibilities to whichever human beings turn out in fact to live in the future. If one has any responsibilities to human beings whose interests one can significantly affect, then one has these responsibilities to any such human beings who happen to live in future times, whatever their numbers and identities. The key question is, what kinds of such widespread responsibilities might be relevant?

Discussions of the ethics of climate change have tended to assume that the primary, if not the only, responsibilities are responsibilities of distributive justice: what we owe to members of future generations are such duties as doing our fair share to solve any common problems and not consuming more than our fair share of common resources such as the capacity of the atmosphere to absorb GHGs without untoward effects. I next want to explore these suggested responsibilities a bit in light of what we have seen to be the nature of the date of the technological transition.

One way of trying to conceive the issue of responsibility is the following. All human beings potentially share some responsibility generally for dealing with climate change and, specifically, for preventing unjustifiable delays in the date of the technological transition, that is, for avoiding the creation of unnecessary dangers for people in the future. Plainly, these specific responsibilities need to be assigned in accord with some allocative principles, such as ability to contribute to the solution or past contribution to the problem. Thus, it will emerge that some people who are unable to contribute to the solution and made no contribution to the problem bear no actual responsibility, while other people bear heavy responsibility on one or both of these (or other) grounds, and so on.[16] In practice, not everyone can reasonably be assigned any responsibility, but in theory everyone is a candidate for bearing responsibility, depending on how the principled assignment of specific responsibilities works out. This conception can lead to the following general picture of the responsibilities.

All the people who are alive now or will live prior to whatever turns out in the end to be the date of the technological transition constitute the general pool of persons eligible to bear some degree of responsibility for when the date will in fact be; clearly, one might also include people in the past, who may or may not already have failed in their responsibilities, as part of the pool of responsibility, but matters are complicated enough without them. The fundamental issue then becomes whether one is carrying out one's fair share of responsibility, if any, given everyone in the general pool, given the total responsibility, and given the allocative principles for the assignment of responsibility. If one fails to carry out one's responsibility, one acts unfairly toward the others in the pool of shared responsibility, who consequently may to some degree—this is a difficult, contested issue—be required to add to their own share of responsibility some portion of the unfulfilled responsibilities of others like oneself who are slackers.[17] Thus, if those of us in this generation fail to carry out our responsibility of preventing avoidable delays in the date of the technological transition, we are guilty of unfairness toward at least some members of future generations.

What I want to suggest, however, is that while this picture of failures of responsibility as unfairness is not strictly inaccurate as a portrayal of a piece of the moral problem, it wildly understates the seriousness of failures on the part of our generation and immediately succeeding ones. I will point out two major reasons why the seriousness of a failure to act now is misleadingly minimized on the usual picture focusing on fairness, sketched above. What is wrong with this standard picture?

First, the usual picture of responsibility is distortingly static. Implicitly, the suggestion is that a fixed amount of effort is necessary to bring about a relatively early date of technological transition and that if we now are not doing our part, others will unfairly need to do more or yet others who otherwise would not have needed to do anything—for example, distant generations who would have inherited a safe environment if we had carried out our responsibilities—will have to take up burdens then in order to make up for our failure now. The strong implication is that all that changes if we fail is that someone else needs to pick up after us: we shirk our responsibilities and so others have to carry them instead of us.

But the identity of those carrying responsibilities is not all that changes, nor is it the most important change. For, as we have seen, the date at which atmospheric accumulations of GHGs cease to expand determines the maximum absolute severity of the resulting weather and other surface problems: the later the date of technological transition, the worse the climate change (more likely than not, although not for sure). The initial picture that understates the problem suggests, in effect, that if the date of technological transition will occur after six generations if the present generation fulfilled its responsibilities, then it will occur after seven generations if we fail to act. An extra generation, the seventh, is unnecessarily and unfairly burdened if we drop the ball now. The unfair burden, however, might be the least of it for the seventh generation. For another effect is that the problem of climate change that would, let us say, have reached a level of severity #6 in the sixth generation would reach a level of severity #7 in the seventh generation.

What will be the difference between severity level #6 and severity level #7? I wish I knew. Conceivably, not a lot—perhaps level #7 is just a bit worse than #6 but not significantly different. On the other hand, level #7 could be dramatically worse if some threshold is passed during the transition between these two levels that would not ever have been passed if the planetary deterioration had stopped at level #6. Perhaps at level #7 species extinctions begin to cascade, perhaps the global human pandemic comes, perhaps the Younger Dryas-like reversal from rapid warming to rapid cooling is triggered, and so on. Obviously these 'levels' are merely an illustrative abstraction, and I do not know exactly what they might mean. Yet the point is clear: to delay is to play with fire (and ice). At some point things will probably become truly nasty. Maybe the nasty one is level #13, and the difference between levels #6 and #7 is unremarkable. Maybe the nasty one is level #3, and the passage beyond #6 to #7 will by then be immaterial. We do not know, and we are very unlikely ever to know very far in advance.

Nevertheless, a good policy is readily evident. We do not need more information in order to know a wise way to act, which means that, contrary to the assertions of defenders of current US obstructionism, uncertainty is no excuse for inaction. Suppose you know that you are walking through a fog toward a cliff, but you do not know how many steps lie between you and the cliff—can you think of a good policy? Yes: stop as soon as you can. Now, I realize that energy policy, which is the key to climate change, is not that simple, especially since in the real world 'stopping' would have large costs too. And I am not advocating 'stopping', whatever that could actually mean: we need to move forward but in a direction that does not lead toward a cliff. Knowing which other direction that is, is not a simple matter. This is why vigorous well-financed research on alternatives to fossil fuel is urgent. What carries over from the analogy with heading for the cliff in the fog is that those such as George W. Bush who say 'Nothing needs to change yet' are being at least as simplistic as someone who says 'Just stop,' and they are flagrantly tempting fate. A 'few' years in the date of technological transition could make a spectacular difference if in those years some point of no return was passed that would otherwise never have been passed. We are not simply strolling in the fog—we are playing poker as we stroll in the fog; and we do not even know what stakes we are playing for, although the stakes could be very, very high, and if our gamble loses, our great-great-grandchildren will pay.

Under this much uncertainty it is perfectly reasonable to pay attention to the costs as well as the benefits of advancing the date of technological transition. It would not be reasonable to try to make the date as soon as possible no matter what the cost. The 'fog' of uncertainty prevents us from knowing whether the potential train crash is just around the corner or much farther down the line, either of which would make what we do largely irrelevant, or somewhere between immediate and distant, which could make what we do utterly crucial. Assuming the worst case would be extremely—and, I would think, excessively—expensive in light of all the other urgent matters also requiring attention.[18] Thus, for example, it would obviously not be reasonable to divert all the funds that might otherwise go toward curing cancer and AIDS into slowing climate change because these diseases are scourges too, and their harms are quite clear, even if the value of any particular line of research in those cases is far from obvious, as it is in the instance of non-fossil energy sources. But, in symmetry, it is also not reasonable to devote only relatively trivial amounts to research on arbitrarily chosen alternatives to fossil fuel, as the USA is now doing. The assumption of the best case, which would be the only way to try to make non-corrupt sense of the lackadaisical passivity of the Bush administration, would be at least as extreme and unwarranted as assuming the worst case.[19] US policy could be far less dismissive without even coming close to overreacting.

Whatever is exactly the right approach to such a case in which there is strict uncertainty (that is, no calculable probabilities of alternative outcomes but some disastrous outcomes definitely possible on the basis of fairly well understood planetary mechanisms), the point remains that if the present generation continues to fiddle around the edges of the problem rather than take a grip on its responsibilities, the moral failure will not consist only or primarily of unfairness to those to whom our burdens are then shifted. Much more important, we will be responsible for allowing the consequences of climate change to become worse—to reach a more extreme point—than they would have been had we acted with some seriousness. Far worse or only a little worse? George W. Bush and Richard Cheney do not know any more than you and I do, and they find it much more difficult to be open-minded about whether to challenge entrenched fossil fuel interests.

The second reason why the usual picture is misleading insofar as it suggests that a moral failure by our generation would be only unfairness—not that unfairness is not already a serious and fundamental moral failure—is that in a historical process such as rapid climate change it is impossible to do later all of what it is possible to do now. Suppose that in fact, although we do not yet know this fact, the species on the surface of the planet, including humans, can on the whole handle the effects of one further doubling of the atmospheric concentration of carbon dioxide reasonably well, but that a further redoubling will produce cascades of extinctions including the extinction of food plants of great value to humans. Burning all remaining fossil fuel will take us well beyond a further redoubling of atmospheric carbon.[20] Determined action now can prevent that further redoubling, but action after a certain point in time (specifically, after a certain additional proportion of the carbon in the remaining fossil fuel has been injected by combustion into the atmosphere) will simply come too late to make a significant difference. The atmospheric commitment will at that time have already been made.

This is because carbon dioxide has a long atmospheric residence time, averaging around a century. Once a certain amount of carbon dioxide is in the atmosphere, we know of no way of removing it. And any given level of GHGs in the atmosphere creates a commitment to consequent climate changes on the surface. Suppose the people in generation #4—we are #1—discover the fact that a further redoubling of atmospheric carbon is going to immiserate some societies and generally impose great strains upon humanity. They are nevertheless doomed to suffer this fate if either of two things has happened. First, obviously, if enough additional fossil fuel has already been burned by us and the intermediate generations to produce the atmospheric commitment to the surface changes, it is physically impossible for them to stave off the changes (without some miraculous stratosphere-cleansing technology not remotely in prospect). More agonizing, even if the fatal amount of fossil fuel has not yet been burned, but generation #4 has been left with a world economy dependent upon fossil fuel because no good alternative energy sources have yet been developed, it is politically impossible for them at that date to stave off disaster. They cannot simply 'stop' the world economy by 'turning off' energy consumption—then practically everyone would die of deprivation. So, they themselves may produce the fatal emissions because they still have no more alternative than we now have because we were content to leave them with no alternative to fossil fuel energy sources and our successors followed our bad example. In a way, they would hang themselves, but of course only because we had not prepared the way for them to have any alternative to fossil fuel use other than starving. By the time of their generation it is too late to be doing the research and development that we could have done at earlier times (that is, now). Many choices in history are irreversible. Either a technology is ready when it is needed or it is not; if it is not, it cannot be used because it is not there. We could perhaps see to it that the safer technology is available by then, but currently the US government is not even trying.

VULNERABILITY AND BETRAYAL

If I were a desperate member of that later generation, I think I would be furious at our generation and the short-sighted and self-centred do-nothing-ism of the USA, Australian, Canadian, and other laggard governments of the early twenty-first century, not to mention the belligerent obstructionism of the Saudis and some of the other governments sitting contentedly on oil that they are absolutely determined to sell. It would be a good thing that one cannot harm one's ancestors, other than by trashing their reputations. They might well view us with the contempt we have for forebears who were slave owners or pirates. This is not how I was hoping to be remembered: as a good-for-nothing great-great-grandfather who wallowed in comfort and convenience to such an extent that no viable options remained.

Suppose that either out of greed and corruption, or simply out of indifference and self-centredness, we cannot be bothered to move aggressively to replace fossil fuels with alternative sources of energy before scarcity and price rises force later generations to do it—we fail to spend even as much on research and development of alternative energy as, say, we spend on a peripheral boondoggle such as ballistic missile defences, which would be irrelevant to almost all the worst threats the USA faces even if the technology could someday pass realistic tests.[21] Our moral offence, it seems to me, goes well beyond unfairness. It constitutes the infliction of harm, a violation of what is arguably the most fundamental moral principle of all: Do no harm.

Why is there no concerted research and development initiative on alternative energy? The morally most acceptable explanation would be forgivable ignorance. Climate change is difficult to grasp: much of the evidence comes from sophisticated models inaccessible even to well-educated and intelligent people who are not specialists, and there are few telegenic disasters from climate change evident yet.[22] If it cannot be seen on TV, it is not happening—this is the American ontology. Climate change is too pervasive to point to, and many perfectly decent people do not understand what the fuss is all about.

The morally most outrageous explanation is greed and cover-up. Those whose wealth depends on pumping all the oil and digging all the coal have a lot to lose if the political decision is ever made to leave the stuff in the ground; many of them intend to see that this political decision is never made and indeed that no one who would make it ever gains significant political power (by winning a US presidential election, for example). For some of these people, the advocates of alternative energy are simply the enemy, and this is a war. It would be naive to expect such selfish people to be moved by concern for the welfare of other people, not to mention other generations. They cannot be persuaded or moved to empathize—they can only be outsmarted and out-manoeuvred. At present, they are winning the struggle and dominate what passes for energy policy in the USA: pump more oil and dig more coal.

In between these extremes is what I hope is the vast majority of people who actually care about the environment in general but are not quite sure where ozone depletion ends and climate change begins and for now do not feel safe invading an Interstate in anything other than a tank-like SUV (or even an absurd Hummer). We in this group are less wise and less compassionate, especially toward those distant in space and time, than we might ideally be, but we do not really want to hurt anyone who is not threatening us. For us, I hope, it might matter if our failure to do much about climate change would do genuine and serious harm to people who are utterly at our mercy.

And assuming again that we are generation #1, this seems to be the situation of the people in generation #7, who are utterly and asymmetrically vulnerable to us. Their very existence is in our hands and the hands of the intermediate generations; if we unleashed a massive nuclear winter (less likely for now than it once was) or failed to control some virulently contagious and fatal epidemic, the people of generation #7 might never live. And the quality of the lives of whoever are born in that generation is under our control to a profound degree, in completely familiar ways. Whether they can enjoy beautiful forests and great universities depends upon whether we leave them any—a single gener-ation cannot grow a magnificent forest (although they can plant one) or suddenly throw together a great university. These things take time: if one generation is to have them, earlier generations must see to it. There is no express route.

In some of these cases, perhaps, if we do nothing, we fail to provide a benefit we might have provided to future generations, but we thereby do them no wrong. Suppose there were no decent university in our state, and we did nothing to create one. Future generations might be unhappy with us because we did not provide this benefit, but I cannot see that we would actually have done them any harm. And they could always start one if they thought it was important, although it would take more than their own lifetime for it to flourish as an outstanding and enduring institution. It could be their gift to the generations that succeeded them.

A failure to take action to put a floor under how bad climate change can become seems to me to be a much worse failure than a failure to give a gift that one might well have given but was under no obligation to give. Suppose that every generation after ours will do whatever it ought to do about climate change in the circumstances that it then faces (perhaps because the damage will have become more obvious as time passes). Then how bad climate change becomes at its worst turns on how much we do now. There may be harms that will occur only if we do nothing because only if we do nothing will climate change become severe enough to cause those harms.

What if an intermediate generation, inspired perhaps by contempt for our generation, did twice what it could reasonably be expected to do and tried to make up for our failure? Generation #1 (us) does nothing, but generation #4 does twice what it could be expected to do in order to make up for our failing—might generation #7 turn out then to be just as well off as if generation #1 had done its share? We are here engaging in abstract speculation of a possibly not very reliable kind, but here is what I can make of it. It is of course conceivable that one share of effort each by generations #2 and #3, plus two shares by #4, would add up to the same thing as one share by each of the four generations. If the task were to build a stone wall by adding individual stones, four shares of effort supplied by three generations ought to produce the same result as four shares by four generations. Let us say that each share of effort contributes 2 feet of height to the whole wall; either way an 8-foot wall results. Suppose the need were for a wall too tall to be jumped by mounted marauders, the requisite height for security was 6 feet, and the marauders were going to attack early in the fourth generation. The fact that generation #1 had done no work would mean that at the beginning of generation #4 the wall would only be 4 feet tall, when it would have been 6 feet tall if generation #1 had done its job. So the marauders would conquer generation #4 before they could get very far with their double effort of wall building, which would have taken the wall to 8 feet at a later point in time. Thus, even with something as simple and cumulative as adding stones to a wall, earlier omissions can have irreversible effects. The attack of the marauders constitutes a critical threshold, and on that day the wall either will or will not be tall enough to stop them. Safety depends on how much has already been done by the crucial date.

And, almost needless to say, irretrievable effects are far more likely in the case of climate change. Return to the example mentioned earlier: perhaps the effects of a doubling of the atmospheric concentration of carbon dioxide are manageable, but the quadrupling (and more) that would result from the combustion of all the fossil fuel exploitable at a profit to those who control it will have much more severe effects. Then it is critical whether the date of technological transition, the date when the atmospheric accumulation ceases to expand, occurs before or after the concentration has quadrupled. Suppose that if this generation launched a serious initiative on alternative energy, it would be very likely that the research and development could be completed in time for widespread adoption of alternative sources well before enough of the vast remaining supplies of cheap fossil fuel had been burned to cause the atmospheric concentration of carbon dioxide to quadruple. But suppose that if serious research and development did not begin until the generation after us, the concentration would quadruple before the eventually emerging alternative forms of energy have replaced enough of the fossil fuels.

How should this generation's failure to act be evaluated? 'They could have helped, but they didn't'? 'They unfairly left their share of the effort to be done by some succeeding generation'? Unfortunately, it seems incomparably worse than those assessments: They made the choice that determined how bad climate change became at its worst, and their choice resulted in its becoming worse than it would have if they had chosen differently. They were not for the most part evil people (although they complacently tolerated corrupt political leaders), but they were simply preoccupied with their own comfort and convenience, not very imaginative about human history over the long run, and not particularly sensitive to the plight of strangers distant in time. They did not mean to do any harm, but in fact they inflicted severe damage on their own descendants. A sad chapter in human history—so much opportunity lost while a tiny clique with financial interests in fossil fuels amassed short-term profit. Will this be our legacy?

Now, the 'two' reasons I have given why a failure to act is worse than an unfair shirking of responsibility—that delay is likely to magnify severity (to make the worst worse) and that historical choices can be irreversible—are essentially the same point: The irretrievability of lost historical opportunities matters in this case because the opportunity that is now being lost is to prevent climate change from becoming as extreme as it will otherwise probably become. I have simply highlighted two facets of one very hard rock.

I have also highlighted the responsibility of the present generation, noting that even if all other generations were to do their part after we had failed to do ours, our failure might well set the bottom limit on how bad things finally become. Naturally if we did our part and one or more succeeding generations failed to do theirs, the depths of the disaster might be at least as bad as or worse than if we had not evaded our responsibility. So, why pick on us? For one thing, we are the only ones available to be picked on, although I hope to leave behind a book provoking future generations as well! More seriously, one is only responsible for what one can in fact affect. We cannot control what future generations do, but the broader public might be able to wrest control of what our generation does from those narrow interests who now dominate it. *Climate policy is energy policy*, and changes in energy policy affect the value of the holdings of some of the wealthiest firms and individuals in the world— they will not surrender their grip on the political power that protects their wealth without a prolonged and dirty fight. But ordinary decent people do outnumber them, so if democracy could be made to work, there would be a little hope.[23] Secondly, as already noted, there is the bittersweet possibility that, as the problems become worse, they will become more visible. Succeeding generations may sadly need less imagination than we do to understand the seriousness of the situation, so they may be a little more likely to act because they are more frightened. In sum, there is no guarantee that if we act, all will be well, but there is a high probability that if we do not act, the best that will be possible will be worse than it relatively easily could have been.

Finally, I have said nothing about how to encourage the alternative energy sources needed to supplant fossil fuel. Technological change is not well understood, although many understand it better than I. That simply throwing public money at a problem does not solve it is amply demonstrated by the tens of billions poured into the Strategic Defense Initiative (SDI), now born again as Ballistic Missile Defenses (BMD), which can only succeed in rigged or farcically easy tests. Perhaps a 'Manhattan Project' for alternative energy would be as bad an idea as the SDI/BMD. One of the reasons for profound doubt about the Kyoto Protocol is the extent to which its various 'flexibility mechanisms', such as the Clean Development Mechanism, create financial incentives to disperse throughout the Third World the same fossil fuel-based technology that brought us climate change in the first place, and contain no strong incentives to use alternative energy. This reflects the extent to which Kyoto was designed to please dominant interests in the USA, although the current US administration dismissed it contemptuously anyway.[24] Yet local governments, state governments, universities, and the private sector need not wait for the federal government to stop favouring fossil fuel; they could provide the initiative and vision absent in Washington.

I defer to others who are wiser in practical matters on exactly how to proceed. But now is the time for thoughtful but determined action to prevent the sale and burning of all the vast remaining cheap fossil fuel, an economic choice that bids fair to become the most short-sighted 'bargain' in human history.

NOTES

1. I first stumbled my way reluctantly into this conclusion in 'Avoidable necessity: global warming, international fairness, and alternative energy', this volume. My fundamental analysis of the issues of distributive justice is 'Subsistence emissions and luxury emissions', this volume. A later summary overview, with some modifications, is Shue (2002).
2. See Houghton et al. (2001). For a lucid brief account, see the chapter by Mahlman in *Perspectives on Climate Change: Science, Economics, Politics, Ethics*, ed. Walter Sinnott-Armstrong and Richard B. Howarth (Amsterdam and San Diego: Elsevier, 2005).
3. See Paterson (2000), Rutledge (2005), and Ness (2005).
4. Fairer processes for the allocation of emissions, such as the one proposed in *Perspectives on Climate Change: Science, Economics, Politics, Ethics* by Dale Jamieson, would increase the incentive for the worst emitters to pursue alternative sources of energy. The importance of the incentive structure created by emissions allocations was clearly set out in the classic article by Grubb (1995). A moral

philosopher who has recently made a serious attempt to incorporate considerations of incentives is Traxler (2002). A critique of Traxler's proposal appears in
Gardiner (2004). Gardiner provides an excellent comprehensive interdisciplinary
overview of the ethics, economics, and science.

5. See 'A legacy of danger: the Kyoto Protocol and future generations', this volume.
My formulation is inspired by Michael Grubb's distinction between dynamic
efficiency and static efficiency—see Grubb (1998, p. 2) and Grubb, Chapuis, and
Duong (1995).

6. It is net increases that matter, naturally. A theoretical alternative would be to
increase sinks for carbon dioxide faster than emissions of carbon dioxide increase,
but this is in practice impossible. Many places would now benefit from reforest-
ation, for example, but the limits on land to serve as carbon sinks will be reached
long before the limits of the human demand for additional energy. Various exotic
engineering solutions are imaginable, but none is yet feasible.

7. How much can be recycled in the short term changes somewhat with changes in
the total accumulation—see the chapter by Mahlman in *Perspectives on Climate
Change: Science, Economics, Politics, Ethics*. The change is in the direction helpful
to humans but is far too small to save us.

8. See, for a remarkably accessible and engaging account, Alley (2000). Also see
Weart (2005).

9. See United States, National Academy of Sciences, National Research Council,
Committee on Abrupt Climate Change (2002, p. v.) On the Younger Dryas,
specifically, see pp. 24–36. Also see Vellinga and Wood (2002).

10. See, for example, McCarthy (2004a, b).

11. See Gagosian (2003). Especially alarming are data indicating a decades-long
decline in the salinity crucial to driving the ocean circulation that allows the
Gulf Stream to warm New England and Western Europe—see Dickson et al.
(2002).

12. Two popular explanations of the underlying mechanisms are Broecker and
Denton (1990) and Alley (2004).

13. See Kasting (1998, p. 18).

14. Yes, they could try to preserve their cultures in some now deserted part of
Australia if the xenophobic Australian immigration policy were changed to permit
them to enter.

15. I refer to the work done, and stimulated, by Derek Parfit.

16. For the argument that all the plausible allocative principles converge on the same
agents in the case of climate change, see 'Global environment and international
inequality', this volume. Two splendid recent discussions of the assignment of
responsibility are Miller (2001) and Green (2002).

17. On this issue, see Kutz (2000) and Murphy (2000).

18. For the Pentagon's usual worst-case thinking, see Townsend and Harris (2004).
Also see Schwartz and Randall (2003).

19. I would myself bet on the corruption explanation, that is, that this administration
is under the control of the oil interests out of which Bush and Cheney come (and
to which they will likely return). The Bush administration continues to cover up
scientific data on climate change; see Revkin and Seelye (2003).

20. See Kasting (1998).
21. The double standard involved in ignoring promising energy technology while throwing billions at unpromising military technology is astounding! And profoundly irrational.
22. Yet warnings accessible to the general public abound. See, for example, Regalado (2003).
23. See Eckersley (2004).
24. I am not endorsing the Kyoto Protocol, precisely because it needlessly pits today's poor against tomorrow's poor in order to avoid inconveniencing the rich at any time—see 'A legacy of danger: the Kyoto Protocol and future generations', this volume. But the current Bush administration not only high-handedly rejected the protocol but sneered at the process of trying to move beyond it, preferring what has become its customary unilateralism. For a general overview of the situation regarding the Kyoto Protocol, see Grubb et al. (2003). For recent empirical findings on effects on the USA, see Parmesan and Galbraith (2004).

REFERENCES

Alley, R. B. (2000) *The two-mile time machine: Ice cores, abrupt climate change, and our future.* Princeton, NJ: Princeton University Press.

Alley, R. B. (2004) 'Abrupt climate change', *Scientific American*, 292, 62–9.

Broecker, W. S., & Denton, G. H. (1990) 'What drives glacial cycles?', *Scientific American*, 262, 49–56.

Dickson, B., Yashayaev, I., Meincke, J., Turrell, B., Dye, S., and Holfort, J. (2002) 'Rapid freshening of the deep north Atlantic Ocean over the past four decades', *Nature*, 416, 832–7.

Eckersley, R. (2004) *The green state: Rethinking democracy and sovereignty.* Cambridge, MA: MIT Press.

Gagosian, R. B. (2003) *Abrupt climate change? Should we be worried?* Woods Hole, MA: Woods Hole Oceanographic Institution, 2003. Available: <http://www.whoi.edu/institutes/occi/hottopics_climate change.html>.

Gardiner, S. M. (2004) 'Ethics & global climate change', *Ethics*, 114(3), 555–600.

Green, M. (2002) 'Institutional responsibility for global problems', *Philosophical Topics*, 30(2), 1–28.

Grubb, M. (1995) 'Seeking fair weather: Ethics and the international debate on climate change', *International Affairs*, 71(3), 463–96.

Grubb, M. (1998) *Corrupting the climate? Economic theory and the politics of Kyoto.* Valedictory lecture. London: Royal Institute of International Affairs.

Grubb, M., Chapuis, T., and Duong, M. H. (1995) 'The economics of changing course: Implications of adaptability and inertia for optimal climate policy', *Energy Policy*, 23 (4/5), 417–32.

Grubb, M., Brewer, T., Müller, B., Drexhage, J., Hamilton, K., Sugiyama, T., and Aiba, T. (2003) *A strategic assessment of the Kyoto–Marrakech system: Synthesis report.*

Sustainable Development Programme Briefing Paper no. 6. London: Royal Institute of International Affairs. Available: <http://www.riia.org; www.iisd.org>.

Houghton, J. T., Ding, Y., Griggs, D. J., Noguer, M., van der Linden, P. J., Dai, X., Maskell, K., and Johnson, C. A. (eds) (2001) *Climate change 2001: The scientific basis. Contribution of working group I to the third assessment report of the Intergovernmental Panel on Climate Change.* Cambridge: Cambridge University Press. Available: <http://www.ipcc.ch>.

Kasting, J. F. (1998) 'The carbon cycle, climate, and the long-term effects of fossil fuel burning', *Consequences: The Nature & Implications of Environmental Change*, 4(1), 15–27. Available: <http://www.gcrio.org/CONSEQUENCES/vol4no 1/carboncycle. html>.

Kutz, C. (2000) *Complicity: Ethics and law for a collective age.* Cambridge & New York: Cambridge University Press.

McCarthy, M. (2004a) 'Disaster at sea: Global warming hits UK birds', *Independent*, 30 July, 1, 7. Available: <http://www.independent.co.uk>.

McCarthy, M. (2004b) 'A giant ecosystem that has functioned for millions of years has begun to break down', *Independent*, 30 July, 7. Available: <http://www.independent. co.uk>.

Miller, D. (2001) 'Distributing responsibilities', *Journal of Political Philosophy*, 9(4), 453–71.

Murphy, L. (2000) *Moral demands in nonideal theory.* Oxford & New York: Oxford University Press.

Ness, E. (2005) 'Detroit is still stuck in reverse', *Onearth*, 26(4), 22–31. Available: <http://www.nrdc.org/onearth>.

Parmesan, C., and Galbraith, H. (2004) *Observed impacts of global climate change in the U.S.* Arlington, VA: Pew Center on Global Climate Change.

Paterson, M. (2000) 'Car culture and global environmental politics', *Review of International Studies*, 26(2), 253–70.

Regalado, A. (2003) 'Panel shifts stance on global warming', *Wall Street Journal*, 17 December, A2.

Revkin, A. C., and Seelye, K. Q. (2003) 'Report by E.P.A. leaves out data on climate change', *New York Times*, 19 June, A1,A20.

Rutledge, I. (2005) *Addicted to oil: America's relentless drive for energy security.* London: I.B. Tauris.

Schwartz, P., & Randall, D. (2003) 'An abrupt climate change scenario and its implications for United States national security'. Available: <http://www.gbn.org/ ArticleDisplayServlet.srv? aid=26231>.

Shue, H. (1993) 'Subsistence emissions and luxury emissions', this volume.

Shue, H. (1995) 'Avoidable necessity: global warming, international fairness, and alternative energy', this volume.

Shue, H. (1999) 'Global environment and international inequality', this volume.

Shue, H. (2002) 'Equity', in *Encyclopedia of global environmental change* (vol. 5, pp. 279–83). Chichester, UK: John Wiley & Sons.

Shue, H. (2004) 'A legacy of danger: the Kyoto Protocol and future generations', this volume.

Townsend, M., and Harris, P. (2004) 'Now the Pentagon tells Bush: Climate change will destroy us', *Observer*, 22 February. Available: <http://www.observer.guardian. co.uk>.

Traxler, M. (2002) 'Fair chore division for climate change', *Social Theory and Practice*, 28(1), 101–34.

United States, National Academy of Sciences, National Research Council, Committee on Abrupt Climate Change. (2002) *Abrupt climate change: Inevitable surprises.* Washington, DC: National Academy Press.

Vellinga, M., and Wood, R. A. (2002) 'Global climatic impacts of a collapse of the Atlantic thermohaline circulation', *Climatic Change*, 54(3), 251–67.

Weart, S. (2005) 'Rapid climate change'. American Institute of Physics' *The discovery of global warming.* Website: <http://www.aip.org/history/climate/pdf/rapid.pdf>.

13

Making exceptions*

Contingencies make us nervous, necessities comfort us. As long as we can deal exclusively in concepts and possibilities, we can ignore facts and probabilities and need not worry whether we have 'the empirics' wrong. Why is so much of what we 'practical philosophers' do of no interest to people of practical affairs? One of our weaknesses, from their point of view, is this preoccupation with mere conceivability and relative disinterest in practical possibility. This weakness shows up most unrelentingly in our tendency to discuss only imaginary cases, which can help us to judge what is non-contradictorily conceivable but cannot answer many other questions relevant to practice, most especially not when to make exceptions to general principles.

As long as one deals only in imaginary cases, which currently seems to be an addictive habit among philosophers (and several other kinds of academic theorists), one needs to know only about concepts—one does not need to try to figure out what usually happens and why that is. One can have views on the justice of arrangements for dealing with climate change without understanding any of the science of climate change and views on killing in war without the remotest understanding of what combat is actually like. Could a use of nuclear weapons possibly be justified? And how seriously should we take the answer to the previous question, if it is affirmative? Does it matter whether that conceivable justified use is remotely likely to be one of the actual uses? One might think not if one read only philosophers.

Great dangers lie in the opposite direction as well, of course: philosopher as amateur strategist, amateur physician, amateur economist, amateur climatologist, and so on. Nevertheless, I want to try to show in what follows why I think it is better to run the risks of dabbling and dilettantism—and flat-out factual error—than to float in the comfortable clouds of abstraction amongst the imaginary cases. One of my central examples is of one of the times I got the relevant facts wrong (torture).

* 'Making Exceptions', Henry Shue, *Journal of Applied Philosophy* 26:4, Wiley, (2009) 307–22. This chapter has endnotes rather than footnotes.

Two features of practical judgements make stripped-down hypotheticals, largely preoccupied with conceptual coherence, unhelpful or misleading. First, practical judgements need to be all-things-considered judgements. The practical agent—certainly the practical political agent—wants to know what to do here and now. She must act, so she needs to arrive at what specifically is to be done, from among the options available. Actions taken have, for instance, economic costs and political costs—literal costs as well as 'opportunity costs', alternative options forgone. Such costs are often morally salient. One way of thinking about what would be the moral thing to do is to think what one would do if one's action were cost free; this may yield what, in one sense, is the ideal solution—the best possible solution where, say, money is no object and where one need not deal with obtuse, prejudiced, or ill-informed political opponents or constituents.

It is often observed that even political philosophers tend to write as if they were a single impartial spectator with full authority and power to execute her own wishes—someone whose fiat simply became fact—and not, say, one member of a voting body who must somehow patch together a coalition with at least one more person in it than the opposition coalition. So our recommendations tend to take the form: Here is what I would do if I ran the world. This may—or may not—be interesting, but it is usually not the best method for answering the question: What is the best thing we can do here and now, given our budget, given our political institutions, given the recent history, and given all of whatever else is actually given (which is in itself a difficult empirical judgement to make)? This is another longer way of saying that the practical person needs an all-things-considered judgement: given that if we spend the money here, we cannot spend it there; given that if we alienate these people, we cannot also alienate those (or we will not have enough support to form the coalition to make this happen); and so on.

I do not assume that action must be conservative, incremental, or palliative. I simply mean: the proposed action must be possible to carry out in fact—we cannot simply 'imagine a tin opener'.[1] And I do not assume a substantive moral view containing none of what would ordinarily be called moral absolutes—everything as compromise and coalition. On the contrary, I now think, for example, that one should absolutely never torture, as I will explain presently. Here I am saying only that one must recognize that every position has morally relevant costs: some planes may crash because one did not find out about bombs one could have found out about if one had been willing to torture enough people. But my position is not: one would ideally never torture if an absolute ban on torture had no costs. My position is: one should never torture in spite of the fact that refraining may have high costs. This position may be mistaken, but it does provide clear guidance about what to do here and now: do not torture anyone. I do not want to pursue the issues about torture in their own right quite yet, however. For now I only want to illustrate the fact that, to put it crudely, rejecting the ideal need not mean rejecting the absolute, that is, the exception-less rule.

Second, practical judgements need to decide whether in the case at hand one should follow the general rule or make an exception. This is a major reason why judgements need to be all things considered, so the importance of deciding about exceptions is worth emphasis. Many of—not, of course, all—the difficult decisions are about whether to follow the general rule or to make an exception. One has answered this question if one has specified an exception-less rule, but naturally that will be unusual—torture is a rare, even if not unique, case. Given that we would not normally supply military assistance to a regime that tortures, what about Egypt, given the other human values at stake in the political role that Egypt plays in its region? Answering the question whether to make an exception all the more strongly forces one to consider all the morally relevant factors, including whatever considerations militate in favour of making the exception to the general moral rule, whether or not one should in the end make it.

The insufficiency of imaginary examples is not mysterious. Hypotheticals are useful for assessing the significance of a single factor: here is the case with the factor in question and here is the case without that same factor—how much difference does that factor make? Hypotheticals are usually streamlined in order to focus on one factor at a time. The many other potentially relevant factors are not usually co-varied in various permutations with the factor in question, but omitted in order to single out the factor highlighted. The merit of imaginary cases is their simplicity. In principle, of course, imaginary cases could be every bit as complex as real cases, but then one would do just as well to examine the real case in the first place. One could have a series of imaginary cases that gradually regained the full complexity of real cases, varying one factor at a time, then pairs of factors, trios of factors, and so on. But this would amount to discussing the real case and considering which factors matter, which is precisely what I am suggesting is best done.

For example, someone will say: 'Let's consider whether torture can ever be justified. What if someone is on the way to kill someone else and the only way we can prevent the killing is to inflict a severe pain on one of the legs of the would-be killer for one hour? Wouldn't this be a case of justified torture, establishing that the prohibition against torture is not absolute?' My own view is that this would be an example of the use of what the police and the military call a non-lethal weapon, which is greatly to be preferred, other things equal, to the use of a lethal weapon. It does not seem to me to be the use of torture at all because so many of the features characteristic of the most widely used torture are missing: the person on whom the pain is inflicted is not under the complete and indefinite control of the state, she is not being humiliated and demeaned, she does not have grounds to fear that the pain will be inflicted repeatedly no matter how she responds, she is not having her will completely broken, and on and on—most of what is most disturbing about torture is missing from the example of a one-hour pain in the leg. The result of all these omissions is that our concluding that such a use of a mildly painful, non-lethal weapon was permissible would tell us nothing at all about standard cases of torture, even if we could agree that this was a peripheral case of torture (and not the use of a mildly painful non-lethal weapon).

We could of course begin with the one-hour, non-degrading pain in the leg and start adding one by one the features that, I would suggest, are central to torture—naturally specifically what they are is contentious. One might think of this as a method of addition: start simple and gradually add complicating features. If one really followed this series of imagined cases all the way through to the full complexity of the real case, there is no reason why it could not be useful. But one could also just begin from the real case, and I would like now to explore in slightly more depth in the body of this chapter how considering possible exceptions drives us beyond imaginary cases to real ones.

A TRIO OF PROPOSED EXCEPTIONS

A fascinating trio of cases, for example, are preventive military attacks, the mitigation of climate change, and the use of torture. In all three cases, the central contention is that because of what is at stake—what stands to be lost if the allegedly appropriate action is not taken—one ought to take a kind of action that one would not ordinarily take: one ought to make an exception because this is, in effect, an emergency. Yet, apart from involving proposed exceptions, the three have little else in common.[2] One source of differences among cases is the kind of consideration that is alleged to be the exceptional factor and the direction in which it is said to push matters. But in the cases of preventive military attack and torture, the exception would be to a prohibition: one ought, it is argued, to do what is ordinarily prohibited, while in the case of climate change the exception would be to what it takes to satisfy a require-ment: one ought, it is argued, to do far more than would normally be enough to fulfil one's obligations to future generations. Further, by my lights, the proposed exception is fully justified in the case of climate change, sometimes justified in the case of preventive attack, but always unjustified in the case of torture. Thus, the answers, as well as the questions, are different. Both the similarities and the differences seem to make it worth looking at each case briefly in order to see how the morally salient 'extra' complications come in and why they cannot be dodged if one is to offer any practically useful comment, beginning with torture and preventive attack, and concluding with climate change.

Here I shall not be presenting anything remotely resembling full arguments about any of the three cases. Instead, drawing on sources containing fuller argument, I shall focus exclusively on the contrasting ways in which the exceptional needs to be handled in the respective cases.

The cases of preventive attack and torture initially appear to have a highly similar structure.[3] A very strong and basic prohibition excludes almost from consideration, much less adoption, a particularly despicable act in each of the two cases: never launch a lethal attack against anyone who has not forfeited his right not to be killed by committing some particularly vicious act, such as launching a lethal attack against you or others; and never torture anyone in any circumstances whatsoever. One version of the basic contrary contention in each case is that as wrong as violations of these prohibitions may be, the stakes are so high in certain instances that a violation is justified in just these instances. When one immerses oneself in the empirical details, however, it turns out that a preventive military strike need not be in violation of the relevant prohibition, although torture certainly is; this muddies the comparison while illustrating why social realities are often pivotal. For one of the main reasons for being extremely cautious about authorization of preventive attack is nevertheless an essentially sociological factor concerning the bureaucratic embodiment of any readiness to perform the acts in question, as I belatedly realized is also the case with torture; this makes the comparison fruitful. Yet, in the end, the salient features of the social institutions necessary to each respectively turn out to be crucially different if one examines them in enough detail; this makes the comparison treacherous. The simple but important methodological point is that little of this is apparent until one plunges into what philosophers tend to dismiss as 'the empirical details', that is, how things actually work.

Let us begin with torture. Clearly torture is morally wrong—no one seriously suggests otherwise. The question is whether, given how very wrong it is, it can ever be justified to engage in it nevertheless. This would be the exception: one would commit the exceptional act of torture in certain exceptional circumstances. What are the circumstances? Other people and I have written ad nauseam about what has unfortunately come to symbolize the issue, the ticking bomb hypothetical, which is now the subject of at least two books.[4] So here I want to paint in the broadest possible strokes. The effort to provide a case of justified torture portrays a time bomb whose clock has been set running by a culprit whom the authorities have luckily captured. If, but only if, the culprit is tortured will she reveal the location of the bomb in time for it to be neutralized before it explodes. Time is of the essence, and the stakes are high, because the bomb will take the lives of numerous innocent people if it explodes. The thrust of my 'Torture' in 1978 was to emphasize the multiple dimensions along which the hypothetical is an idealization compared to ordinary cases of torture and to argue, in effect, that to conclude that ordinary torture is justified is to generalize across dissimilar cases—from ideal case to ordinary case. For example, among the ideal features built into the ideal case are (1) that the authorities have the right person, (2) that the person reveals information accurately and promptly enough for the bomb to be defused, and—most important—(3) that this is indeed exceptional behaviour, in that the political authorities do not habitually round up and torture 'suspicious' people in case they might know and reveal something useful.

The analogous perfect case for preventive military attack is the terrorist cell with genetically modified smallpox virus that is immune to known vaccines and, once released, would launch a lethal and uncontrollable pandemic of deadly and highly contagious disease. Time is of the essence here too because intelligence agencies have located the central supply of the virus for now, but once vials are dispersed among the network of terrorist cells in various different cities, there will be no practical possibility of tracking it all; and the stakes are extraordinarily high, possibly involving literally billions of lives (incomparably higher stakes than in the standard ticking bomb case for torture). Once again, many features are ideal: (1) the intelligence agencies have targeted the right people (this is not the mythical WMD in Iraq 2003), (2) the cruise missiles will hit the correct location and destroy all the virus without any escaping, and (3) this strike is indeed exceptional behaviour, in that the special forces or others are not generally going around the world secretly targeting 'suspicious' groups of people and killing them with no accountability.[5]

In reality, however, differences as well exist between the implications of the hypothetical accounts of the perfect time for torture and of the perfect time for preventive attack. One crucial principle that a preventive attack appears to violate is the prohibition against lethal attack on people who have done nothing to forfeit their right not to be killed. The contention defending the position that they have committed no rights-forfeiting wrong would be that the terrorist cell (granting for the sake of argument that they are terrorists—and that we know what we mean by 'terrorist') may have been stockpiling deadly weapons but that this is no different from the conduct of the US and the USSR during the Cold War when weapons of mass destruction capable of killing at least hundreds of millions were amassed on both sides. Arms races have much to be said against them, but international rules that permit arms races may be preferable to international rules that permit preventive strikes against arms accumulations when the arms may otherwise never be used and may in fact have been acquired for the sake of deterrence. Thus, the acquisition of arms, as distinguished from their use, ought not to be considered a wrong—in any case, not a wrong of such magnitude as to constitute a forfeiture of the right not to be attacked. So those conducting the arms acquisitions are not to be considered to have forfeited their right not to be killed. Such people are engaging in what passes for normal behaviour in the kind of world in which we actually live where arms racing is common.

If such a moral defence as the one just summarized were correct, a preventive attack on anyone acquiring weapons but not having attacked with them (or being imminently going to attack with them) would be a wrongful attack on people who had not done enough to have forfeited their right against such lethal attack. The preventive attack would be wrong (but might, as torture might, be justifiable in spite of being wrong). The neatness of the parallel between preventive attack and torture is, however, disrupted if a preventive attack would nevertheless not be wrong and would, in particular, not be a violation of the prohibition on attacking those who have not committed a wrong sufficiently serious to have forfeited their right against lethal attack.

Here is one reply to the defence of WMD acquisition. Weapons of mass destruction are different. They provide ways to commit a serious enough wrong other than by attacking (or being imminently going to attack), namely by demonstrably intending unconditionally to attack and having acquired a kind of weapon of mass destruction that would inflict vast numbers of deaths and could not be defended against once it had been released by the attacker. The firm and unconditional intention to inflict vast numbers of deaths, plus the acquisition of a capacity to do so against which there is no defence, can plausibly, if still controversially, be construed as the commission of a wrong against those who are the intended victims sufficient to forfeit the right not to be attacked preventively by the intended victims. In such a case a preventive attack can reasonably be understood as not only defensive but even also retaliatory: a retaliation against the formation of the firm unconditional intention to attack and against the implementation of the intention as far as the stage of acquiring decisively effective means to carry it out.[6]

Now, many questions about the contentions just offered would need to be answered in a full defence of preventive military strikes.[7] Here I simply want to acknowledge that I consider it not to be obvious that a preventive attack against terrorists fully intending to use a weapon of mass destruction—that is, a weapon that will almost immediately kill vast numbers and cannot be defended against after it is launched—is not a defensive attack justified by a wrong committed in forming an unconditional intention to use weapons of mass destruction and acquiring the weapons with which to carry it out.[8] The preventive attack may be the last effective defensive measure that it is possible for the intended victims to take prior to actually being victimized.[9] In such a case a preventive military attack against such people would not need to be defended as an exception to the prohibition on attacking those who have not forfeited their right not to be killed, for these people would have forfeited their right even though they had not yet launched their attack and even though their attack was not imminent. This preventive attack would not violate the prohibition against attacking those who have done no wrong and so would not need to be defended as a justified exception to that prohibition. Preventive attacks are of course exceptions to the general practice of not attacking people who have not attacked you.

If this is correct, torture can be defended only as a wrong that can in exceptional circumstances nevertheless be justified, but preventive military attack need not always be considered a wrong. Preventive military attack is still justifiable only in exceptional circumstances—when aimed at people with an unconditional intention to use WMD that they have acquired for this purpose—but the exceptional circumstances make it a rightful action, not a justifiable wrong, as torture would be if it were indeed ever justified. This is an important lack of parallel. This difference notwithstanding, other parallels between torture and preventive attack make pursuit of this comparison fruitful for our purposes.

Let me next return to torture and explain very briefly why I think it is impossible ever to justify it even as an exceptional wrong; then we will come back to preventive attack. From the mistake that I now think that I made in 1978 about torture we can see the great value of a fuller empirical grasp of relevant social institutions. The kind of reason that is decisive against torture—the nature of underlying social institutions—turns out also to be the basis for an objection against preventive attack that is serious, but perhaps less serious, than in the instance of torture.

One can quite convincingly maintain that even if torture would be justified in the rarefied circumstances of the ticking bomb hypothetical, this conclusion that torture was justified in that kind of case provides no guidance for the very different kinds of cases that are ordinarily faced in real life where, for example, one is usually not sure one is about to torture the right person and one is not sure that he will not deceive, dissociate, collapse, die, or otherwise manage to avoid divulging the crucial information in good time. To infer that what is appropriate in a case of type A is also appropriate in a case of type B, where types A and B have such significant relevant differences as those just listed, is to make a mistaken, groundless generalization. But of course the next move is for someone to ask: But what if a real case were relevantly similar to the ideal case? The ticking bomb scenario may be rare, but it is not impossible.[10] What if we have a real instance of the perfect time to torture? How could we not be permitted to torture then, when the theoretically justifiable exception had arisen in fact?

Obviously serious epistemic problems arise about how one would actually know what the ideal case assumes one knows: that one has the right person, that the information will be divulged in time, and so on. In 'Torture' I expressed considerable scepticism about any perfect cases actually turning up, but could then see no way honestly to avoid conceding that if someone reasonably thought that the perfect time for torture had arrived, it would be permissible for him to go ahead and conduct the torture. If it turned out that he was correct in his beliefs about the situation and he did in fact prevent an awful catastrophe from occurring, we should, I suggested, not punish him for having tortured but instead offer assistance to both torturer and victim to restore their integrity from the degradation both would have (justifiably although wrongly) experienced. The good result would naturally have done nothing to reduce the degradation inherent in torture, but the suffering of the degradation on the part of the conscientious torturer might be viewed as roughly analogous to the suffering of mutilation or death in a justified war. The subject of the torture would also have suffered degradation that no human being deserves to suffer; that he had brought it upon himself should give us no solace. The conscientious torturer would in effect have suffered a crippled soul in the service of those whom he had saved from catastrophe, perhaps as just warriors suffer crippled bodies in the service of those whom they protect.

I now consider this concession to have been misguided (and not a little romanticized), as I have recently explained in 'Torture in Dreamland'. In the perfect case for torture, while torture is rare because restricted to such appropriate cases, the torture is perfectly successful: suddenly someone with no experience or training, who has never tortured anyone before, quickly extracts vital information from someone dedicated to withholding that very information. This is a sociological fantasy:

> We have abstracted from the social basis—the institutional context—necessary for the practice of torture. For torture is a practice. Practitioners who do not practice will not be very good at what they do. Who are we imagining that they will practice on? Practitioners without the best equipment will also not be very good. Where will they obtain their 'cutting edge' equipment? How will they test it in order to be sure it will work when the catastrophe looms? . . .

> The moderate position on torture is an impractical abstraction—it is torture in dreamland. The only operationally feasible positions are toward the extremes. . . .

> If the conscientious offender is to be an effective and competent torturer, or to have them on call, he must . . . be the tip of a bureaucratic iceberg of institutionalized torture . . . One can imagine rare torture, but one cannot institutionalize rare torture. The suggestion of rare torture has no place in the real world of politics. It is an optimistic thought with no social embodiment.[11]

I had hoped in 1978 that between the extremes of no torture and widespread torture there might be the 'moderate position' of just a little torture (because—one of the other dangers of practical ethics—I had wanted not to sound utopian but pragmatic, by taking a moderate stance).

In fact, there simply is no middle ground in the case of torture. Torture takes skill, dispositions, and knowledge that are gained only from experience.[12] Natural instincts must be suppressed, unnatural ones nurtured; this is all part of the degradation of the torturer. So our only choices are: no torture at all or the cadres of trained torturers that one now finds in the USA, among other nations.[13] Because of the degradation with which torture pollutes human civilization I think we must reject the state torture bureaucracies that now flourish with the support of our taxes and accept that there may be life-saving information that will not be extracted, with the result that some of us may die when torture might have saved us, but civilized social life may as a result survive.[14]

Whatever the correct moral judgement on the substantive point, the methodological point is that one cannot morally evaluate torture without knowing what torture is—what it is as an embodied social institution, not merely a thought. What I did not fully appreciate in 1978 is that torture is an institutionalized practice with a culture, expert teachers, innovative students, equipment testing, technique improvement, international communication, plus corrupt medical doctors who collaborate, and corrupt lawyers (such as those around Bush and Cheney) who cover up and deny.[15] Torture is a contagiously corrupt profession that blights the integrity of medicine, law, politics, and the other professions it infects. To think that these are 'mere facts' is to think that one can make wise moral assessments of practices without knowing how the practices work—indeed, without really knowing what the practices are. This is intellectually and morally irresponsible.[16] Aristotle may have been mistaken in thinking that one needs to be old to do ethics, but one certainly needs not to be naive, however many or few years that takes.

Now, as I have already indicated, I think some of the most powerful considerations against preventive military attacks, even attacks against terrorists with weapons of mass destruction, parallel the argument above against torture: one cannot have rare, effective torture because effective torture is a social institution with a bureaucratic life of its own. If torture is tolerated, it will be employed frequently, precisely as it is in fact now, so the option of rare, effective torture turns out simply not to be on the menu. The main dangers concerning preventive attacks also turn on the social institutions implicated. These too naturally have longer stories than can be told here—I will again try quickly to sketch some outlines.

One of the direct parallels between preventive attack and torture is obviously the assumption of having the right target to strike and the assumption of having the right person to torture. Precision-guided munitions are much more likely than 'dumb' munitions to hit what they are aimed at. This does not, unfortunately, mean that mistakes are less likely to be made, but only that different kinds of mistakes are now more likely, especially mistakes based on faulty intelligence such as the now infamous repeated bombings of wedding parties in Afghanistan. Outcomes increasingly depend on who decides at what to aim and on the quality of the information on which they rely.[17] Hypothetical cases generally build in perfect knowledge.[18] Transferring this assumption into the real case translates into assuming a highly competent intelligence agency to locate any target, such as vials of genetically modified smallpox.[19]

Foreign intelligence agencies are in fact notoriously unsuccessful, missing much more than they catch. (James Bond is a fictional character!) Whether one considers the low rate of success to be incompetence or misfortune depends on what one believes is practically possible. Whatever the explanation, one must expect false negatives and false positives. For the sake of one's own interest, one worries about the false negatives. The false positives would, if preventive strikes are carried out, constitute state murders of innocent people. For the sake of avoiding the commission of murderous wrongs, one must worry about the false positives (infamously, Iraq 2003).[20]

Many of the philosophical moves here are predictable. One can argue that the prohibition on the killing of people who have committed no wrong sufficient to forfeit their right not to be killed is sufficiently stringent that one should avoid running any significant risk of committing such murders, with the implication that one could permissibly launch a preventive military attack (or any other military attack, for that matter) only if one had specific and strong reason to trust that one's beliefs about one's targets were extremely well founded. Against this it can be argued that the supreme duty of the state is to protect people—at least its own and possibly others—against murder, certainly including terroristic murder, and that a state that failed to act on strong and solid, even if not irrefutable, evidence that preparations were well advanced for a devastating assault on large numbers of innocents would be a contemptible failure at its primary mission. This can be construed as the risking of the taking of life versus the risking of the failure by responsible agents to protect life, which leads into familiar argumentative pathways. In some respects this is like the familiar problem of how many students the campus police should themselves risk killing in order to try to prevent the disturbed gunman from killing even more, except that in our case, unlike the cessation of firing when the gunman is killed, any contagion released would not stop spreading when those who released it were dead (so that it is crucial that the virus release never begins).

In both the case of torture and the case of preventive attack, a critical factor is the nature of a social institution. But the institutions in question vary, and the more specific problems are quite different. For torture the key is that given the way torture works, it is impossible, not logically, but psychologically and sociologically, that there could be effective but only rare torture. In sum, the minimum moral price of any effective torture is very high. For preventive attack one key is whether intelligence gathering—or some other method of information assessment[21]—can be sufficiently reliable that it is not irresponsible to take human lives on the basis of it. The historical record provides much reason for scepticism about our capacity to gather evidence that must be definitive about, among other things, the intentions of people in other cultures who are acquiring weapons, but this evidence does not seem to me to be as decisive as the evidence that torture will never be effective but extremely rare.

For example, one can easily imagine a highly trustworthy, highly trained double agent with fluent language skills and a deep knowledge of the culture who had infiltrated a terrorist cell and observed for herself that a scientist collaborating with the cell had genetically modified a virus and manufactured an ample supply of it now stored in sealed containers, heard repeatedly for herself confidential and sincere conversations establishing that the members of the cell were irrevocably dedicated to releasing the virus in ten European ports and ten North American ports, and had confirmed that arrangements were in place for twenty couriers to board twenty separate ocean-going vessels within the next week to disperse into the twenty cities. The double agent could identify the current location of all the virus with a laser pointer, which missiles could follow to the precise site.

Now obviously this is a paradigm ideal spy, not a typical American or British agent. Should this imagined case be taken any more seriously than the imagined conscientious and competent torturer who almost immediately extracts vital information from the torture subject in spite of never, or almost never, having attempted to torture anyone before? Why isn't the hypothetical ideal spy just as mythical as the hypothetical ideal torturer? This is not the place to discuss this comparison in much detail. It seems, however, that while the ideal torturer borders on the miraculous, because it is so difficult to conceive how this highly skilful practitioner with no practice comes into being, the ideal spy could be the result of the expenditure of vast amounts of money for language teaching and cultural socialization and for the salaries (and life insurance policies) of large enough numbers of spies to create a reasonably high probability of having the right person in the right spot. Clearly, while there are quite separate reasons why one might have doubts about the advisability of the maintenance of such extensive spying operations—comparable, presumably, to the massive Cold War spying establishments—it is not especially difficult to see what it would take; and no miracles are required, although some good luck about being in the right place at the right time would, as always, not hurt. In my judgement, the ideal torturer sounds like fantasy, but the ideal spy mainly sounds fabulously expensive (and of course a sad diversion of resources that could have had more productive uses, although the diversion is arguably the responsibility of terrorists, not terrorist hunters). There are great social costs to having a huge intelligence establishment, as there were during the Cold War, but, in my judgement again, they seem not as corrosive as a cadre of torturers.[22] (More could be said of course about how to compare a bureaucracy for routinized deception and the violation of trust, to a bureaucracy for the routinized infliction of pain and humiliation for the sake of coercion—neither is a very pretty picture.)

Assume the ideal spy is not a fantasy, and that the spy agency is not also a torture agency such as the CIA, however likely in practice. I have been told, although I cannot document, that planes of the US Navy sometimes operate under rules of engagement that prohibit them from attacking from the air any site that is not clearly and totally military without an observer on the ground who can mark the target with a laser. One could certainly have a similarly restrictive rule about preventive military attacks: no attacks without an observer on the ground at the time marking what she herself knows beyond any reasonable doubt is an appropriate target. It would seem to me that such a rule would show appropriate respect for the value of human life, both the lives of innocents who might otherwise be struck by an air attack meant for terrorists and the lives of the innocents who would otherwise be killed by the terrorists' virus. With a bit more elaboration not appropriate here we could have a clear case of a morally justified preventive military attack.

The purpose here of course is methodological. However misguided my substantive judgement about the attack being justified may be, what does seem clear is that one cannot offer any practical ethical guidance without the incorporation of a considerable degree of empirical information including less than firm hypotheses about how institutions do, must, or can function. Can one be justified in attacking preventively? Only if intentions can be clearly established. Can intentions be clearly established? Only if there is direct and long-term personal contact with the agents. Can such contact be arranged? Only if intelligence services use human agents capable of assessing intentions in a specific cultural context. Can that be arranged? Only if agents are intelligent, well trained, courageous, and numerous. And so on. Our understanding of the moral issues guides us: one surely cannot kill anyone who has so far done no one else any harm unless it is abundantly clear that he firmly intends to do great harm and has made active and effective preparations to do so. We would need highly reliable information about such matters. Is there some feasible social institution that might be able to generate it? I am guessing there could be.

All this would of course establish only that a preventive attack could be morally permissible to the extent of satisfying the set of epistemic requirements discussed. Whether military solutions, even if permissible, are the best available solutions could be decided only after far broader considerations. Here there is space only to mention two further issues, but explore neither.

First is another deeply institutional issue. Thanks mainly to the Charter of the United Nations we now have a relatively entrenched norm against the first use of force. Nations are to employ military force only in response to an attack (which may include imminent attack, the occasion for genuine pre-emption as distinguished from preventive attack) or when authorized by the Security Council. Rightly or wrongly, one factor that is—or, prior to 2003 anyway, was—eroding the norm against first use of force is the growing acceptability of humanitarian intervention, although one can certainly argue about whether responding to, say, an attempted genocide is aptly thought of as a first use of force. If preventive attacks—even preventive attacks narrowly constrained to attacks on thoroughly confirmed weapons of mass destruction in the hands of terrorists—also gained additional acceptance, this might further erode the norm against first use of force. So the norm and practice of no first use of force would itself come under at least some additional pressure. Even granting that preventive attack was justifiable in some particular instances, would we want to have a general rule allowing it, especially if that general rule threatened to undermine the general rule against first use of force? This raises familiar issues about generalization, the relation of acts and practices, and so on; I want only to flag it. It could be that some individual preventive attacks would in themselves be morally permissible but that it would be unwise, for various reasons, to allow a general practice to develop.[23]

Second, while all these funds and energies mentioned above were going into the training of spies, what would be being done to bridge cultural, social, religious, and other gaps, the bridging of which might reduce the sources of terrorism over the longer term? Longer-term improvements do not protect against short-term deadly threats, but short-term solutions equally do not usually eliminate sources of longer-term dangers. Nothing so far said here begins to settle the issue of how much resources to devote to, respectively, short-term response and long-term prevention. But I hope the methodological point is clear: we must examine institutional presuppositions and implications if we are to be able to offer any judgements about what to do.

In any case, I would like to turn finally to what may be the greatest long-term danger of all: climate change. Consideration of what to do about climate change involves many issues of international justice, but it also raises questions of intergenerational justice, at which I want to glance here. Most philosophical work on intergenerational justice has concerned the just savings rate: to what extent, if any, ought the current generation to save for future generations, bearing in mind that future generations will likely, it has been assumed, be better off than the current generation in any case (because they will inherit progressive changes even where the motivation for the changes was nothing to do with their welfare but with the welfare of those initiating the changes).

As soon as one learns much of anything about the likely effects of climate change on humans, it becomes clear that it is not particularly likely that future generations will in fact be better off than the current generation and, even if much less than the worst occurs, it is entirely possible that our own generation will turn out to have been the best off in human history, with a decline in human welfare from here on because, as we undermine the environmental conditions for productive human economies, especially agricultural economies, water supplies and food crops may be progressively distorted and/or undermined.[24] The question, then, sharply reverses from 'How much benefit should we aim to provide to future generations?' to 'How much harm are we permitted to cause them while pursuing what we would in the past have taken to be ordinary activities of our own?'

Our new-found empirical understanding of our planet's climate, which has mushroomed over the last quarter of a century, radically transforms the nature of the issue of intergenerational justice. So far, however, we have simply noted a general reversal of fortune that eliminates good options and leaves mostly a range of bad ones—no basis so far for any assertions of exceptional duties, which is the point on which I want to compare climate change to torture and preventive attack. Further argument depends upon particular empirical hypotheses, and specifically upon evidence about how bad the options in the possible range are. It turns out that some of the likely options are quite bad and some of the entirely possible options are catastrophic.

Four morally salient features of climate change are now clear; I summarize them in ascending order of severity:

1. In failing to stop accelerating climate change, we are not failing to provide assistance to future generations, but inflicting harm on them— we are violating on a massive scale the principle: Do no harm.

2. In continuing to fail to stop accelerating climate change, we are inflicting harm upon additional generations beyond those who are already at this time destined to face more adverse conditions of life.

3. In continuing to fail to stop accelerating climate change, we are not simply continuing to make the conditions of life for future generations worse, but creating opportunities for the crossing of thresholds beyond which climate change would feed upon itself through positive feedbacks that would not have occurred if we had acted sooner and would become severely worse.

4. In continuing to fail to stop accelerating climate change, we are not simply creating opportunities for the crossing of thresholds beyond which climate change would become severely worse, but creating opportunities for the crossing of thresholds beyond which climate change will become catastrophically worse.[25]

As long as any one of these four propositions is correct—not to mention all four—we are engaged in one of the most massive inflictions of harm by one group of humans on other groups of humans in history. The 'empirical situation', which is to say, what is happening, is extraordinary and unprecedented. What is the appropriate response to these exceptional circumstances?

Viewing the situation as an opportunity rather than a burden, we are an absolutely pivotal generation in a highly strategic position within human history. A change in the trajectory of our energy policy could be a profoundly valuable legacy to whoever lives in the future. Plainly, it is reasonable to expect us to do more than people in ordinary circumstances would be expected to do with regard to future generations. The pressing questions are of course: 'How much more than normal?' and 'What are the appropriate measures?'

One of the empirical complications is that while climate change is indubitably already under way and some harms are now inevitable—some in fact already occurring—many harms, including the most terrifying (encapsulated in propositions 3 and 4 above) are uncertain in the technical sense that their probabilities are incalculable. Some people argue that it is unreasonable to expect even the most well off of the current generation to make definite sacrifices for the sake of preventing indefinite future harms, even ones that will be disastrous or catastrophic if they occur. Normally we discount the magnitude of a possible disaster by its probability, when we know the probability. When its probability is incalculable—the occurrence of the disaster is uncertain and thus may never eventuate—we tend in effect to discount it informally even though we have no probability by which to discount it formally. This is to say that we do not take the possible disaster especially seriously, and we do not go to very substantial lengths to prevent it. However, the empirical fact that the climate disasters would be disasters that we are complicit in bringing about, and very likely in making worse than necessary, seems to me to cast the matter in a whole new light.

Obviously these are complicated matters, but I have presented extensive arguments elsewhere for the following conclusions.[26] One ought to try urgently to make a possible outcome progressively more unlikely until the marginal costs of further efforts become excessive, irrespective of the outcome's precise prior probability, which may not be known in any case, when the circumstances include three features: (1) *massive loss*—the magnitude of the possible losses is massive; (2) *threshold likelihood*—the likelihood of the losses is significant, even if no precise probability can be specified, because (a) the mechanism by which the losses would occur is well understood, and (b) the conditions for the functioning of the mechanism are accumulating; and (3) *non-excessive costs*—the costs of prevention are not excessive (a) in light of the magnitude of the possible losses and (b) even considering the other important demands on our resources. These three features jointly constitute a sufficient set for prompt and robust action to be required. We know that our actions now are opening the doors to some terrible outcomes; we ought to re-close as many of these doors as we can.

Now this is perhaps enough for our purposes about climate change, although what has been said here is extremely superficial. If even one of the four hypotheses that I have presented about the deterioration in the future conditions of human life we may be causing is correct, then we face genuinely exceptional circumstances—quite extraordinarily exceptional circumstances. The exceptional losses would be suffered by whichever humans live after us, whatever their identities.[27] What precisely follows about our duties would be considerably longer in the telling, but the methodological point, it seems to me, is again clear. We could not have come this far, and can surely go little further, in our understanding of the direction in which we ought to move without an understanding of the physical and political dynamics of climate change. The abstract accounts of intergenerational justice we inherited focused on the wrong problem: How much to save? Generally, how much to help? That is in fact not the issue now. The issue is: How far ought we to go to reduce the harm we are likely otherwise to do by leading what we have come to assume are our ordinary lives? We learn this by delving into 'the empirical details', not this time (as in the cases of torture and preventive attack) institutional underpinnings, but physical underpinnings in the dynamics of the planet.

Unlike the cases of torture and preventive war we are not in the case of climate change considering making an exception to a prohibition. Instead we are considering going beyond a normal requirement. The similarity is that we have once again not brought our ready-made abstractions to the situation and asked how they might apply. We have immersed ourselves in the situation and tried to be open to its morally salient features. Our moral theories guide us by telling us which kinds of features would be salient if they were there, but we must also look and see if they are here and now.[28]

NOTES

1. Old joke: physicist and economist are stranded on desert island with a carton of tinned food. Physicist struggles to drop a rock on a tin that will split it open or to arrange for the sun to melt a hole in a tin. Economist simply says: 'Imagine a tin opener' (in the spirit of: 'Assume full information').
2. Perhaps therefore juxtaposing them serves little purpose—we shall see!
3. Speaking strictly autobiographically, I was for a long time guided in my thinking about preventive attack by analogy with the analysis I had earlier given of torture—until I saw some crucial differences, to be sketched below.
4. See Henry Shue, 'Torture', *Philosophy & Public Affairs* 7, 2 (1978): 124–43; and 'Torture in dreamland: Disposing of the ticking bomb', *Case Western Reserve Journal of International Law* 37, 2 and 3 (2006): 231–9. <http://www.case.edu/orgs/jil/archives/vol37no2and3/Shue.pdf>. Books by others include Bob Brecher, *Torture and the Ticking Bomb* (Oxford: Blackwell, 2007), and Yuval Ginbar, *Why Not Torture Terrorists? Moral, Practical, and Legal Aspects of the 'Ticking Bomb' Justification for Torture* (Oxford: Oxford University Press, 2008). A brief decisive treatment is in David Luban, 'Unthinking the ticking bomb' in C. R. Beitz and R. Goodin (eds) *Global Basic Rights* (Oxford: Oxford University Press, 2009). Also see Matthew Alexander with John R. Bruning, *How To Break A Terrorist: The U.S. Interrogators Who Used Brains, Not Brutality, To Take Down The Deadliest Man in Iraq* (London: Free Press, 2008).

5. At present US special forces are in fact secretly going around the world killing suspicious people with no apparent public accountability—see Eric Schmitt and Mark Mazzetti, 'Secret Order Lets U.S. Raid Al Qaeda', *New York Times*, 10 November 2008. <http://www.nytimes.com/2008/11/10/Washington/10mili tary/html?pagewanted=all>. (accessed 24 November 2008).

6. On its being retaliatory, see Suzanne Uniacke, 'On getting one's retaliation in first', in H. Shue and D. Rodin (eds) *Preemption: Military Action and Moral Justification* (Oxford: Oxford University Press, 2007), pp. 72–4. Could Soviet and US possession of WMD during the Cold War be condemned on the same grounds as WMD possession by terrorists? This depends, I think, on whether Soviet and American intentions to use their WMD could be made out to be conditional on the behaviour of their respective adversaries in a way in which terrorist intentions are not conditional on the behaviour of their adversaries. This is a complex question, which I cannot pursue here. On the general issue whether this distinction between conditional and unconditional intentions holds, see the chapters by Rodin and Luban in Shue and Rodin (eds), op. cit.

7. This issue is discussed in most of the chapters in Shue and Rodin (eds) op. cit.

8. My misgivings about the justifiability of preventive strikes are laid out in 'What would a justified military attack look like?' in Shue and Rodin (eds) op. cit., pp. 222–46.

9. On the right to take the last effective defensive measure, see Zahler Bryan, 'Initiating force: A re-consideration of the justifications', MPhil thesis (Oxford, 2007).

10. See Oren Gross, 'The Prohibition on Torture and the Limits of the Law', in S. Levinson (ed.) *Torture: A Collection* (Oxford: Oxford University Press, 2004), p. 234.

11. Shue 2006 op. cit., pp. 237–8.

12. The ticking bomb hypothetical is thus, in Onora O'Neill's helpful contrast, an abstraction as well as an idealization—see 'Ethical reasoning and ideological pluralism', *Ethics* 98, 4 (1988): 711–12. There is no possible social embodiment of rare torture. The choices are torture routines or torture abolition.

13. Jane Mayer, *The Dark Side: The Inside Story of How the War on Terror Turned into a War on American Ideals* (New York: Doubleday, 2008).

14. Henry Shue, 'The debate on torture: Response', *Dissent* 50, 3 (2003): 90–1.

15. Philippe Sands, *Torture Team: Deception, Cruelty and the Compromise of Law* (London: Allen Lane, 2008).

16. Second worst is simply to peddle abstract principles and 'leave the applications to the empirical experts'. Such bare-principle peddlers belong to the I-will-buy-the-can-of-paint school of ethics: 'Here, you apply it.' But the so-called 'application' is what takes the subtle and balanced moral assessment that philosophers are supposed to do well. How are the 'empirical experts' supposed to know, for example, which features of the phenomenon in question are the morally salient features?

17. The knowledge that the munitions are more accurate may also encourage more attempts, including riskier attempts undertaken in the false confidence that the eye of the needle can always be threaded. Consider the dozens of failed attempts to assassinate Saddam Hussein with missiles at the beginning of the assault on Iraq in 2003.

18. As already indicated, economists and other model-building social scientists are probably guilty even more often than philosophers on this score of lazily assuming perfect knowledge.

19. Shue, 'What would a justified military attack look like?' in Shue and Rodin (eds) op. cit., pp. 230–2.

20. For a troubling bureaucratic explanation by a former intelligence officer of why one ought to expect many more false positives than false negatives, see Greg Thielmann, 'Intelligence in preventive military strategy', in W. W. Keller and G. R. Mitchell (eds) *Hitting First: Preventive Force in U.S. Security Strategy* (Pittsburg, PA: University of Pittsburgh Press, 2006), pp. 156–9.

21. For an important alternative proposal about how to improve decision-making about preventive strikes, see Allen Buchanan and Robert O. Keohane, 'The preventive use of force: A cosmopolitan institutional proposal', *Ethics & International Affairs* 18, 1 (2004): 1–22; and Allen Buchanan, 'Institutionalizing the just war', *Philosophy & Public Affairs* 34, 1 (2006): 2–38. For critiques, see Steven Lee, 'A moral critique of the cosmopolitan institutional proposal', *Ethics & International Affairs* 19, 2 (2005): 99–107; and Shue, 'What would a justified military attack look like?' in Shue and Rodin (eds) op. cit., pp. 235–8.

22. This makes the heroic assumption that the spying and the torturing are separate. In fact, in the USA at present much, not most, of the torture is conducted by the CIA—see Alfred W. McCoy, *A Question of Torture: CIA Interrogation, From the Cold War to the War on Terror* (New York: Metropolitan Books, 2006).

23. See Allen Buchanan, 'Justifying Preventive War', in Shue and Rodin (eds) op. cit., pp. 128–32; and David Luban op. cit., 'Appendix 2: Buchanan's Objection to Generalization Arguments', pp. 199–201.

24. I am obviously not persuaded of hypotheses of indefinite substitutability as applied to 'eco-system services'.

25. See 'Deadly delays, saving opportunities: creating a more dangerous world?', this volume. For earlier formulations, see 'Responsibility to future generations and the technological transition', this volume; and 'A legacy of danger: The Kyoto Protocol and future generations', this volume.

26. This paragraph draws on 'Deadly delays, saving opportunities: creating a more dangerous world?', this volume.

27. The so-called 'non-identity problem' is irrelevant.

28. 2014: It was an exaggeration to say that we must 'eliminate completely the use of fossil fuel if the temperature is not to exceed 2°C' (303, 308, and note 35 on 306). We must eliminate completely *net additions* to the atmospheric concentration of CO_2. The atmospheric concentration would not rise if carbon emissions on the surface were once again, as they were in 1850 and before, small enough to be absorbed by vegetation and the oceans. Carbon absorption by vegetation is basically fine. Carbon absorption by the oceans is basically causing acidification, which is treated as a problem separate from climate change, but it is severe, a cause of species extinctions, dangerous to the human food supply, and probably irreversible. So for all practical purposes we must cease to rely on fossil fuels as a primary source of energy because their use must be radically reduced, but not literally eliminated. Coal cannot be a major source of electricity, and petroleum cannot be a main energy source for transportation.

14

Deadly delays, saving opportunities: creating
a more dangerous world?*

> Will there really be a 'morning'?
> Is there such a thing as 'Day'?
> Could I see it from the mountains
> If I were as tall as they?
>   Emily Dickinson

We now know that anthropogenic emissions of greenhouse gases (GHGs) are
interfering with the planet's climate system in ways that are likely to lead to
dangerous threats to human life (not to mention non-human life)[1] and that
are likely to compromise the fundamental well-being of people who live at a
later time.[2] We have not understood this for very long—for most of my life, for
example, we were basically clueless about climate. Our recently acquired
knowledge means that decisions about climate policy are no longer properly
understood as decisions entirely about *preferences of ours* but also crucially
about the *vulnerabilities of others*—not about the question 'How much would
we like to spend to slow climate change?' but about 'How little are we in
decency permitted to spend in light of the difficulties and the risks of difficulties to which we are likely otherwise to expose people, people already living
and people yet to live?' For we now realize that the carbon-centred energy
regime under which we live is modifying the human habitat, creating a more
dangerous world for the living and for posterity. Our technologically primitive
energy regime based on setting fire to fossil fuels is storing up, in the planet's
radically altering atmosphere, sources of added threat for people who are

* 'Deadly Delays, Saving Opportunities: Creating A More Dangerous World?', in *Climate Ethics*, ed. Stephen M. Gardiner, Simon Caney, Dale Jamieson, and Henry Shue (Oxford University Press, 2010), 146–62. This chapter has endnotes rather than footnotes.

The gestation of this chapter has been long and painful, and I have been assisted in wrestling with various versions of it by responsive audiences at San Diego State University, University of Washington, Cornell University, University of Oslo, University of Oxford, New York University Law School, University of Edinburgh, and the University of Tennessee. In the last round, I have benefited especially from comments by Wilfred Beckerman and Nicole Hassoun.

vulnerable to us and cannot protect themselves against the consequences of our decisions for the circumstances in which they will have to live—most notably, whichever people inherit the worn-and-torn planet we vacate.[3]

LARGE DISASTERS, CONSIDERABLE LIKELIHOODS

As we academics love to note, matters are, of course, complicated. Let's look at a few of the complications, concentrating on some concerning risk.[4] Mostly, we are talking about risks because, although we know strikingly much more about the planetary climate system than we did a generation ago, much is still unknown and unpredictable. I will offer three comments about risk. The third comment is the crucial one and makes a strong claim about a specific type of risk, with three distinctive features. After illustrating the three features with the effects of a possible bird flu epidemic, I then argue somewhat more fully that the three features are also jointly characteristic of the effects of climate change, with strong implications for how we should regard our recently discovered complicity in producing climate change and thereby worsening the circumstances to which whoever succeeds us will need to adapt. Then I will consider the specific implications for what it is most essential and urgent to do.

The first point to be made about risk and climate change, however, is that not everything is uncertain, for two reasons. One is simply that some threatening changes in the climate are already occurring, as practically every informed person acknowledges. I would not have the scientific knowledge to sort through very many specifics, but clearly, for example, patterns of rainfall and storm intensity have already changed somewhat, resulting in both flooding and drought.[5] The other reason is that unless virtually all human understanding about the climate were completely misguided, other changes are practically certain to occur; for example, sea level will surely rise significantly.[6] If nothing else, the volume of the water would increase from the rise in temperature that has already straightforwardly been measured. And other factors are converging on sea-level rise, such as amazingly rapid melting of Arctic and Greenland ice that both directly increases the amount of water in the ocean, when the melting ice was previously on land (i.e. was an ice sheet, not an ice shelf),[7] and indirectly warms the planet by reducing albedo through elimination of the reflectivity of the snow. Some island nations in the South Pacific are already well into the process of being submerged by rising sea levels. Nothing in my argument to follow turns on how much is already in fact happening, since I will mostly be discussing risk of future events, but it is simply factually misleading to talk as if climate change is all risk only and nothing untoward is happening yet.

Now, what about the risks, which virtually anyone will acknowledge are fortunately still most of the problem? The second point to note about risk is that it is highly significant morally whether one is choosing a risk for oneself or imposing it, conditionally or unconditionally, on others. A certain level of risk may be a reasonable one for me to choose for myself but not a reasonable one for me to impose on others. Therefore, even if the level of risk from climate change imposed on future generations were the same as the risk for us—of course, it is not remotely the same—it might still be unreasonable for us to impose it on them, even if it were not unreasonable for us to choose it for ourselves. I am free to choose to mountain climb, but I could not reasonably propose that an experience of mountain climbing be a requirement for graduation from university, because mountain climbing is too dangerous to require of others generally. That we are imposing risks that others will inherit at birth is extremely important.

Risk is most often explained as the product of magnitude and probability. The magnitude is a measure of the seriousness of the loss risked, and the probability is a measure of the likelihood of that loss occurring. Some corporate and government opponents of vigorous action to slow climate change, especially coal and oil interests, have made much of alleged 'uncertainty'. In many cases they have purposely distorted the science and wildly exaggerated the extent of our current ignorance; Steve Vanderheiden has aptly characterized this intentional smoke blowing as 'manufactured uncertainty'.[8] The tobacco companies always claimed that the connections between smoking and bad health were uncertain; the coal and oil companies claim the connections between carbon combustion and bad climate are uncertain. Neither connection is uncertain. But what I want to show here is that there is a crucial kind of cases in which a considerable degree of uncertainty does not matter even if we are appropriately uncertain without having been tricked by industry and government propaganda.

I will defend the suggestion—this is the third, and chief, point about risk—that there are cases in which one can reasonably, and indeed ought to, ignore entirely questions of probability beyond a certain minimal level of likelihood. These are cases with three features: (1) *massive loss*: the magnitude of the possible losses is massive; (2) *threshold likelihood*: the likelihood of the losses is significant, even if no precise probability can be specified, because (a) the mechanism by which the losses would occur is well understood, and (b) the conditions for the functioning of the mechanism are accumulating; and (3) *non-excessive costs*: the costs of prevention are not excessive (a) in light of the magnitude of the possible losses and (b) even considering the other important demands on our resources.[9] Where these three features are all present, one ought to try urgently to make the outcome progressively more unlikely until the marginal costs of further efforts become excessive, irrespective of the outcome's precise prior probability, which may not be known in any case.

We know that our actions now are opening the doors to some terrible outcomes; we ought to reclose as many of these doors as we can. The suggestion, then, is that these three features jointly constitute a sufficient set for prompt and robust action to be required.[10] When all three conditions are present, action ought to be taken urgently and vigorously. Doing nothing but calling for further research is morally irresponsible, I will now argue. Obviously, further research is also good provided that it is not a substitute for effective action.

Basically, the argument is that because the magnitude of particular losses is so serious, the only acceptable probability is as close as possible to zero, provided this reduction in likelihood can be achieved at a cost that is not inordinate. Some losses would be utterly intolerable, especially 'losses' involving massive deprivations of necessities to which all people, regardless of individual identity, have rights simply as human beings. This applies to (a) some cases in which the probability is known and small but still significant and (b) some cases in which the probability cannot be calculated but can be known to be significant (because the relevant mechanism is understood and the conditions for its functioning are appearing). Only the latter would be a case of uncertainty in the technical sense, that is, an event with no calculable probability. Obviously, several aspects of this argument each need separate discussion.

I begin with a preliminary reminder about uncertainty. That something is uncertain in the technical sense, that is, has no calculable probability, in no way suggests that its objective probability, if known, would be small.[11] There is a grand illusion here: if we cannot see what the probability is, it must be small. Perhaps we assume a visual metaphor: we cannot see the probability because it is too small to see, so it must be really tiny. This inference is totally groundless. If all we know is that the probability cannot be calculated, then we do not know anything about what it is; if we do not know anything about what it is, then we do not know whether it is small or large.

However, we might have independent evidence that a likelihood is either small or large, without being able to calculate the probability. Cases of type (b) just mentioned above are such cases in which we cannot calculate a probability but know on other grounds that the likelihood is significant. The point now is that the simple fact that the probability is uncertain does not entail that it is small. Thinking so would be like thinking that if you are not sure where a city is located, the city must be small. We may simply be totally overlooking an entire dimension of a problem that will turn out to be huge. Things can be invisible for reasons other than being small. Some probabilities unknown at one time turn out later to be very large. Often the universe has major surprises for us, some very unpleasant.

Next we turn to cases illustrating the three features that I think require us to push the probability as close as we can to zero, whatever exactly it is now, given that it is significant. One example of the kind of case I have in mind is the reasons for the measures now being taken to prevent a possible bird flu pandemic.[12] In a bird flu pandemic, (1) the losses would be massive; (2) the likelihood of occurrence is significant even if it cannot be calculated because the mechanism of occurrence is well understood and conditions for its functioning have appeared; and (3) the costs of prevention, while far from negligible, are (extremely) moderate in light of (a) the possible losses and (b) even the other legitimate demands on resources.

First, the losses could be massive. Tens of millions of people died from the 1918 flu epidemic that helped to end World War I.[13] Now that we have enhanced globalization, including rapid movements of large numbers of people for great distances, it is entirely possible that deaths from a flu pandemic would be in the hundreds of millions of people—a modern, global plague.

And second, we understand the mechanisms by which this would happen and can see conditions favourable for the working of the mechanism arising. This second one is the 'anti-paranoia' requirement, designed to narrow the range of possibilities on which we need to act. By requiring a clear mechanism we avoid reacting similarly to every imaginable threat. If all the oxygen on earth burst into flame, that, too, would be a disaster, but we do not know of any way that could happen. The specification of a clear mechanism is the central contributor to our conviction that the probability is significant in spite of our not being able to calculate it.

Human flu is highly contagious, and the active bird flu has already rapidly mutated several times. Nothing naturally prevents a mutation into a form directly transmissible from human to human—it is the precise probability of this occurring that is unknown. Once the mutated flu was passing directly among humans, it would move quickly if no directly applicable vaccine had been prepared in sufficient quantity in advance of the outbreak of the pandemic. Vaccine has a production time of months using current technology; this is the problem of lead time, which is monumentally important in the case of climate change. It would take a very long time, depending on how many labs were manufacturing vaccine, to produce, say, one billion doses, which would still leave five out of six humans unprotected, providing only enough vaccine for a population the size of either China or India.

Meanwhile, the virus might mutate again, making the vaccine already produced until that time ineffective. So actually, the best argument for doing nothing to prevent a pandemic would be the fatalistic argument that it was impossible to stay ahead of the virus. But it is not known to be impossible—that, too, is uncertain—so I think we should try, as to some degree we are, because, third, preparing facilities for the manufacture of vaccine in large quantities, while expensive, is not prohibitively so.[14] It would be difficult to imagine a better investment of public funds than subsidizing this manufacturing capacity.

What I want to emphasize is that no precise probability of the pandemic plays any role in the argument whatsoever, primarily because the magnitude of the possible loss is so great—tens of millions or hundreds of millions of human lives. Another probability that would matter would be a virtual certainty that attempts at prevention would fail, making the funds spent on expanded production of vaccine a waste; even so, unless the costs were astronomical—on the order of a perpetual boondoggle such as the dysfunctional US ballistic missile defences, for example—would it even begin to seem unreasonable to try. Extra manufacturing capacity for flu vaccine would cost only a tiny fraction of what is currently being wasted on misguided military systems.

Obviously, what I am next going to suggest is that a number of phenomena that could result from climate change, especially climate change allowed to build up even more momentum before anything serious is done to slow it, are like the possible flu pandemic in having the three key features: (1) the possible losses are massive; (2) while the precise probability of these losses occurring is unknown, their likelihood is significant because the mechanism by which they would occur is well understood and conditions for its functioning are falling into place, and (3) the costs of preventing these losses are not excessive—at least for now—in light of the magnitude of the possible losses, even taking into account the other important current demands on resources.

The three features must apply to each potential loss that is given weight in deciding what to do, and, of course, that each feature applies is an empirical claim that must be established with detailed scientific argument. My only hope here is to formulate a reasonable set of criteria; I lack the knowledge to make all of the various cases that the criteria are in fact satisfied. So I will merely briefly indicate the kind of empirical cases that need to be spelled out. We can count on the rapidly developing science to spell them out.

How might massive losses arise from climate change? For example, among ecosystems, agricultural systems are especially touchy.[15] Crops for humans need to be edible, which basically means they need to be just right. It cannot be too hot or too cold, too wet or too dry. If they are under-ripe, they cannot be eaten; if they are overripe, they cannot be eaten. If the rain comes too soon, they parch later; if the rain comes too late, they have already shrivelled or will rot. Farmers already gamble on the weather. Climate change is long-term weather change. Gambling on climate change is raising the odds greatly against the already wagering farmers, who keep us alive, when they are lucky.[16]

Generally speaking, if the weather changes faster than the crops can adapt, there is trouble, that is, shortage of food. Severe shortage in one place tends to mean higher prices in other places, if those whose own agriculture failed have enough money to import food. The famine can be exported, but it cannot be made to evaporate. As Amartya Sen demonstrated in *Poverty and Famines*, those with high incomes bid up the price of food, and those with low incomes starve.[17] So in the case of climate change, too, (1) the potential human losses could be massive. This is for many reasons, but a lethal one is disruption of food supplies, causing volatile food prices. Others include the need for massive relocations of population from low-lying shores inundated by rising sea levels.

(2) The mechanisms leading from burning fossil fuel, above all, to the climate changes are increasingly well understood.[18] Those connecting the climatic changes in turn to human misery were already well understood, since necessities as elemental as food and shelter are directly assaulted by the physical phenomena constituting climate change, such as more intense storms and atypical weather.

(3) The costs of prevention are moderate, although far from insignificant.[19] First come the 'no-regrets' measures that eliminate current costly energy waste and thereby improve living standards and reduce dependence on Middle Eastern dictatorships such as Saudi Arabia which are lightning rods for terrorism and entice heedless Western politicians into needless wars and bloated military budgets. Much of what we need to give up next after the economically and politically profitable, no-regrets reductions are frivolous preferences, life-shortening luxuries, and pointless indulgences. What we must give up after those depends on how long we continue to make the problem worse by continuing to derive our energy from fossil fuel before we begin to make it better by switching to alternative sources.[20] Plainly, delay will not make the necessary transition less painful—it will only shift it off us and onto others.

THE CREATION OF A MORE DANGEROUS WORLD

So far we have only a quick overview of the case of climate change, and now we need to look at selected aspects a little more thoroughly. The nature of what I have so far been vaguely referring to as 'massive losses' can be specified more precisely. I will examine four aspects of danger, acknowledging three and setting aside the fourth.

Creating danger

First, and most significant, failing to deal with climate change constitutes not only failing to protect future generations but inflicting adversity on them by making their circumstances more difficult and dangerous than they would have been without as much climate change, and more difficult and dangerous than circumstances are now for us.[21] If the current climate change were a naturally occurring problem, like some effects of human ageing, and we did nothing to deal with it, we would leave future generations facing a problem that was only as severe when we bequeathed it as when we inherited it. We would have failed to provide protection—done nothing to make their lives less dangerous. That would be blameworthy, but what we would be guilty of would be a 'sin of omission': neglecting to provide protection for subsistence rights that was ours to give if we had chosen to bother.[22]

Failing to deal with our climate change is not like that, because the current climate change is not naturally occurring. Political choices about energy policy are causing climate change. At some points in the planet's history, climate change has occurred naturally, but the climate change happening now is, as the scientists say, anthropogenic: people are causing it, by bringing about the emission of increasing amounts of greenhouse gases such as the CO_2 from the burning of fossil fuels in car engines and electricity generating plants. Human activities are undermining the environmental conditions to which human beings have successfully adapted, making the environmental conditions for future generations more threatening for them than the present conditions are for us. 'Doing nothing' about climate change in the sense of simply continuing business as usual is—far from actually doing nothing—continuing to change the environmental conditions that future generations will face for the worse. To persist in the activities that make climate change worse, and thereby make living conditions for future generations worse, is not merely to decline to provide protection. It is to inflict danger, and to inflict it on people who are vulnerable to us and to whom we are invulnerable.[23] The relationship is entirely asymmetric: they are at our mercy, but we are out of their reach. Causation runs through time in only one direction. Lucky for us.

Endangering additional generations

Second, failing to deal with climate change constitutes inflicting danger on additional generations that could have been spared. It is not only that future generations that are already fated to be adversely affected by the GHGs that have already been injected into the atmosphere by previous generations since the spread of the industrial revolution will face more adverse conditions of life than if we had managed to get a grip on our fossil fuel consumption. Yet later generations, the great-great-great-grandchildren rather than the grandchildren, that might have been spared this problem if it had been solved sooner, will suffer from it. Suppose that if our generation did whatever it ought to do to stop accelerating climate change, the effects of climate change would have become manageable by some Generation L. If we do not do what we ought, and everything else remains the same, then at the very least the next generation, Generation M, will suffer from climate change. So, besides making life more treacherous for every generation from A to L, we would have inflicted completely avoidable problems on Generation M (and doubtless others), which would have been free of these problems if we had restrained our environmentally damaging activities, assuming only that tackling the problem sooner means solving it sooner.

This assumption would not be straightforwardly true if, say, some technology needed to mature before it could be successfully applied to climate change and attempts to employ it sooner would be futile. If we had reason to believe this was the situation, however, we would have no basis for merely increasing fossil fuel consumption as usual. First, we could, instead of attempting to use the immature technology before it was ready, be seriously investing in research on improving the technology or on alternative technologies, rather than simply indulging in our own high-emissions consumption. Our investment now might allow an intermediate generation still to implement the by then mature technology in time to save Generation M. Second, and more important, we do not need to develop any new technologies in order simply to cease wasteful and frivolous uses of fossil fuels and to defeat short-sighted politicians who block policies that would make the wasteful pay and that would create disincentives for excessive emissions. Time passes while the problem remains untackled, so additional generations will suffer. But this is not the worst.

Creating additional dangers

Third, failing to deal with climate change constitutes not simply continuing to make the environment for human life more threatening but unnecessarily creating opportunities for it to become significantly more dangerous by feeding upon itself through positive feedbacks that would otherwise not have occurred—creating opportunities for the danger to escalate one or more levels. We have hardly scratched the surface of the seriousness of continued delay in facing the challenge of climate change. Climate change is dynamic. It involves many poorly understood feedbacks, negative as well as positive. It is conceivable that a continued worsening will trigger a negative feedback, such as an increase in the kinds of clouds that reflect sun waves back away from the earth, that will actually improve the situation for humans. Unknowns remain. But some of the best understood and most likely feedbacks are positive, compounding the problem. For example, if emissions of CO_2 cause the Arctic tundra to thaw, as they appear well on the way to doing, the thawing tundra will release vast amounts of methane (CH_4), which is a far more powerful GHG per unit than CO_2 and will make climate change significantly more severe than it would have been if the tundra had not thawed.[24]

The opportunities we create for net positive feedbacks to occur may not be taken, or the positive feedback may somehow be more than cancelled out by some now only more dimly foreseeable negative feedback. But it still seems wrong to create the opportunity for the positive feedback for no good reason. If I play Russian roulette with your head for my amusement as you doze and the hammer of the revolver falls on an empty chamber, I will have done you no physical harm. But I will have seriously wronged you by subjecting you to that unnecessary risk. We do no wrong when we unavoidably inflict risks on future generations, or even perhaps if we have compelling reasons for doing so where it would be avoidable. But we do wrong them if we subject them to opportunities for matters to worsen severely for no good reason except that we could not be bothered to change our comfortable habits and that the owners of the coal and oil reserves are greedy for maximum return. We can be justified in imposing a risk on others when the harm to ourselves from avoiding the risk to them would be severe—perhaps even if it would only be significant—but not when avoiding the imposition of the risk on them would cause us only mild inconvenience, or even serious but manageable difficulty, or leave us merely rich, not super rich.

The fourth aspect of danger is the most fearsome. For completeness, I need to mention it, but I will not rely on it in my argument.

Creating desperate dangers

Fourth, failing to deal with climate change constitutes not only unnecessarily creating opportunities for the planetary environment to become significantly worse for humans (and other living things) but unnecessarily creating opportunities for it to become catastrophically worse. It is not merely that (1) we make living conditions more dangerous for some generations that already will suffer from climate change and that (2) we make conditions dangerous for one or more generations that could have been secure from the threats of climate change and that (3) we create opportunities for the environment to degenerate severely. Worse still, (4) we could contribute to turning severe problems into literally insoluble problems. Or, of course, possibly not—this would, once again, be a question of the justifiability of avoidably imposing risks of adversity on defenceless others.

Unnecessarily imposing a risk of uncontrollable change—change that the people subject to it could neither steer nor stop—would be much like creating, for no good reason, a highly contagious fatal disease and leaving it behind without a cure for future generations to contend with.

Various mechanisms for runaway climate change are well understood and have in fact operated in the past. A runaway climate is certainly possible in the future because it has been actual in the past. A general category employed by scientists is abrupt climate change, which can be defined as 'a large-scale change in the climate system that takes place over a few decades or less, persists (or is anticipated to persist) for at least a few decades, and causes substantial disruptions in human and natural systems'.[25] We know, for example, that rapid warming can lead to abrupt cooling, because it did in the Younger Dryas roughly 10,000 years ago and, as we know from astoundingly informative ice cores, several times far earlier.[26] So there is no doubt that something devastating to humans could happen if climate change crosses a threshold that we can cause it to cross or prevent it from crossing.

The 2007 report from Working Group I of the IPCC, however, is sceptical about abrupt climate change in the current century: 'Abrupt climate changes, such as the collapse of the West Antarctic Ice Sheet, the rapid loss of the Greenland Ice Sheet or large-scale changes of ocean circulation systems, are not considered likely to occur in the 21st century, based on currently available model results. However, the occurrence of such changes becomes increasingly more likely as the perturbation of the climate system progresses.'[27] While we should, I think, take little comfort from the fact that our own century might be safe from the most extreme possibilities, if the report is correct in its judgement, the possibility of desperate danger does not, then, fully satisfy my second condition, threshold likelihood. Although we understand various mechanisms that could lead to runaway climate change, we do not yet have strong reason to believe that the conditions in which those mechanisms operate are coming together—at least, not yet. So I return to the previous point: creating additional but non-catastrophic danger by creating opportunities for positive feedbacks to cause climate change to escalate one or more levels above where it is already destined to go.

And the possibility of such severe danger, even short of desperate danger, is more than enough to concern us. The ones who need to worry about severe climate change are the most vulnerable, including children yet to be born, who may reap the whirlwind if we sow the wind. Those who will suffer most, if anyone does, will be people with absolutely no past role in causing the problem and with no other kind of responsibility for it (and other species, most with no capacity for morally responsible action but full capacity for suffering and frustration). This would put the kind of wrong done by the avoidable precipitation of severe climate change, it seems to me, in the general moral category of the infliction of damage or the risk of damage on the innocent and the defenceless. This is far worse than simply neglecting to protect rights, as wrong as that is, and is more like recklessly dropping bombs without knowing or caring whom they might hit. Can someone seriously argue that we are not morally responsible for avoiding the wreaking of such havoc?

And—feature three, once more—the human costs of preventing climate change from becoming severe could be modest, if well managed and begun promptly.[28] Much of our current GHG emissions serve worthy, even essential or admirable, goals. But substantial portions of it result from thoughtlessness, laziness, and wastefulness; and much serves purposes that are opulent, frivolous, or pointless.[29] I do not want to sound like a puritan; perhaps we are all free to engage in a certain amount of frivolity and pointless joy—at least, if we do no serious harm to others. On the other hand, much commends a life of simplicity, although I will not press that point here.[30] The main point here is that frivolous and pointless GHG emissions, far from being harmless, may be storing up threatening problems for whoever lives in future generations. There is low-emissions frivolity and there is high-emissions frivolity. I take no position here on low-emissions frivolity. High-emissions frivolity is another matter: it can be a serious threat to many other living things.

The overall picture, then, is that for the sake of benefits to ourselves that are, even if not forbidden, utterly insignificant, we are inflicting on whoever comes after us an unknown but substantial risk of a significantly more dangerous world—a dangerous world that would be to no minor extent our own creation: collateral damage from the primitive energy regime now fuelling our lifestyle, not intended but no longer unforeseen. Even collateral damage in war is required to be proportional to the achievement of something important through a necessary action. To what present necessity would severe adversity on the part of successive generations of humans who succeed us be proportional?

PROPORTIONALITY AND RELATIVITY

Judgements about proportionality—especially proportionality between incommensurable values such as qualities of human life and quantities of financial costs—cannot be precise.[31] I want to emphasize, however, the presence of two relativities in, but the absence of a third from, the proposed set of three jointly sufficient conditions for prompt and robust action. As already mentioned, the third factor, non-excessive financial cost, is obviously not independent of the first factor, magnitude of human losses. What would be an excessive cost for preventing relatively smaller human losses might not be excessive for preventing relatively larger human losses. I take this to be the plainest of common sense. We cannot quantify very usefully, I think, but we can rank: a cost that would be excessive for preventing one additional destructive Atlantic hurricane per year might not be excessive for preventing the flooding of scores of the world's major cities by rising sea levels. Reasonable expenditure is obviously relative to the seriousness of the losses prevented.

I am tempted to say that no cost would be excessive for avoiding severe climate change that could lead to distortions of agriculture and yield additional starvation by way of global food price fluctuations.[32] Such undercutting of the food system would be a monumental human tragedy. But the fact is that, as rich as we humans are in 2010, our financial resources are finite, so costs must, second, also be assessed in light of other legitimate current demands on resources. Right now, on the order of eighteen million people are dying each year of readily remediable chronic poverty for want of relatively small sums of money and related institutional changes.[33] One could not sanely claim that unlimited sums should be devoted to blocking the possibility of future severe climate change if that entailed that one would, in consequence, refuse to spend what it would take to eliminate current severe poverty. This specific dilemma, however, is totally false: the budget for climate change does not need to be deducted from the budget for chronic poverty. It could be deducted from the budget for misguided military adventures.[34] Nevertheless, the point remains: at some level, expenditures on even the avoidance of dangerous climate change could be excessive, compared not to folly but to legitimate alternative uses. So in principle, what count as proportionate expenditures on the mitigation of climate change designed to stabilize it at a less dangerous level must be conceded to be relative not only to the losses that could occur if the expenditure is not made on prevention of climate change but also to the losses that would occur if, as is now far from being the case, the climate expenditure had to be taken away from other genuinely urgent matters. Therefore, the third condition within the sufficient set needed to be stated in a way that makes reasonable costs relative to the extent of the human losses that are the subject of the first condition and relative to other real—as opposed to politically manufactured—emergencies.[35]

What the costs do not need to be relative to, however, is a possible additional consideration: the probability of the massive losses. The second of the three features required for the set sufficient for action, threshold likelihood, has been formulated in order to deal with likelihood by relying on a threshold, not relying directly on probability. This is where I recommend diverging fundamentally from the standard manner of dealing with risk, which normally multiples the magnitude of possible losses by the probability of the losses occurring. My single most crucial claim here is that we ought not to discount huge possible losses by their probability when the likelihood of their occurrence is above some threshold level.[36] We cannot spend vast sums to prevent every catastrophe that is simply conceivable or barely possible. The likelihood must rise above a minimum threshold, as I have repeatedly emphasized. In the case of climate change, I believe this threshold is passed when (a) a mechanism and (b) emerging conditions for its working have been established. This is surely not the only basis on which the threshold can be satisfied.[37] But the essential point is that once the threshold is passed, one takes vigorous action until—third feature—the costs of doing so become excessive. In sum, reasonable costs of action are relative to how massive the possible losses are if the expenditures are not made and to how great the losses are if the expenditures are diverted from other important uses, but reasonable costs of action are not directly relative to the probability of those losses occurring when the possible losses are massive and their likelihood is above a minimal threshold. One does not discount by the probability; one checks to see whether there is a significant likelihood, based on solid evidence, that massive losses may occur. If so, one takes preventive action.

If it is certain that one person will die, one can say that the probability of a death is 1 and the magnitude of death is 1; on the usual way of calculating risk, $1 \times 1 = 1$. If 1,000,000 people might die, but the probability is known to be 0.000001, the usual calculation of risk is: $1,000,000 \times 0.000001 = 1$. Arithmetically, the two risks are equal. I have my doubts about whether we ought to respond to a one-in-a-million chance that one million people will die in the same manner that we respond to the certainty that one person will die; I am inclined to think that we should do much more in the case of the possible deaths of one million. This, however, is familiar and contentious territory. What I am claiming here is the following: (1) if we know that one million people might die, and we know that the likelihood is significant, then we should take action to prevent the million deaths until the costs of those actions are clearly excessive; and (2) one way we know that the likelihood is significant is when (a) we understand the mechanism by which the deaths are likely to occur and (b) we have begun to create the conditions that lead the mechanism to function.[38]

Much more needs to be worked out before we can judge with clarity how vigorous, expensive, and urgent our efforts ought to be to reduce our chances of making our own descendants miserable. For now, however, we are in absolutely no danger of overshooting and simply need to make a serious beginning. And we have seen that our responsibilities for the climate change we are producing are of a different, more demanding, kind from the responsibilities conventionally assumed, even by those who acknowledge our responsibility. Two aspects above all are clear: (1) we are called upon not only to provide security for the members of humanity who live later but also to refrain from causing them dangers; and (2) even if the worst does not eventuate, the lesser dangers we may cause are quite sufficient to ground responsibility for robust action now.[39]

THE MOST ESSENTIAL PRECAUTION

What specifically should we do? Here is where the science really matters. The single most important fact about climate change will be the historic peak level of atmospheric concentration of greenhouse gases, and what is crucial to where the concentration peaks is the percentage of the carbon now safely sequestered underground in the form of coal and oil that are extracted and injected into the atmosphere as CO_2. Of course, other GHGs matter as well. However, if we burn all of the fossil fuel under the surface of the earth, the atmospheric concentration of CO_2 will quadruple.[40] Business as usual is misleadingly packaged for PR purposes as the 'preservation of diversity' in energy sources.[41]

Either the carbon under the planet's surface is injected into the air through burning or not. It can be kept out of the atmosphere either by being left where it is now under the ground or the sea or by being burned only after effective carbon-sequestration techniques are developed. The opposition of interests is sharp: what is good for those who want all of the carbon extracted and burned with or without effective sequestration is bad for the climate and for the other 99.999 per cent of humanity. And waiting for the price to rise until fossil fuels become non-competitive greatly risks—as far as I can see, guarantees—that too much carbon will already have been injected into the planet's layer of GHGs before the price rises high enough to cut demand. The friends of fossil fuel—the carbon peddlers—have joined the enemies of humanity.

That is a strong statement. The grounds for it are the underlying science, the physical dynamics of climate. Climate change is driven by the atmospheric concentration of GHGs; this is what determines how much radiation is trapped on the planet. The atmospheric concentration is driven by annual emissions in excess of those compatible with the climate humans evolved in adaptation to—call that the sustainable rate of GHG emissions. Every year that the annual rate of emissions is larger than the sustainable rate, the atmospheric concentration grows. That is the stinger: every year that we fail to bring carbon emissions (and other GHG emissions) down to a sustainable level, the atmospheric concentration expands and more heat is trapped inside it. The atmospheric concentration has been expanding now for a century and a half. In recent years, it has been ballooning faster almost every year: the rate of increase is increasing.[42] 'It is *very likely* that the average rates of increases in CO_2, as well as in the combined radiative forcing from CO_2, CH_4, and N_2O concentration increases, have been at least five times faster over the period from 1960 to 1999 than over any other forty-year period during the past two millennia prior to the industrial era.'[43]

Even if the rate of increase were not increasing, as it is, the underlying arithmetic would be inexorable. The relation between unsustainable annual emissions and the atmospheric concentration is roughly like the relation between annual budget deficits and the national debt, or like the relation between annual population growth and size of total population at stabilization. The longer it takes a country to go down to replacement levels of fertility—the more years of growth in population—the larger the population size when the country stops growing. The more years of budget deficits, the larger the national debt is when the budget is finally balanced. In the best circumstances imaginable, the more years of unsustainable emissions, the higher the atmospheric concentration of GHGs when the concentration stabilizes—if it ever does. If we do not stop until we have pumped all of the oil and dug all of the coal, we will have the largest possible level of carbon dioxide concentrated in the atmosphere. And as far as we can tell, the larger the atmospheric concentration, the greater the disruption of the climate to which humans were adapted. At a minimum, we create a risk of greater disruption.

Matters are worse in two respects. We face political inertia and physical inertia. One cannot change the energy regime overnight because the super rich who own and distribute the fossil fuels have powerful political friends and articulate intellectual defenders.[44] Politics guarantees that high carbon emissions will continue for some time. That is bad enough. But the physical problem of lead time, analogous to the cultivation period for flu vaccine, is almost unimaginably daunting. In general, the whole planetary mechanism of atmosphere, oceans, and surface-level weather has enormous inertia overall once it is moving in a particular direction.[45] This is not the kind of dynamic process that gets reversed in a hurry. But the worst news may be specifically about CO_2, the most important GHG:

A[n atmospheric] lifetime for CO_2 cannot be defined . . . The behaviour of CO_2 is completely different from the trace gases with well-defined lifetimes. Stabilisation of CO_2 emissions at current levels would result in a continuous increase of atmospheric CO_2 over the 21st century and beyond . . . In fact, only in the case of essentially complete elimination of emissions can the atmospheric concentration of CO_2 ultimately be stabilised at a constant level . . . More specifically, the rate of emission of CO_2 currently greatly exceeds its rate of removal, and the slow and incomplete removal implies that small to moderate reductions in its emissions would not result in the stabilisation of CO_2 concentrations, but rather would only reduce the rate of its growth in coming decades. A 10% reduction in CO_2 emissions would be expected to reduce the growth rate by 10%, while a 30% reduction in emissions would similarly reduce the growth rate of atmospheric CO_2 concentrations by 30%.[46]

I repeat the critical finding: 'only in the case of essentially complete elimination of emissions can the atmospheric concentration of CO_2 ultimately be stabilised at a constant level'. It is, therefore, urgent to move aggressively now to cut CO_2 emissions sharply.

This science has strong implications for how we think about policy toward climate change. We need to ask: 'What must we do now to keep the total atmospheric concentration below a dangerous level?' not 'By how much would we like to reduce our emissions?' We need to focus on the target, which is lowering the risk of great danger, and reason back along the means–ends connections to what we must do now.

And the costs of lowering the risk of severe threat can be affordable if action begins soon enough. The longer we wait to start, the more it is likely to cost and the more abrupt the reductions in emissions would later have to be in order to keep the atmospheric concentration below a dangerous level.[47] How much we will need to tighten our belts depends on how rapid the transition to alternative energy is. Defenders of the carbon status quo say that to reduce emissions as much as scientists suggest would decimate the economy by depriving it of energy. But that is only if the economy continues to be dependent on fossil fuel. The economy can remain vibrant, while we avoid potential danger, as long as its energy source is not fossil fuel. The key is to move away from fossil fuel sooner, not later, before price rises force a switch. We need to get down to sustainable levels of annual GHG emissions, not when oil or—heaven help the future—coal 'runs out' and not when its price rises too high, but as soon as possible, leaving as much carbon as possible in the ground, where it is harmless, or burning it only after we understand how to sequester the CO_2 for a very long time.[48]

We have been considering *imposing* risks on the vulnerable of the future. One natural objection would take the line: 'What do you mean, "imposing"? Will not future generations be able to make choices for themselves?' Well, they will choose from the range of options we leave them.[49] Here are two vital factors they cannot choose because these will have been determined by earlier generations like us: (1) the size of the atmospheric concentration of GHGs already present (and unlikely to decline significantly during the succeeding century insofar as CO_2 is a factor) and (2) the dominant energy regime. If they inherit, say, an atmospheric concentration triple pre-industrial revolution levels and a still entrenched fossil fuel regime such as the one we labour under now—still digging that coal and pumping that oil—the people of the future are screwed. We would have been complicit in the imposition of a range of choice containing no good options in two ways: (1) we would have made the atmospheric concentration larger than it needed ever to become, and (2) we would have cooperated in the maintenance of an antiquated and corrosive fossil fuel regime, and the high-casualty foreign policies serving it, that humans need to escape from.

We do not have a long time left to do the job, even now. If we cannot soon reverse the political inertia of favouritism toward fossil fuel interests, the date of technological transition—that is, the date when the atmospheric concentration of GHGs ceases to increase—recedes into the future, and the level at which the atmospheric concentration finally stabilizes grows meanwhile like a planetary cancer, condemning more and more people to environmental danger and potentially undermining the ecological preconditions for sustainable human economies.[50]

If we in the present allow the continuing acceleration of a steady deterioration in the climate, the generation of today's students—or shockingly, even my own generation—could turn out to have had it as good as it gets. For the well-being and security of humans, history could be all downhill from here. Philosophers and economists used to think of the problem of intergenerational justice as the problem of the just savings rate; the danger was that we might shortchange ourselves by saving or investing too much of our own resources for the sake of people in the future because each future generation would in any case—it was assumed—be better off than the previous one. So we needed to discount the value of benefits to people in the future. The spectre of climate change means, by contrast, that we may be confronting the issue of the just deterioration rate. How much worse off than the previous generation can we permit the next to be? And will we allow the deterioration to continue until critical thresholds for human security are passed? Economic sustainability has ecological preconditions (unless one makes the assumption of literally infinite substitutability, which is not unusual among conventional economists but fantastic nevertheless if extended to the entire environment, including climate).

One way to characterize in moral terms the choice to run a genuine risk of massive loss for those who follow us is that it would be the voluntary and knowing infliction of a grievous wrong. We would have chosen to leave open the possibility of great distress, or even disaster, when, at relatively little cost to ourselves, we could have closed off that possibility. We could have protected people in the future against threats to their well-being; instead, we would have increased the threats and left them vulnerable to threats they likely cannot handle. Yet, however appropriate this first moral characterization, most people do not respond well to being threatened with a guilty label, and there is no need to try to lay a guilt trip on our own generation.

OPPORTUNITY FOR A LEGACY OF SECURITY

A much more positive moral characterization of the situation we now face is equally appropriate: thanks to the remarkable ingenuity of the scientists of the present day, invaluable understanding of the dynamics of the planetary climate system has been gained that places us in the position to provide vital protection to people in the future who would very likely otherwise find it impossible to protect themselves. Apart from blind technological optimism, we have no grounds for expecting that humans in the next century would have the capacity to protect themselves if we do nothing toward that purpose. But we have the capacity to leave them a legacy of security instead of a legacy of danger.[51]

The spectacular opportunity opened to us by our new understanding of the climate—most important, the realization that we must not allow much more of the carbon under the soil and the sea to be injected into the atmosphere, and certainly not all of it—is that we can protect future generations by keeping as much as possible of the remaining fossil fuel right where it is now. Bottom line: Do not leave your descendants—and more important, the descendants of much poorer people, such as most people in Africa—in avoidable danger.

Instead, provide them with security. Create an energy regime that will leave as much as possible of the remaining sequestered carbon out of circulation.

We can have all we need economically, and much of what we want but do not need, while promptly moving away from burning fossil fuels to alternative energy sources. No vital interests are at stake in the choice among energy sources for those of us who do not own coal and oil. But many vital interests are at stake for those in the future whose fates are vulnerable to our choices. We can leave them social institutions that will protect them—in particular, a cleaned-up energy regime that does not vomit GHGs into the sky. An energy regime not based on fossil fuels will make the worst effects of climate change that are now increasingly likely once again nearly impossible. Let us seize the opportunity to bequeath this magnificent gift of protection against vulnerability.

NOTES

1. 'Between a quarter and a third of the world's wildlife has been lost since 1970, according to data compiled by the Zoological Society of London.' See 'Wildlife Populations "Plummeting"', *BBC News*, 16 May 2008. Available at <http://new svote.bbc.co.uk/mpapps/pagetools/print/news.bbc.co.uk/2/hi/uk_news/ 7403989>. For an attempt to calculate the purely economic value of the loss of biodiversity now to be expected, see Pavan Sukhdev, *The Economics of Ecosystems & Biodiversity (TEEB): An Interim Report* (European Communities, 2008). This is being described as the *Stern Review* (see note 20 below) for biodiversity and is available online at <http://ec.europa.eu/environment/nature/biodiversity/econom ics/pdf/tceb_report.pdf>. Also see <http://ec.europa.eu/environment/nature/bio diversity/economics/index_en.htm>.
2. In accepting Article 2 of the Framework Convention on Climate Change in 1992, world leaders committed themselves to achieve 'stabilization of greenhouse gas concentrations in the atmosphere at a level that would prevent dangerous an-thropogenic interference with the climate system'. From <http://unfccc.int/essen tial_background/convention/background/items/1349.php>. So far, they have failed to stabilize concentrations at all.
3. Many philosophers are preoccupied with what is known as the non-identity problem, which was formulated by Thomas Schwartz in two contemporaneous pieces, 'Obligations to Posterity', in *Obligations to Future Generations*, ed. R. I. Sikora and Brian Barry (Philadelphia: Temple University Press, 1978), pp. 3–13; and 'Welfare Judgments and Future Generations', *Theory and Decision* 11 (1979): 181–94; and by Derek Parfit, *Reasons and Persons* (New York: Oxford University Press, 1984). As far as I can see, individual non-identity has no implications at all for what we ought to do. At most, it has some implications for how we explain our moral judgements.

4. How to think about the imposition of risk is an exceedingly difficult question. One valuable collection is Paul Slovic, ed., *The Perception of Risk* (London and Sterling, VA: Earthscan, 2000).

5. Andrew C. Revkin, 'New Climate Report Foresees Big Changes', *New York Times*, 28 May 2008; and US Climate Change Science Program, *Weather and Climate Extremes in a Changing Climate*, Final Report, Synthesis and Assessment Product 3.3 (June 2008). Available at <http://www.climatescience.gov/Library/sap/sap3-3/final-report/default.htm>.

6. 'There is strong evidence that global sea level gradually rose in the 20th century and is currently rising at an increased rate, after a period of little change between AD 0 and AD 1900. Sea level is projected to rise at an even greater rate in this century.' Nathaniel L. Bindoff, Jürgen Willebrand, Vincenzo Artale, et al., 'Observations: Oceanic Climate Change and Sea Level', in *Climate Change 2007: The Physical Science Basis, Contribution of Working Group I to the Fourth Assessment Report of the Intergovernmental Panel on Climate Change*, ed. Susan Solomon, Dahe Qin, Martin Manning, et al. (Cambridge and New York: Cambridge University Press, 2007), pp. 408–14, at p. 409.

7. The chunk of Antarctic ice that collapsed in February–March 2008 was a piece of an already floating ice shelf, the Wilkins ice shelf, and will not contribute to sea-level rise. See Associated Press, 'Chunk of Antarctic Ice Collapses', *New York Times*, 26 March 2008.

8. See Steve Vanderheiden, *Atmospheric Justice: A Political Theory of Climate Change* (New York: Oxford University Press, 2008), pp. 192–202. He documents some of the intentional deceptions of US citizens about the science for which the Bush–Cheney White House was notorious.

9. Obviously, the third feature is not independent of the first but proportional to it. My three features fit generally within what Neil A. Manson has called 'the three-part structure of the precautionary principle', consisting of a damage condition, a knowledge condition, and a remedy; see his penetrating article 'Formulating the Precautionary Principle', *Environmental Ethics* 24 (2002): 263–74. My suggestion is also, I think, a variant of what Stephen M. Gardiner very fruitfully isolates as a 'core precautionary principle'; see 'A Core Precautionary Principle', *Journal of Political Philosophy* 14 (2006): 33–60. Whether either Manson or Gardiner would accept my specific suggestion is, of course, another matter. I will for the most part ignore the more general philosophical issues that they perceptively consider, although I do not believe I run foul of anything they have established.

10. The suggestion is thus parallel to the suggestion of a sufficient set of conditions for states to take responsibility for the effects of their actions on affected people outside the territory they govern in Henry Shue, 'Eroding sovereignty', this volume. In both cases, the idea is that one has responsibility to protect against what might otherwise be the effects of one's own actions the people who would be vulnerable and unable to protect themselves, in the one case across space (including national boundaries) and in the other case across time.

11. I realize that not everyone accepts that there are objective probabilities.

12. I am not hinting at any connection between climate change and bird flu; this is a comparison.

13. Gina Kolata, *Flu: The Story of the Great Influenza Pandemic of 1918 and the Search for the Virus that Caused It* (New York: Farrar, Straus and Giroux, 1999).

14. The conduct of the US occupation of Iraq currently costs $16 billion per month (the annual budget of the United Nations). Is one more likely to be killed by a terrorist based in Iraq or by bird flu?

15. Stephen II. Schneider, Serguei Semenov, Anand Patwardhan, et al., 'Assessing Key Vulnerabilities and the Risk from Climate Change', in *Climate Change 2007: Impacts, Adaptation and Vulnerability, Contribution of Working Group II to the Fourth Assessment Report of the Intergovernmental Panel on Climate Change*, ed. Martin Parry, Osvaldo Canziani, Jean Palutikof, et al. (Cambridge: Cambridge University Press, 2007), pp. 779–810. And see generally Cynthia Rosenzweig and Daniel Hillel, *Climate Change and the Global Harvest: Potential Impacts of the Greenhouse Effect on Agriculture* (New York: Oxford University Press, 1998), and *Climate Variability and the Global Harvest: Impacts of El Niño and Other Oscillations on Agro-ecosystems* (New York: Oxford University Press, 2008).

16. See, for example, David Fogarty, 'Warming Globe to Test Farmers' Adaptability', *International Herald Tribune*, 5 May 2008, p. 13.

17. Amartya K. Sen, *Poverty and Famines: An Essay on Entitlement and Deprivation* (New York: Oxford University Press, 1983).

18. Hervé Le Treut, Richard Somerville, Ulrich Cubasch, et al., 'Historical Overview of Climate Change Science', in *Climate Change 2007: The Physical Science Basis, Contribution of Working Group I to the Fourth Assessment Report of the Intergovernmental Panel on Climate Change*, ed. Susan Solomon, Dahe Qin, Martin Manning, et al. (Cambridge and New York: Cambridge University Press, 2007), pp. 93–127. Also see Andrew C. Revkin, 'Strong Action Urged to Curb Warming', *New York Times*, 11 June 2008.

19. For evidence that large reductions in greenhouse gas emissions by the United States are possible at low cost, see Jon Creyts, Anton Derkach, Scott Nyquist, et al., *Reducing U.S. Greenhouse Gas Emissions: How Much at What Cost?* US Greenhouse Gas Abatement Mapping Initiative (Chicago: McKinsey, 2007).

20. For one argument that the costs of mitigation, or abatement, will only go up, and steeply, see Nicholas Stern, *The Economics of Climate Change: The Stern Review* (Cambridge: Cambridge University Press, 2007).

21. This is a different formulation of the general thesis I advanced in 'Climate', this volume.

22. For one argument for subsistence rights, see Henry Shue, *Basic Rights: Subsistence, Affluence, and U.S. Foreign Policy*, 2nd edn (Princeton, NJ, and Oxford.: Princeton University Press, 1996).

23. See 'Responsibility to future generations and the technological transition', this volume.

24. See 'Permafrost Threatened by Rapid Retreat of Arctic Sea Ice, NCAR/NSIDC Study Finds', *National Snow and Ice Data Center Media Advisory*, 10 June 2008.

25. John P. McGeehin, John Barron, David M. Anderson, and David Verardo, *Abrupt Climate Change* (Washington, DC: US Climate Change Science Program, 2008),

p. 1. Also see Hans Joachim Schnellnhuber et al., eds, *Avoiding Dangerous Climate Change* (Cambridge: Cambridge University Press, 2006).

26. See Richard B. Alley, *The Two-Mile Time Machine: Ice Cores, Abrupt Climate Change, and Our Future* (Princeton, NJ: Princeton University Press, 2000); US National Academy of Sciences, National Research Council, Committee on Abrupt Climate Change, *Abrupt Climate Change: Inevitable Surprises* (Washington, DC: National Academy Press, 2002), pp. v, 24–36; M. Vellinga and R. A. Wood, 'Global Climatic Impacts of a Collapse of the Atlantic Thermohaline Circulation', *Climatic Change* 54.3 (2002): 251–67; and Eystein Jansen, Jonathan Overpeck, Keith R. Briffa, et al., 'Palaeoclimate', in *Climate Change 2007: The Physical Science Basis, Contribution of Working Group I to the Fourth Assessment Report of the Intergovernmental Panel on Climate Change*, ed. Susan Solomon, Dahe Qin, Martin Manning, et al. (Cambridge and New York: Cambridge University Press, 2007), pp. 433–97.

27. Gerald A. Meehl, Thomas F. Stocker, William D. Collins, et al., 'Global Climate Projections', in *Climate Change 2007: The Physical Science Basis, Contribution of Working Group I to the Fourth Assessment Report of the Intergovernmental Panel on Climate Change*, ed. Susan Solomon, Dahe Qin, Martin Manning, et al. (Cambridge and New York: Cambridge University Press, 2007), p. 818. They add: 'Catastrophic scenarios suggesting the beginning of an ice age triggered by a shutdown of the MOC [meridional overturning circulation] are thus mere speculations' (p. 818). Compare Juliet Eilperin, 'Debate on Climate Shifts to Issue of Irreparable Change: Some Experts on Global Warming Foresee "Tipping Point" When It Is Too Late to Act', *Washington Post,* 29 January 2006, A1.

28. See Creyts, Derkach, Nyquist, et al., *Reducing U.S. Greenhouse Gas Emissions.*

29. A distinction between 'survival emissions' and 'luxury emissions' was advocated in Anil Agarwal and Sunita Narain, *Global Warming in an Unequal World: A Case of Environmental Colonialism* (New Delhi: Centre for Science and Environment, 1991), p. 5. I pursued their suggestion in 'Subsistence emissions and luxury emissions', this volume. I did not think of these two kinds of emissions as exhaustive: many emissions are neither subsistence nor luxury. Steve Vanderheiden treats 'survival emissions' and 'luxury emissions' as exhaustive categories in *Atmospheric Justice*, pp. 67–73, 242–3.

30. For provocative reflections on the value of a simpler life, see Duane Elgin, *Voluntary Simplicity: Toward a Way of Life That Is Outwardly Simple, Inwardly Rich*, rev. edn (New York: William Morrow, 1993); Wallace Kaufman, *Coming Out of the Woods: The Solitary Life of a Maverick Naturalist* (Cambridge, MA: Perseus, 2000); and David E. Shi, *The Simple Life: Plain Living and High Thinking in American Culture* (New York and Oxford: Oxford University Press, 1985).

31. One can adopt a common metric and then quantify, of course, as some economists do, but the choice of common metric is determined largely by convenience, and the assignments of relative value are highly arbitrary. How many dollars is a decline in the quality of nutrition in Myanmar in 2410 worth now?

32. Nicole Hassoun knows how tempted, and I am grateful to her for discussions of the issues underlying the three conditions.

33. See Thomas Pogge, 'Severe Poverty as a Human Rights Violation', in *Freedom from Poverty as a Human Right*, ed. Thomas Pogge (Oxford: Oxford University Press, 2007), pp. 11–53.

34. As I write, the Bush–Cheney 'war of choice'—the totally unnecessary, ill-considered, and egregiously counterproductive plunge into the invasion and occupation of Iraq— is costing, as already mentioned, $16 billion per month in operating costs. The elimination of such murderous folly would free up vast sums. One year of the budget for the Iraq occupation would take care of chronic poverty plus several years of vigorous action on climate change. For fuller consideration of some of the issues underlying the precipitate rush into war in Iraq, see Henry Shue and David Rodin, eds., *Preemption: Military Action and Moral Justification* (Oxford: Oxford University Press, 2007).

35. Bryan G. Norton makes this general point in defence of his preferred guide, the safe minimum standard (SMS): 'save the resource, provided the costs of doing so are bearable'; see *Sustainability: A Philosophy of Adaptive Ecosystem Management* (Chicago: University of Chicago Press, 2005), p. 346. Similarly, he builds affordability into his statement of the precautionary principle: 'take affordable steps to avoid catastrophe tomorrow' (p. 352). 'Bearability' and 'affordability' depend, for Norton, on which other extremely important matters also require resources.

36. Weapons of mass destruction (WMD) in Iraq could have fitted the same principle if, for example, competent UN inspectors had been finding substantial evidence of their existence. See Shue and Rodin, *Preemption*. It is clear, however, that this eagerly launched war had entirely different purposes and that the case based on WMD was politically concocted as a pretext; see Mark Danner, *The Secret Way to War: The Downing Street Memo and the Iraq War's Buried History* (New York: New York Review Books, 2006). It is profoundly ironic, and deeply evil, that the Bush–Cheney administration, while suppressing incontrovertible evidence of climate change gathered by thousands of the world's best scientists—see Vanderheiden, *Atmospheric Justice*, ch. 6—was manufacturing phony evidence to justify a war that is consuming many times the resources needed to deal with the real problem they vigorously covered up. Bare politically motivated assertion was accepted as the justification for what has become the longest war in American history, while a powerful body of scientific support for action to mitigate climate change was simply denied. See Mark Mazzetti and Scott Shane, 'Bush Overstated Evidence on Iraq, Senators Report', *New York Times*, 6 June 2008; and US Senate, 110th Congress, 2nd Session, Select Committee on Intelligence, 'Report on Intelligence Activities Relating to Iraq' (June 2008). Even NASA's press releases were for years distorted by political appointees; see Andrew C. Revkin, 'NASA Office Is Criticized on Climate Reports', *New York Times*, 3 June 2008.

37. I would think that one of the most productive avenues to explore is what other grounds might satisfy the second condition. The generic form of the second condition is that the likelihood of the losses specified in the first condition must be above some minimum threshold. The specific form this condition takes in the cases of bird flu and climate change is mechanism and emerging conditions for functioning of the mechanism. But the generic condition could well be instantiated by other specific forms of threshold.

38. One million deaths is, of course, simply an example. There are other kinds of 'massive losses' besides deaths, such as calamitous declines in standards of living or the undermining of civilization. If no one died but everyone had to live like cavemen, that would be a massive loss.

39. A splendid survey and analysis of the literature on the moral case is Stephen M. Gardiner, 'Ethics and Global Climate Change', *Ethics* 114.3 (April 2004): 555–600. I tried to establish that each of three independently persuasive arguments all led in this same general direction in 'Global environment and international inequality', this volume.

40. James F. Kasting, 'The Carbon Cycle, Climate, and the Long-Term Effects of Fossil Fuel Burning', *Consequences: The Nature & Implications of Environmental Change*, vol. 4, no. 1 (1998). Available at <http://www.gcrio.org/CONSEQUENCES/vol4no1/carbcycle.html>.

41. TV ads in 2008 successfully opposed effective action on climate change in the US Senate and were sponsored by the coal and oil lobbies. ExxonMobil ads, for example, say, 'We're going to need them all,' meaning, 'Do not cut back on oil.'

42. See David Adam, 'World Carbon Dioxide Levels Highest for 650000 Years, Says US Report', *Guardian*, 13 May 2008; and US Department of Commerce, National Oceanic and Atmospheric Administration, Earth System Research Laboratory, Global Monitoring Division, 'Trends in Atmospheric Carbon Dioxide—Mauna Loa'. Available at <http://www.esrl.noaa.gov/gmd/ccgg/trends/>.

43. Jansen, Overpeck, Briffa, et al., 'Palaeoclimate', p. 436. Emphasis in original; 'very likely' is used technically in that volume to mean a greater than 90 per cent probability; see p. 23.

44. For the general picture, see Daniel Yergin, *The Prize: The Epic Quest for Oil, Money, & Power* (New York: Simon & Schuster, 1992); and Michael T. Klare, *Resource Wars: The New Landscape of Global Conflict* (New York: Henry Holt, 2001).

45. See Meehl, Stocker, Collins, et al., 'Global Climate Projections', pp. 747–843.

46. Meehl, Stocker, Collins, et al., 'Global Climate Projections', pp. 824–5.

47. Nicholas Stern, 'The Economics of Climate Change', *American Economic Review* 98.2 (May 2008): 1–37.

48. It is not clear that there is any such thing as 'running out', in any case. If coal becomes expensive enough, it may simply join its prettier cousin the diamond as a luxury good.

49. See the analysis of this as the 'domination of posterity' in John Nolt, 'Greenhouse Gas Emission and the Domination of Posterity', in *The Ethics of Global Climate Change*, ed. Denis Arnold (New York: Cambridge University Press, 2011).

50. For the initial development of the conception of the technological transition, see 'Responsibility to future generations', this volume.

51. For an account of how the clean development mechanism of the Kyoto Protocol is failing to deal with the danger, see 'A legacy of danger: the Kyoto Protocol and future generations', this volume.

15

Face reality? After you! A call for leadership on climate change*

In Joseph Heller's comic war novel, *Catch-22*, the catch-22 of the title refers to a supposed military regulation that allowed one to be relieved of military service if one was insane, but further provided that no one who realized he would be better off out of military service could possibly be insane. Humanity's so far leaderless approach to dealing with rapidly accelerating climate change embodies a similar, but profoundly tragic, catch-22 that has, among other twists and contradictions, transmuted justice into paralysis.

Many thought that the natural global leader of the effort to gain control of global climate change would be the United States, with its splendid cadre of scientists and its history of technological innovation. But our politicians have failed to be worthy of our scientists or of the trust we citizens have placed in them. Facing reality appears to be increasingly unpopular among those who pass as our national political leaders. Those who refuse to face reality often find that what they ignore may come back to bite them, and worse, it may hurt others who trust them with their well-being. It is unclear which members of the US Senate have sold their souls to the fossil fuel interests and which have simply closed their minds. But the effect is the same: the facts on the ground—and in the air, water, and ice of the planet—are racing further and further ahead of the faltering US political efforts to respond to them. And the American failure of political leadership is one major factor that is crippling efforts to negotiate multilateral action at the international level.

The United States is not, however, the only laggard nation when it comes to the threat of increasingly rapid climate change. While the fact that many developing as well as developed countries, including the United States, were willing to make voluntary pledges of action in response to the Copenhagen Accord provides a small foothold for hope,[1] the pledges are not even remotely adequate for reaching their own stated goal.[2] Moreover, the fulfilment of many

---

* 'Face Reality? After You! A Call for Leadership on Climate Change', *Ethics & International Affairs*, 25:1 (2011), Carnegie Council, 17–26. This chapter has endnotes.

of those pledges has been made contingent on what others do: the European Union, New Zealand, and Norway, for example, promise to act more aggressively only after 'the achievement of a global and comprehensive agreement'.[3] This is one example of catch-22 absurdity, given that it was the failure at Copenhagen to reach a global and comprehensive agreement that forced the world to settle for the weak Copenhagen Accord under which these highly conditional pledges are now being made. The Copenhagen Accord was supposed to be a substitute for the global agreement not reached, not to be dependent itself on some elusive agreement.

While our feckless leaders reason in circles, the problem itself deepens relentlessly, continuing to exceed the conservative estimates made by the Intergovernmental Panel on Climate Change (IPCC), burdened as it has always been with the need to reach political consensus on its reports. For example, the *New York Times* reports that surprisingly warm water around Greenland is forcing highly reputable scientists to confront the possibility that sea-level rise during this century may be 3 feet or more—far exceeding the 8 inches of rise that occurred during the twentieth century.[4] Such a rise would have dramatic effects on New York, London, Shanghai, and other major cities, not to mention low-lying nations, such as Bangladesh and the Maldives. As the author of the report notes, we do not know the probabilities of various sea-level rises because we cannot be bothered even to fund the necessary scientific research; but the conditions for such a dramatic sea-level rise may be coming together in reality, with or without our knowledge. The climate change deniers want us to keep burning coal and pumping oil until the prices rise so high because of scarcity of supply that we can no longer afford them. At that point the owners of the coal and oil will have made all the profit they can reasonably expect, and more. Meanwhile, our leaders are not worried because the extent and speed of climate change are uncertain; they are uncertain because we do not do the science; and we do not do the science because our leaders are not sufficiently worried.

But what we need is not only more research, however desperately important that is, but also action, even in the face of the remaining uncertainty, which science can meanwhile progressively reduce. 'Uncertainty' is the ever-welcome battle cry of the well-financed professional deniers of climate realities.[5] Beyond what Steve Vanderheiden calls the 'manufactured uncertainty', which simply throws sand into our eyes, uncertainty comes in many forms and degrees. As I have noted elsewhere, the genuine uncertainty in the case of climate change takes a distinctive form with three features:

> (1) *massive loss*: the magnitude of the possible losses is massive; (2) *threshold likelihood*: the likelihood of the losses is significant, even if no precise probability can be specified, because (a) the mechanism by which the losses would occur is well understood, and (b) the conditions for the functioning of the mechanism

are accumulating; and (3) *non-excessive costs*: the costs of prevention are not excessive (a) in light of the magnitude of the possible losses, and (b) even considering the other important demands on our resources. Where these three features are all present, one ought to try urgently to make the outcome progressively more unlikely until the marginal costs of further efforts become excessive, irrespective of the outcome's precise prior probability, which may not be known in any case. We know that our actions now are opening the doors to some terrible outcomes; we ought to re-close as many of these doors as we can. The suggestion, then, is that these three features jointly constitute a sufficient set [of reasons] for requiring prompt and robust action.[6]

Why should action to mitigate climate change be so urgent and vigorous? Beyond the arguments just mentioned, it is helpful to separate analytically the question of why action is urgent now from the question of why we ought to be the ones to act.[7] One of the 'inconvenient truths' that the professional deniers of climate change rarely mention is the recently revised understanding of what the scientists call the 'atmospheric residence time' of carbon dioxide, which is of course the main greenhouse gas produced by the burning of fossil fuels. The atmospheric residence time of a greenhouse gas is the length of time an average molecule remains in the atmosphere once it reaches it. As recently as twenty years ago, we all believed that the atmospheric residence time of carbon dioxide was about a century, which already seemed a very long time for the effects of a seemingly transient emission to last. Now climate scientists realize that the atmospheric residence time of carbon dioxide in particular (different greenhouse gases have different 'lives') is more like a millennium.[8]

From an ethical perspective, this means that even if we, the current generations, were in no way at all responsible or accountable for the acceleration in the speed of climate change, there would be what might be thought of as a Good Samaritan reason why we ought to be the ones to perform urgent action now. This first reason is simply that we are the ones who are here. The Good Samaritan had no 'prior history' with the man in the ditch, and certainly no responsibility for his fate. But the man would remain in the ditch unless someone helped him out—there and then. The accumulation of carbon dioxide in the Earth's atmosphere will continue to increase unless those of us in the here and now cut back our emissions of it. There is no one else to do it: it is our fate to be alive when the problem of climate change has first been understood. Since carbon dioxide remains in the atmosphere for so very long, any addition to the atmospheric accumulation is mostly a net addition: that is, the cumulative total keeps ballooning because the natural removal occurs vastly more slowly than the anthropogenic injection. The best validated climate models all agree that the peak atmospheric concentration of greenhouse gases will determine the extent of the surface disturbances: the larger the maximum concentration, the higher the surface temperature, the more extreme the surface storms, and so on. We could be the ones to set the globe onto a path to a maximum concentration that is lower than it would otherwise be with business as usual—because now we understand the problem, and we are here now. This is a golden opportunity to leave a positive legacy—to be historic global Good Samaritans.

So, as I say, we would have reason to act even if we bore no prior responsibility. But of course, unlike the Good Samaritan, we are not confronting an opportunity to help with a problem that is not of our own making. The evidence that the current climate change is anthropogenic is solid and growing. And it is not merely anthropogenic: much of it originates in the United States in particular. There are two reasons for this, one concerning the past and one concerning the present.

First, it is estimated that during the entire twentieth century the United States produced 30.3 per cent of the entire world's total carbon dioxide emissions from the burning of fossil fuels, closely followed by Europe's 27.7 per cent. These two centres of wealth together contributed over half the global total of greenhouse gases, while, for example, China, India, and the rest of developing Asia, despite being home to far more people, contributed only 12.2 per cent over the same period.[9] Two reasons are often offered to support the standard defence that the West's awesome causal responsibility generates no moral responsibility: (a) we did not know that we were interfering with the dynamics of the planet and so did not intend the damage; and (b) now everyone else, most notably China, is doing it, too.

As to the first point, it is true that until about a quarter century ago we did not fully appreciate the damage we were doing to the earth's delicate balance by so radically modifying the composition of the planet's atmosphere. Perhaps life is generally unfair. Perhaps it was a dirty trick that the planet should have had so much coal and oil when burning the stuff wrecks the planet's atmosphere, and the arrangements are unfair to everyone. But here we are, and the question we now must face is: Given that fossil fuel emissions must be rapidly and radically reduced if global temperatures are not to soar from carbon emissions, and that the prompt transition to alternative fuels will involve burdens on humanity, what is the fairest way to distribute these burdens— the issue of cosmic unfairness aside? Who should more reasonably bear the heaviest burdens: those nations whose Industrial Revolution contributed (unknown to them, admittedly) the greatest percentage of the carbon dioxide emissions and who became wealthy and powerful in the process, or those nations that produced very few emissions (equally not knowing that it mattered) and are only now developing and becoming wealthy and possibly powerful?[10]

As to the second point, the fact that everyone—or, at least, increasingly many nations—is now 'doing it' is simply irrelevant to responsibility for destructive activity in the past. This cannot be a reason for the United States to continue to do nothing now, in the present, at the national level. It is a reason, however, why everyone now ought to contribute appropriately to a solution to the problem. (This, by the way, includes India and, in particular, China—to whom the United States, as well as Australia and others, certainly ought not to be increasing exports of coal that are self-righteously not burned domestically.[11]) I shall say more shortly on the implications of what I mean by 'appropriately'.

Most flagrantly, the United States still has no comprehensive national policy to limit either greenhouse gas emissions in general or carbon dioxide emissions in particular, despite all we now know about climate change and its current and likely future human effects. That we persist in emissions without limits is an appalling political failure. US emissions have continued to increase decade after decade in spite of our progressively clearer understanding of their destructive effects on the global climate. This complete failure to act at the national level has moved beyond fecklessness to recklessness.

The absence of a national US policy on greenhouse gas emissions is thus not a failure to volunteer to help with a problem not of our own making, but an increasingly egregious exacerbation of a problem that we have been a major factor in creating and continue to be a major factor in worsening. The fact that we are not merely declining to be benevolent or charitable, or even simply failing to fulfil a general duty that falls on everyone, but are persistently and increasingly engaging in environmentally destructive behaviour has a fundamental implication for the central catch-22 that I would like to sketch here.

PERVERTING JUSTICE INTO PARALYSIS

Why is the United States doing nothing to slow, or even to stop accelerating, a damaging practice that is likely to impoverish our own offspring and all the people who live after us? Reasons abound. For example, the January 2010 decision of the US Supreme Court in *Citizens United v. Federal Election Commission* flung open the floodgates for secret corporate contributions to the campaigns of congressional candidates who favour fossil fuel interests and who oppose climate change legislation, as well as to groups who lobby against action on climate change.[12] And the teaching of science in many US secondary schools leaves students ill-prepared to understand climate science and to evaluate critically the sugar-coated arguments of the professional deniers. The combination of these two factors—inadequate understanding of the science and well-financed campaigns to undermine the science—produces a public discourse that often does not even focus on the real issues.[13]

The explanation for the US policy failure involves many more factors,[14] but there is a single crucial factor, a bad moral argument, that I think plays a significant role in salving political consciences. This is the central catch-22 that perverts a concern with justice into paralysis. What I have above called the 'appropriate' sharing of the burden has been at the heart of international negotiations on climate change from the very beginning. 'Common but differentiated responsibilities' is the famous phrase from Article 4.1 of the 1992 United Nations Framework Convention on Climate Change.[15] The idea behind the phrase is that all parties have responsibilities, but parties in different situations have different responsibilities. This is fine in the abstract, but it obviously cries out for specification. The political battles have concerned the precise sharing of responsibilities, and those battles, as seen in Copenhagen, for instance, seem to have degenerated into the following impasse: The less wealthy countries say to the more wealthy ones, We will take action only if you provide assistance. And the more wealthy countries say, We will provide assistance only if you take action. 'After you!' repeats each side. Such posturing is generally understandable bargaining behaviour. But if these postures remain frozen, they create the catch-22: the wealthier will not act until the poorer act, and the poorer will not act until the wealthier act.

Both sides tend to defend their bargaining positions as representing nothing worse than an insistence on not doing more than one's fair share until others have done their fair share. In short, it is said to be simply a concern for justice. But this is an inaccurate characterization of the situation for the United States, as well as for a number of other parties. It is one thing to refuse to do *more* than one's own share until others have done, or have agreed to do, at least their fair shares. However, it is an entirely different matter to refuse to do *even* one's own share—that is, to refuse to do anything at all—until others have done or have agreed to do so as well. Ethically, one's minimum obligation is to do at least one's own fair share, irrespective of whether one should ever do more than one's fair share to compensate for the non-compliance of others. This minimum obligation is especially compelling when, as in the instance of climate change, one's share includes ceasing destructive activity that creates a danger for vulnerable others now and in the future.

Sometimes it can be a clever bargaining tactic to threaten to refuse to do *A* unless someone else does *B*. If the tactic works, both *A* and *B* get done, partly thanks to one's willingness to threaten not to do *A*. Whether this tactic is acceptable, however, depends on the concrete features of these abstract *A*s and *B*s, and the concrete situation generally. Let us suppose several people are bleeding to death, that it takes one rescuer per victim to stanch the flow of blood, and that each available rescuer is obligated to rescue one victim. If I am willing to fulfil my duty by rescuing one victim but am worried that no one else will rescue any other victims, I suppose I might threaten to let 'my' victim bleed to death unless other rescuers assist other victims, hoping thereby to be helpful to the other victims by provoking their rescues. But this does not seem a very clever tactic for this situation, and surely I should not carry out my threat to do nothing and allow the victim I could save to die if the others do not respond to my threat and assist 'their' victims.

The situation concerning climate is similar. The national economies of the world are haemorrhaging greenhouse gases, and the flows need to be stanched. Some flows are larger and have been going on for longer than other flows. It would be helpful to have a comprehensive agreement about who is going to do what. But it is not a clever strategy for solving the problem to stubbornly refuse to do anything at all until a comprehensive agreement is reached, especially if it is blindingly obvious that many others are attempting to employ this same strategy, producing paralysis.

What is needed to break the current stalemate is leadership.[16] We need one state to break the paralysis by unilaterally (if necessary) taking action in the hope that the others will respond to its example—and to their own comprehension of the inherent importance of the problem. In this way, one can then say not 'After you,' lest I be treated unfairly, but 'I went first, so now you,' lest you treat me unfairly—still appealing to fairness. This is obviously not a universally effective tactic, either, but it is particularly appropriate when one already has a moral duty to act and it is abundantly clear that one's fair share of the burden is well in excess of anything one has yet contributed.

This is precisely the situation regarding the United States and climate change. The dangers to the planetary environment are now well documented and exceptionally urgent.[17] Every nation has an obligation to do (at least) its own fair share in limiting emissions. The fair share of the United States is patently greater than the effective nothing that it is currently doing at the national level. So the United States should clearly be doing something; and if it acted boldly and decisively, its positive example seems much more likely to be effective than its failed defensive strategy of refusing to act until others do. In fact, some others, notably the European Union, are already doing a little, and many seem ready to act more vigorously if they see some leadership being exercised by one or more major powers.

The need for global leadership is so desperate that the duty to provide it falls on anyone who has the capacity to lead.[18] The Good Samaritan did not perform what philosophers now call a supererogatory act—an act above and beyond the call of duty. Rather, the duty to deal with the man in the ditch fell to whomever came along and was able to help. Ex ante, the Good Samaritan was no different from any other passer-by. He was distinguished in the end only by the fact that he chose to act on a duty that others had chosen to ignore. Whichever nation is capable of leading a change in direction away from fossil fuels has a duty to do so before our carbon emissions send us into our own climatic ditch, and this certainly includes the United States.

Like other nations with high cumulative carbon emissions, the United States already has a historical responsibility to act in order to undo the damage done. In addition, like any nation capable of leading the move away from a reliance on fossil fuel, the United States has a duty to provide vital leadership completely independent from its historical responsibility. It has, then, a double duty regarding the threat to the planet's climate: an underlying specific historical responsibility based on previous and ongoing contributions to the threat, enhanced by an additional responsibility based on its scientific knowledge and financial power to provide desperately needed leadership.

What can individual citizens do to bring about US leadership on climate change? First, work unrelentingly to replace climate reality-denying senators with people who understand the problem. This may require legislatively undoing the effects of *Citizens United v. Federal Election Commission* (another political catch-22?). Climate change is the overriding issue for the twenty-first century, and the country and the world need a US Senate capable of ratifying a reality-based treaty on climate change when one can be negotiated. Second, while the US government (except for the Environmental Protection Agency) has so far evaded action at the national level, action has been and is being taken at other levels, including such regional action as New England's Regional Greenhouse Gas Initiative (RGGI) and the West Coast's Western Climate Initiative, led by well-informed governors and Canadian provincial leaders;[19] and urban initiatives, especially international collaborations, such as the International Carbon Action Partnership, Cities for Climate Protection, and the Clinton Climate Initiative.[20] Time is not on our side: 'In the 10 years from 1995 to 2005 atmospheric CO_2 increased by 19 ppm [parts per million]; the highest average growth rate recorded for any decade since direct atmospheric CO_2 measurements began in the 1950s.'[21]

We cannot wait for ignorant, inattentive, or indifferent leaders, and we cannot afford to fall into poisonous pessimism. We must act where action is possible: and for now this is at the regional, state, and local levels. The good news is that these sub-national actors can reach out internationally to others who are rising to the challenge; for example, the RGGI might be able to engage in emissions trading with the EU and do an end run around Washington. If enough action is taken at all levels of US governance other than the national level, corporate interests may then perhaps lobby to eliminate the resulting 'patchwork' of regulations and be willing to accept a reasonable uniform national initiative. Finally, we must intently face reality, which means that we must listen to those who honestly and assiduously study the problems, not to those who are paid to preserve the carbon energy regime that so greatly contributes to them.[22]

NOTES

1. Daniel Bodansky, 'The Copenhagen Climate Change Conference: A Postmortem', *American Journal of International Law* 104 (2010), pp. 230–40.
2. Joeri Rogelj, Claudine Chen, Julia Nabel, et al., 'Analysis of the Copenhagen Accord Pledges and Its Global Climatic Impacts—A Snapshot of Dissonant Ambitions', *Environmental Research Letters* 5 (2010), pp. 1–9. See also note 16.
3. Lavanya Rajamani, 'The Making and Unmaking of the Copenhagen Accord', *International and Comparative Law Quarterly* 59 (July 2010), p. 837.
4. Justin Gillis, 'As Glaciers Melt, Science Seeks Data on Rising Seas', *New York Times*, 13 November 2010. See also Leslie Kaufman, 'Front-Line City in Virginia Starts Tackling Rise in Sea', *New York Times*, 26 November 2010.
5. See Naomi Oreskes and Erik M. Conway, *Merchants of Doubt: How a Handful of Scientists Obscured the Truth on Issues from Tobacco Smoke to Global Warming* (New York: Bloomsbury Press, 2010); and Steve Vanderheiden, *Atmospheric Justice: A Political Theory of Climate Change* (New York: Oxford University Press, 2008), pp. 38–44.
6. 'Deadly delays, saving opportunities: creating a more dangerous world?', this volume. See also Neil A. Manson, 'Formulating the Precautionary Principle', *Environmental Ethics* 24 (2002), pp. 263–74.
7. I am grateful to Simon Caney for pointing this out.
8. G. A. Meehl, T. F. Stocker, W. D. Collins, et al., 'Global Climate Projections', in S. Solomon, D. Qin, M. Manning, et al., eds, *Climate Change 2007: The Physical Science Basis*, Working Group I Contribution to the Fourth Assessment Report of the Intergovernmental Panel on Climate Change (Cambridge: Cambridge University Press, 2007), p. 824.
9. World Resources Institute, 'Earth Trends, Environmental Information: Contributions to Global Warming: Historic Carbon Dioxide Emissions from Fossil Fuel Combustion, 1900–1999'; available at <http://earthtrends.wri.org/maps_spatial/maps_detail_static.php?map_select=488&theme=3>.
10. This is argued more fully in Henry Shue, 'Historical Responsibility', Technical Briefing for Ad Hoc Working Group on Long-Term Cooperative Action under the Convention (AWG-LCA), SBSTA, UNFCC, Bonn, 4 June 2009; available at <http://unfccc.int/files/meetings/ad_hoc_working_groups/lca/application/pdf/1_shue_rev.pdf>.
11. Elisabeth Rosenthal, 'Nations That Debate Coal Use Export It to Feed China's Need', *New York Times*, 21 November 2010. See also Karl Gerth, *As China Goes, So Goes the World* (New York: Hill and Wang, 2010), esp. ch. 8.
12. *Citizens United v. Federal Election Commission*, 558 U.S. (2010).
13. Journalistic coverage of the science is also poor—see James Painter, *Summoned by Science: Reporting Climate Change at Copenhagen and Beyond* (Oxford: Reuters Institute for the Study of Journalism, 2010).
14. See Stephen M. Gardiner, *A Perfect Moral Storm: The Ethical Tragedy of Climate Change* (New York: Oxford University Press, 2011).
15. 'United Nations Framework Convention on Climate Change'; available at <http://unfccc.int/essential_background/convention/background/items/1349.php>. For thorough legal

analysis, see Lavanya Rajamani, *Differential Treatment in International Environmental Law* (Oxford: Oxford University Press, 2006).

16. See e.g. the central plea for leadership in United Nations Environment Programme, 'The Emissions Gap Report: Are the Copenhagen Accord Pledges Sufficient to Limit Global Warming to 2°C or 1.5°C?' November 2010. I am grateful to Anja Karnein for this source.

17. I attempt to explain more fully why they are urgent in Henry Shue, 'Human rights, climate change, and the trillionth ton', this volume.

18. I was provoked to think about the ethical status of this duty to lead by Anja Karnein.

19. See Leigh Raymond, 'The Emerging Revolution in Emissions Trading Policy', in Barry G. Rabe, ed., *Greenhouse Governance: Addressing Climate Change in America* (Washington, DC: Brookings Institution Press, 2010), pp. 101–25.

20. See Henrik Selin and Stacy D. Vandeveer, 'Multilevel Governance and Transatlantic Climate Change Politics', in Barry G. Rabe, ed., *Greenhouse Governance*, pp. 353–65.

21. P. Forster, V. Ramaswamy, P. Artaxo, et al., 'Changes in Atmospheric Constituents and in Radiative Forcing', in Solomon, Qin, Manning, et al., eds, *Climate Change 2007: The Physical Science* (Cambridge: Cambridge University Press, 2007), p. 137.

22. The Cancún Agreements, negotiated in December 2010, are pitifully weak, and the USA has once again settled for a lowest common denominator that is shamefully far below the minimum urgently needed now.

16

Human rights, climate change, and the trillionth ton*

The desultory, almost leisurely approach of most of the world's national states to climate change reflects no detectable sense of urgency. My question is what, if anything, is wrong with this persistent lack of urgency? My answer is that everything is wrong with it and, in particular, that it constitutes a violation of basic rights as well as a failure to seize a golden opportunity to protect rights. I criticized the outcome of the initial climate conference in Rio de Janeiro in 1982, the Framework Convention on Climate Change, for establishing 'no dates and no dollars: no dates are specified by which emissions are to be reduced by the wealthy states and no dollars are specified with which the wealthy states will assist the poor states to avoid an environmentally dirty development such as our own. The convention is toothless.'[1] The general response to such criticisms was that the convention outcome was a good start.

More than a quarter of a century later, the outcome of the 2009 Copenhagen Conference of the Parties to the Convention is equally toothless, once again containing no dates and no dollars. *The New York Times* described the twelve-paragraph Outcome Document as 'a statement of intention, not a binding pledge to begin taking action'.[2] It contains a vague commitment to the end of keeping the global temperature rise to no greater than 2°C beyond pre-industrial levels but no specification of, much less commitment to, the means necessary to that end.[3] And the USA has with great fanfare vaguely committed

---

* 'Human Rights, Climate Change, and the Trillionth Ton', by Henry Shue, in Denis G. Arnold (ed.), *The Ethics of Global Climate Change* (Copyright Cambridge: Cambridge Univ. Press, 2011), 292–314. Reprinted with Permission. 'Ton' here, denotes 'metric ton'.

[1] 'Subsistence emissions and luxury emissions', this volume.

[2] A. C. Revkin and J. M. Broder, 'A Grudging Accord in Climate Talks', *New York Times,* 20 December 2009.

[3] Para. 1 'recognizes' 'the scientific view that the increase in global temperature should be below 2 degrees Celsius.' Para. 12 requires the assessment of the implementation of the accord to 'include consideration of strengthening the long-term goal . . . including in relation to temperature rises of 1.5 degrees Celsius', whatever this means. See United Nations, Framework

itself to contributing to a fund to assist poorer nations, while providing no specification of how much it might put in the fund, and pitifully pledged laughable emissions cuts equivalent to a reduction of 4 per cent below 1990 levels.[4] Yet, once again, the defence of the outcome has been that it is a good start.

But after nearly a quarter of a century of delay in confronting the fact that we are ourselves creating an onrushing problem for those who come after us, good intentions, a first step, and a good start are not nearly good enough. After more than two decades of denial, the time has passed for perpetual first steps. Indeed, we wrong the people of tomorrow by doggedly persisting in contributing to conditions in which they will be unable to fulfil their basic rights. If we all continue simply to consume fossil fuels like there is no tomorrow, there may indeed be no tomorrow for those who will have to try to live in it. And if we persist in business as usual, we miss a spectacular opportunity to provide protection for people who come after us that they will be in no position to provide for themselves. Here is one general argument.

Rapid climate change places current and future generations in precisely the kind of general circumstances that call for the construction of rights-protecting institutions.[5] While the identities of future individuals are not yet determined and are thus not knowable,[6] we know that as humans they will all be entitled to human rights, including rights that depend on a functioning economic system, which itself depends in turn on a planetary environment within which those future humans can adapt and support themselves. They need, therefore, to inherit from past generations an environment that is neither radically inhospitable nor radically unpredictable. Now, however, we are producing through our carbon-fuelled economic activities increasingly rapid environmental change that currently has no fixed or otherwise

Convention on Climate Change, Conference of the Parties, Fifteenth Session, Agenda item 9, 'Draft Decision -/CP.15', FCCC/CP/2009/L.7 (18 December 2009).

[4] Bill McKibben, 'Heavy Weather in Copenhagen', *New York Review of Books*, 57:4 (11–24 March 2010), 32–4.

[5] For a reasonably accessible broad general introduction to the science of climate change, see J. Houghton, *Global Warming: The Complete Briefing*, 4th edn (Cambridge University Press, 2009). Also see Save the Children UK, *Legacy of Disasters: The Impact of Climate Change on Children* (London: Save the Children UK, 2007).

[6] Many philosophers have been much taken with what is known as the non-identity problem, which was formulated by Thomas Schwartz in two contemporaneous pieces, 'Obligations to Posterity', in R. I. Sikora and B. Barry (eds), *Obligations to Future Generations* (Philadelphia: Temple University Press, 1978), 3–13, and 'Welfare Judgments and Future Generations', *Theory and Decision*, 11 (1979), 181–94; and later by D. Parfit, *Reasons and Persons* (New York: Oxford University Press, 1984). As far as I can see, our current lack of knowledge of particular identities in later centuries has no implications at all for what we ought to do now about future climate change. At most, it has some implications for how we explain our more basic moral judgements. For a different view, see Edward A. Page, *Climate Change, Justice and Future Generations* (Northampton, MA.: Edward Elgar, 2006), 132–60.

predictable limit. The members of future generations of humans are completely vulnerable to the choices made by earlier generations, including ours, which is the very first generation in human history to have acquired the knowledge necessary to understand either the problem of climate change or possible solutions to the danger of excessively rapid change.

A situation in which some humans are utterly at the mercy of other humans, but those others have the capacity to create institutions to protect the vulnerable against the forces against which they cannot protect themselves, is the paradigmatic situation calling for the recognition and institutionalization of rights. Moreover, in the case of climate, the forces against which future humans will otherwise be defenceless are being unleashed by us ourselves, so we bear especially great responsibility for ceasing to make matters worse, compensating those whom we will have harmed by what we have done and will still do, and constructing social institutions, including most especially a regime for dealing with human-induced climate change, to protect those who could not otherwise protect themselves because the causal lead times in matters affecting the climate extend far beyond single generations. The opportunity to replace short-sighted rights-undermining practices reliant on primitive fossil fuel energy technologies with far-sighted rights-protecting institutions resting on alternative energy technologies is historic, unprecedented. We can take steps now to protect rights for those who come after us, steps that it may be too late for anyone to take later.

The purpose of a right is to provide protection for human beings against a threat to which they are vulnerable and against which they may be powerless without such protective action. For this protection to endure, rather than needing constantly to be improvised and reimprovised, the protection needs to take the form of an intergenerational institution. Human rights are an expression of human solidarity; we commit ourselves to each other to try to guarantee to all the urgent interests that some cannot provide for themselves. It is especially sad, as a matter of Western intellectual history, that the criticism that Marx derived from Hegel that individual rights are a kind of juridified or legalized form of the war of all against all, authorizing each of us to battle to be left alone by the others, should have taken as deep a hold among academic theorists as it has.[7] Degenerate systems of rights in which rights are reduced to

---

[7] K. Marx, 'Zur Judenfrage', in K. Marx/F. Engels, *Werke*, Bd. 1 (Berlin: Dietz Verlag, 1972 [1844]), 347–77; translated and abridged but with commentary as 'On the Jewish Question', in J. Waldron (ed.), *Nonsense Upon Stilts: Bentham, Burke and Marx on the Rights of Man* (London: Methuen, 1987), 137–50. Prominent scholars adopting the Marxian criticism include C. Taylor, 'Atomism', in C. Taylor, *Philosophy and the Human Sciences, Philosophical Papers*, vol. 2 (New York: Cambridge University Press, 1985 [1979]), 187–210; and M. A. Glendon, *Rights Talk: The Impoverishment of Political Discourse* (New York: The Free Press, 1991), 47. For elaboration of this point, see H. Shue, 'Thickening Convergence', in D. K. Chatterjee (ed.), *The Ethics of Assistance* (Cambridge University Press, 2004), 217–41.

purely negative rights of non-interference can have such an 'atomistic' form in which each merely tries to protect himself *from* the others, especially if rights are narrowed to rights to negative liberty. But developed systems of human rights, while they contain some protections against those others who are in fact predatory (even then in the form of solidarity among potential victims against the potential predators), are much more about cooperating with others in solidarity to create social institutions and practices that provide protection *to* others when they face dangers they cannot handle on their own.[8]

One question naturally is: Which rights of the people to come are threatened by climate change, and in which particular ways? Fortunately, a strong contribution to answering this question in detail has been made by Simon Caney, who has carefully shown how climate change will specifically threaten at least three rights, the right to life, the right to health, and the right to subsistence.[9] Here I shall simply rely on Caney's arguments about which rights so that I can focus on two other questions as they arise in the context of climate change: Which features do rights-protecting institutions need to have and what specifically are the tasks that need to be performed to protect rights against the threat of rapid climate change?

NECESSARY FEATURES OF RIGHTS-PROTECTING INSTITUTIONS: I, INTERNATIONAL

We cannot of course protect each other against all threats to all rights, and I have referred to the dangers for which protection ought to be provided as 'standard threats'.[10] In an important recent book Charles Beitz has stated the purpose of human rights as 'to protect urgent individual interests against certain predictable dangers ("standard threats") to which they are vulnerable under typical circumstances of life in a modern world order composed of

---

[8] For a fuller discussion see H. Shue, 'Solidarity Among Strangers and the Right to Food', in W. Aiken and H. LaFollette (eds), *World Hunger and Morality*, 2nd edn (Upper Saddle River, NJ: Prentice-Hall, 1996), 113–32.

[9] S. Caney, 'Climate Change, Human Rights and Moral Thresholds', in S. Humphreys (ed.), *Human Rights and Climate Change* (Cambridge University Press, 2009), 69–90. Also see Oxfam International, *Climate Wrongs and Human Rights: Putting People at the Heart of Climate-Change Policy*, Oxfam Briefing Paper 117 (Oxford: Oxfam International, 2008); P. Baer, T. Athanasiou, and S. Kartha, *The Right to Development in a Climate Constrained World: The Greenhouse Development Rights Framework* (Berlin: Heinrich Böll Foundation, 2007); and International Bank for Reconstruction and Development/The World Bank, *Development and Climate Change: World Development Report 2010* (Washington, DC: The World Bank, 2010), Box 1.4, 53 (mentioning 'the right to food, the right to water, and the right to shelter'.

[10] H. Shue, *Basic Rights*, 2nd edn (Princeton University Press, 1996), 29.

states'.[11] This rich characterization contains more elements than can be commented upon here, but I want to begin from Beitz's primary emphasis on human rights as 'matters of international concern'.[12] I shall argue, first, that any institutions to protect the rights threatened by climate change must be international; second, that they must also be intergenerational; and, third, that we must begin to build imaginative international intergenerational institutions immediately.

For better or for worse, the 'modern world order' assigns individual human beings to the control, possible protection, and possible predation of national governments. At the level of theory of the ideal, one can ask whether the best global order would be such an international system of national states; I seriously doubt it. On the other hand, be that as it may, at the level of what is practical for the foreseeable future, one can instead ask how, given the global order we actually have, the human rights of individuals can best be protected. The arrangement that we in fact have is what Beitz calls 'a two-level model' of protection for rights: 'The two levels express a division of labor between states as the bearers of the primary responsibilities to respect and protect human rights and the international community and those acting as its agents as the guarantors of these responsibilities.'[13] I have sometimes referred to these second-level, back-up responsibilities as 'default duties'.[14] When a national government fails to carry out its primary responsibility to protect rights, responsibility defaults to the second level consisting of the remainder of humanity, organized under the other national governments and constituting the remainder of the international community. This is essentially the model or picture underlying, for instance, what has come to be called the 'responsibility to protect' (or R2P).[15]

One good question is why primary responsibility should be assigned to national states; often that question turns into the question of ideal theory: Is a system of states really the best way to organize the globe? This is a question that, as I have said, I leave aside for consideration in some less desperate time, in order to pursue another good question: Why, in the international system we now have, default duties or secondary responsibility should fall to 'the international community'? This is so far a very general and rather grand question, and we want to focus here specifically on the pressing issue of accelerating climate change.[16] An initial part of the answer to the grand, general question is

---

[11] C. R. Beitz, *The Idea of Human Rights* (New York: Oxford University Press, 2009), 109.

[12] C. R. Beitz, *The Idea of Human Rights*.

[13] C. R. Beitz, *The Idea of Human Rights*, 108.

[14] Shue, *Basic Rights*, 2nd edn, 171 ff.

[15] International Commission on Intervention and State Sovereignty, *The Responsibility to Protect* (Ottawa: International Development Research Centre, 2001).

[16] The rate of climate change is itself increasing and appears to be feeding upon itself, that is, generating dominantly positive feedbacks—see P. Forster, V. Ramaswamy, P. Artaxo, et al.,

that it is very difficult to imagine what other agents could effectively step into the breach when a national state is failing to protect its own people other than some agency of the international community (perhaps, as in the case of climate change, an agency of a kind yet to be fully conceived and constructed). Because national states are so extremely jealous of their state sovereignty and their other prerogatives and tend to be armed and dangerous, very few effective options are available for bringing about inside their claimed jurisdiction any change that they view as unnecessary or oppose for any other reason. Action genuinely representative of the international community, even if implemented primarily by one or a few other national states, is one of the only options likely ever to have both the legitimacy and the power to accomplish inside a sovereign state anything not welcomed, much less seriously opposed, by the state in question. Otherwise, when national governments were hostile to rights, the individuals they control would simply have to be abandoned to their fates.

If it is difficult to conceive any effective agency other than some agency of the international community in most cases in which a state has failed to protect the most fundamental rights, it is truly inconceivable that a national state could protect the rights threatened specifically by climate change while acting alone. Every state is a 'failed state' as far as climate is concerned. Climate change is such a deeply global problem that only coordinated international action could possibly deal with it. Climate is itself a set of literally global phenomena driven by transnational systems at various levels such as the jet stream, the Atlantic thermohaline circulation, and El Niño and La Niña. And the greenhouse gases (GHG) that are disturbing the historical climate equilibria, whatever their national origins, themselves mix into a single global atmospheric pool blanketing the entire planet and producing climate disturbances for emitters and non-emitters alike. Climate is one case in which effective national protective measures are simply not possible. Anything China does can be undone by the USA, and vice versa. The futility of uncoordinated national efforts at protection against the effects of climate change is certain. The only conceivable protection of any rights threatened by climate change is protection through concerted action by the international community as a whole.

It is worth noting, on the other hand, that international institutions need not be centralized. Given the failure so far of the Conference of the Parties to the Framework Convention on Climate Change to reach any general

'Changes in Atmospheric Constituents and in Radiative Forcing', in S. Solomon, D. Qin, M. Manning, et al. (eds), *Climate Change 2007: The Physical Science Basis*, Working Group I Contribution to the Fourth Assessment Report of the Intergovernmental Panel on Climate Change (Cambridge University Press, 2007), 129–234; and H. Le Treut, R. Somerville, U. Cubasch, et al., 'Historical Overview of Climate Change Science', in Solomon et al. (eds), *Climate Change 2007: The Physical Science Basis*, 93–127.

agreement about a multilateral treaty that will effectively eliminate carbon emissions in any reasonable period of time, it may well be that more hope lies with international coalitions, and coalitions of coalitions, of agents operating at various different levels than through standard international treaties.[17] If, say, the state of California, the Chinese Ministry of Environmental Protection, Walmart, and the EU could agree on aggressive action, it might not be necessary to wait for ill-informed laggards such as the majority of the US Senate. It would be outrageously tragic if the fate of the whole earth depended on the US Senate. This is a largely non-normative issue at a different level of analysis that could not be pursued here even if I were the right one to pursue it. The important negative caution is simply that action can be international without needing to be based on the kind of treaty that the US Senate so rarely deigns to ratify.

NECESSARY FEATURES OF RIGHTS-PROTECTING INSTITUTIONS: II, INTERGENERATIONAL

It is equally certain that any effective protections for the people of later generations must be begun in prior generations, making climate change an inherently intertemporal/transgenerational, as well as an inherently international, challenge. It is quite simply impossible for the social arrangements necessary to control the severity of climate change to be put into place by the generation who need protection. The overall climate system has extraordinary momentum; it is difficult to divert it from whatever path it is on, but once it has been diverted, as it now has been since the fossil fuel-fed Industrial Revolution, it is then equally difficult to divert it yet again from that new path over any period of time less than a century. Above all, this is because of the atmospheric residence time of the most dangerous GHG, carbon dioxide (CO_2). Only a few years ago scientists believed that the average atmospheric residence time of a molecule of CO_2 was roughly a century.[18] They have now realized that it is so much longer than a century that it is currently incalculable: 'A[n atmospheric] lifetime for CO_2 cannot be defined . . . The behaviour of CO_2 is completely different from the trace gases with well-defined lifetimes. Stabilisation of CO_2 emissions at current levels would result in a continuous increase of atmospheric CO_2 over the 21st century and beyond . . . In fact, only

[17] I am grateful to Scott Moore for raising this issue.
[18] This was already a great oversimplification—for a more sophisticated approach, see F. Joos, M. Bruno, R. Fink, et al., 'An Efficient and Accurate Representation of Complex Oceanic and Biospheric Models of Anthropogenic Carbon Uptake', *Tellus*, 48B (1996), 397–417.

in the case of essentially complete elimination of emissions can the atmospheric concentration of CO_2 ultimately be stabilised at a constant level... More specifically, the rate of emission of CO_2 currently greatly exceeds its rate of removal, and the slow and incomplete removal implies that small to moderate reductions in its emissions would not result in the stabilisation of CO_2 concentrations, but rather would only reduce the rate of its growth in coming decades.'[19]

The practical implication of this rather mind-boggling conclusion is this: Once CO_2 is emitted into the atmosphere, it stays there over any period of time of interest to humans; and once more CO_2 is emitted, more stays there. Or in other words, at whatever level of CO_2 the atmospheric concentration peaks, that concentration will stay for a long, long time—multiple generations at a bare minimum. This makes the duration of climate change like few other problems, except perhaps the generation of nuclear waste, which is also extraordinarily persistent, and the manufacture of the most persistent toxic chemicals. And of course the pervasiveness of climate change is incomparably greater than nuclear waste or any toxics about which we so far know. While it is always good for rights-protecting institutions to be enduring, for them to deal specifically with the dangers of climate change no other option is possible.

When I was a small boy in rural Virginia in the 1940s, travelling evangelists would pitch their tent for a week in our county and attempt to convert us (although most of us thought we had signed up locally). One of their standard ploys was to try to terrorize us by preaching on the final evening about the 'unforgivable sin': It was essential to convert before you committed it—later would be too late because on this one there was no going back. I used to lie awake after returning home from the tent meeting, worrying that I might have committed the 'unforgivable sin' already without having realized it at the time, since the evangelist's account of it was as vague as it was ominous, and so be eternally damned before I had even gotten a good start on life (or had much fun). Adolescence, of course, brought other worries, and I gave up on the 'unforgivable sin', coming to doubt that there was any such thing. Now, however, I sometimes think the atmospheric scientists may have figured out what the 'unforgivable sin' is after all: emitting so much CO_2 that uncountable generations face a severely disrupted and worsening climate that blights their lives from the beginning! The penalty is not quite the promised eternal damnation, but bad enough, and, worse, the penalty falls not on the unforgiven sinners/emitters but on their innocent descendants, dooming them from the start.

---

[19] G. A. Meehl, T. F. Stocker, W. D. Collins, et al., 'Global Climate Projections', in Solomon et al. (eds), *Climate Change 2007: The Physical Science Basis*, 824–5.

NECESSARY FEATURES OF RIGHTS-PROTECTING INSTITUTIONS: III, IMMEDIATE

Philosophers and normative theorists have an unfortunate tendency often to begin offering prescriptions for dealing with problems before appreciating in sufficient depth the specific features of the particular problem. We ought to consider further the implications of the recent IPCC finding, quoted in the third paragraph above, that in the case of CO_2, in effect, what goes up does not come down any time soon. Common sense suggests that although we may be constantly adding CO_2 to the atmosphere, what is there already must somehow be drifting off somewhere else or otherwise be being neutralized. It is correct that, on the scale of centuries, the molecules of CO_2 decay, with the carbon and oxygen atoms entering new combinations. But the fact that the decomposition occurs only over such an extended time frame means that for all practical purposes—that is, as far as human interests are concerned—the concentration of CO_2 is simply expanding indefinitely, now that we are ejecting so much more from fossil fuel burning. It is absolutely critical how large the concentration is when it peaks—when it reaches its largest size—and it will not peak as long as we keep adding to it, because all additions amount practically to net additions.

Therefore, we need an immediate exit strategy from fossil fuel consumption and institutions to begin implementing it. I quote the alarming crucial finding again: 'In fact, only in the case of essentially complete elimination of emissions can the atmospheric concentration of CO_2 ultimately be stabilised at a constant level.' This is a stunning fact about the atmospheric chemistry of our planet, which makes continued use of fossil fuel (by far the primary source of the CO_2 emissions) seem much like indulgence in an 'unforgivable sin'. And this is why the proposal by the Obama administration to cut annual GHG emissions flows by only 4 per cent below 1990 flows is such a sad gesture—this would continue rapidly to increase the size of atmospheric stocks of GHG. As a physicist friend familiar with the US election campaign of 1992 recently put it, 'It's the stocks, Stupid!'

Indeed, these recent empirical findings support three momentous judgements critical to the nature and extent of our responsibility to terminate fossil fuel usage. First, if we do not act to prevent future climate change from becoming more severe, we have no grounds for confidence that anyone else ever can—there can be no later reversal of additional deteriorations, for a long time, with any known technology.[20] We know of no acceptable way for

---

[20] One can of course fantasize about unknown technological fixes. Some of the more popular ones, such as geo-engineering, are morally highly dubious—see S. M. Gardiner, 'Is "Arming the Future" with Geoengineering Really the Lesser Evil?', in Gardiner et al. (eds), *Climate Ethics*, 284–311.

subsequent generations to reverse whatever damage our actions will unleash upon them. In order to bring about any mitigation of climate change a century from now, vigorous action needs to be taken now. Second, we in any case ought to be the ones to act, in part, because we are the ones currently making matters (irreversibly) worse for everyone to come in the foreseeable future. We ought to take the vigorous action. Third, thanks to their hard-won understanding of the effects of reliance on carbon fuels, our scientists have opened the door to our bequeathing to those who follow us a priceless legacy: institutions that will nurture alternative energy sources that do not progressively undermine the environment to which the human species and other contemporary species are successfully adapted. Vigorous action now, by us, can produce invaluable results in the protection of threatened rights. What must we do?

NECESSARY TASKS FOR RIGHTS-PROTECTING INSTITUTIONS: I, 'DO NO HARM'

Like increasing numbers of scientists, I personally believe that there is a strong case for choosing the minimum goal mentioned in the Copenhagen Outcome Document of limiting warming to 2°C above pre-industrial levels.[21] The World Bank accepts that even a rise of 2°C would include these results: 'Between 100 million and 400 million more people could be at risk of hunger. And 1 billion to 2 billion more people may no longer have enough water to meet their needs.'[22] Others who do not accept this particular goal may of course view the following as a hypothetical example with an arbitrary goal; the general logic of the position remains, whichever goal one selects from within the range of what could plausibly be genuinely believed to be likely to avoid dangerous interference with the climate system.[23] The same logic holds provided any firm limit on emissions is adopted.

---

[21] Indeed, much can be said for a more ambitious goal of a global temperature rise of no more than 1.5°C above pre-industrial levels, which would require constraining atmospheric carbon to 350 ppm—hence the name of the website: <http://350.org>. The World Bank has endorsed the goal of preventing temperature rises from exceeding 2°C—see International Bank for Reconstruction and Development, *Development and Climate Change*, 3.

[22] International Bank for Reconstruction and Development, *Development and Climate Change*, 5. Both projections are taken from the 2007 IPCC report.

[23] For well-informed speculations about what the world would look like after temperature rises of various amounts up to 6°C, see M. Lynas, *Six Degrees: Our Future on a Hotter Planet* (Washington, DC: *National Geographic*, 2008). For a more technical account, see G. W. Yohe, R. D. Lasco, Q. K. Ahmad, et al., 'Perspectives on Climate Change and Sustainability', in M. Parry, O. Canziani, J. Palutikof, et al. (eds), *Climate Change 2007: Impacts, Adaptation and Vulnerability*, Working Group II Contribution to the Fourth Assessment Report of the Intergovernmental

If we want to limit global warming, for example, then, to 2°C above pre-industrial levels, we must avoid emitting the trillionth metric ton of carbon (Tt C) to be confident of having even a 50 per cent chance of meeting our target.[24] We have already emitted 0.5 Tt C and are therefore already committed to 1°C of warming. 'Having taken 250 years to burn the first half-trillion tonnes of carbon, we look set, on current trends, to burn the next half trillion in less than 40.'[25] Recent research suggests that the most helpful way to conceive our challenge, if we want to avoid warming of more than 2°C, is as the challenge of humanity's remaining within a total cumulative carbon budget of 1 Tt C, although of course it may turn out that this cumulative cap needs to be revised as scientific understanding progresses.[26] Total cumulative emissions of carbon must not surpass 1 Tt C, or global average surface temperature will, as likely as not, rise more than 2°C above pre-industrial levels, due to CO_2 alone. As shorthand, then, we can view our challenge as staying within a cumulative carbon budget of 1 Tt C, or, in short, avoiding the trillionth metric ton.[27] We have only forty years—about half a normal affluent lifetime—to eliminate completely the use of fossil fuel if the temperature rise is not to exceed 2°C.[28] Before today's college students retire, humanity must either have perfected carbon sequestration or stopped digging coal and pumping oil, the remaining reserves of which hang over us like a sledgehammer.

Panel on Climate Change (Cambridge University Press, 2007), Tables 20.8 and 20.9, 828–9. It is obvious that even a 3°C rise would be intolerable, with greater rises being increasingly catastrophic.

[24] M. R. Allen, D.J. Frame, C. Huntingford, et al., 'Warming Caused by Cumulative Carbon Emissions Towards the Trillionth Tonne', *Nature*, 458 (30 April 2009), 1163–6.

[25] M. Allen, D. Frame, K. Frieler, et al., 'The Exit Strategy', *Nature Reports Climate Change*, 3 (May 2009), 57. Atmospheric physicists at the University of Oxford maintain a website on which a ticking meter shows the blinding speed of the global release of carbon from fossil fuels at <http://trillionthton.org/>.

[26] Three factors might cause us to revise the cap: (1) we might make revisions regarding our conception of 'dangerous anthropogenic interference in the climate system', and change the 2°C target; (2) we might choose a larger or smaller cap to reflect risk preferences regarding the costs of abatements and climate change damages; or (3), as scientific uncertainty resolves, we might find we need to revise the cap even while holding the target and risk preference fixed. I am grateful to my colleague David Frame for these points and other helpful suggestions.

[27] Obviously if our goal were to be more ambitious than remaining below 2°C above pre-industrial levels, we would need to adopt an even tighter budget. At this point what is crucial is that we move aggressively downward in our carbon emissions; the ultimate target can be, and surely will be, adjusted as time goes by.

[28] I despaired of the capacity of US institutions of governance to rise to the challenge even before the January 2010 decision of the US Supreme Court in *Citizens United* vs *Federal Election Commission* (558 U.S. [2010]) opened the floodgates for the use of unlimited corporate funds against political candidates who support any legislation aimed at the necessary limitation of the use of fossil fuel. Now it is difficult to imagine effective political action in the USA against such wealthy entrenched power as the oil and coal industries, but legal action could conceivably work—see next note.

Accepting that the fundamental specific challenge has the shape of a need to stay within a cumulative budget of carbon emissions, which one must accept if one adopts any ceiling on temperature rises, has radical implications for how we ought to behave. Because we, together with all the generations who follow in our wake, must stay inside a single limit, carbon emissions are zero-sum across generations. That carbon emissions are *zero-sum across all emitters throughout foreseeable time* is profoundly important for the nature of our responsibilities for our actions regarding the rights of people to come. Every metric ton of carbon emissions for which one person is responsible is one less metric ton of carbon emissions available for all the other persons who will live during the foreseeable future. We are in direct competition for a scarce resource with our own great-grandchildren, and everyone else's great-grandchildren. Every time I fly across the Atlantic is one less time that anyone else can fly across the Atlantic in a plane burning fossil fuel. Our challenge is intergenerationally zero-sum.

This is obviously far from making the problem unique. The consumption of any non-renewable resource is intergenerationally zero-sum. Any unit of it that I consume is one less unit for everyone else across time to consume. We have simply learned recently that on the timescales that matter to humans, the planet's capacity to deal with carbon without rises in surface temperature is one of the non-renewables, even if over several centuries the atmospheric carbon will break down.

Business-as-usual consumption of carbon fuel is, then, doubly dangerous for the people to come. First, carbon emissions contribute to making climate change more severe. Second, carbon emissions contribute to using up whatever quota is set on behalf of humanity for total cumulative carbon emissions; e.g. the remaining 0.5 Tt C if we want to have a 50 per cent chance of avoiding a temperature rise of more than $2°C$. These two are each separately compelling reasons to institutionalize the initial exit strategy from fossil fuel as soon as humanly possible and assign responsibilities accordingly. The strategy will naturally need, as already indicated, to be adjusted as events develop and better understanding of climate dynamics grows. But these double dangers make vigorous action, far more robust than anything now contemplated by conventional politicians, urgent.

The main argument usually given for inaction is uncertainty. What if, in spite of the growing sophistication and increasing validation of climate science, the dangers of climate change have somehow been exaggerated? I have dealt with this complex issue at length elsewhere, where the upshot was what amounted to a modified formulation of the 'precautionary principle': one ought to ignore entirely questions of probability beyond a certain minimal level of likelihood in cases with 'three features: (1) *massive loss*: the magnitude of the possible losses is massive; (2) *threshold likelihood*: the likelihood of the losses is significant, even if no precise probability can be specified, because (a) the mechanism by which the losses would occur is well understood, and (b) the conditions for the functioning of the mechanism are accumulating; and (3) *non-excessive costs*: the costs of prevention are not excessive (a) in light of the magnitude of the possible losses and (b) even considering the other important demands on our resources'.[29] The danger of climate change clearly has these three features, so uncertainty ought not to prevent us from taking action that does not have excessive costs, with excessive measured only by the magnitude of the possible losses, not discounted by their probability.

The specification of the appropriate bearers of the duties to protect any particular right can often be the intellectually most challenging aspect of spelling out the form that protection for the right might best take, as Beitz rightly emphasizes.[30] The fact that continuing carbon emissions by us exacerbate both the double dangers is also the first of three powerful reasons why we clearly ought to begin to curtail carbon emissions sharply now. Business as usual is not inaction—it is the knowing infliction of more straitened circumstances of life on those to come. One of the most compelling principles for the assignment of responsibility is the principle that the persons inflicting a harm must stop—first, do no harm—and must, if possible, compensate for the damage they themselves have already done. Our generation cannot plead ignorance of the implications of our choice of energy source. The harm consists of the creation of circumstances in which the fulfilment of such rights as those to life, health, and subsistence comes under increasing threat.[31] Those circumstances will evidently arise no later than the emission of the trillionth metric ton.[32] The best way to begin to protect the rights of future generations is to stop threatening them—to cease contributing to conditions that may make enjoyment of those rights impossible. If we do not act vigorously now, we will inflict avoidable harm on people who live in our polluted wake. While

---

[29] 'Deadly delays, saving opportunities: creating a more dangerous world?', this volume. The fossil fuel industries continue to do their utmost to confuse the general public by the propagation of what has been aptly called 'manufactured uncertainty'—see S. Vanderheiden, *Atmospheric Justice: A Political Theory of Climate Change* (New York: Oxford University Press, 2008), 192–202, for documentation of some of the notorious intentional deceptions during the Bush–Cheney Administration. However, lawsuits based on damage caused by climate change may be beginning to worry the fossil fuel industry—see J. Schwartz, 'Courts as Battlefields in Climate Fights', *New York Times*, 26 January 2010. In *Native Village of Kivalina v. ExxonMobil Corp.* it is alleged that 'the industry conspired "to suppress the awareness of the link" between emissions and climate change through "front groups, fake citizens organizations and bogus scientific bodies"'.

[30] See especially Beitz, *Human Rights*, ch. 7.

[31] Harm need not be done to an individual identifiable in advance.

[32] As indicated in note 20, harm undoubtedly begins at lower rises in temperature—indeed, it has already begun today. Many species will be driven to extinction by radical changes in habitat, for example. Setting the bar this low is merely a concession to the reality of the political power backing fossil fuel use. Strictly speaking, I am advocating merely 'Do no unavoidable harm,' which is a pale copy of the Hippocratic principle.

I believe that this first reason is by itself entirely sufficient, an additional, although more complex, reason further strengthens the case for major responsibility for urgent action falling upon the generations now alive, the first to begin to understand climate dangers.

NECESSARY TASKS FOR RIGHTS-PROTECTING INSTITUTIONS: II, FAIRNESS

We have noted already that continuation of the consumption of fossil fuels not only contributes to greater severity of climate change but also continues to use up whatever the remaining quota of 'tolerable' additional carbon emissions may be—'tolerable' in the extremely weak sense of not necessarily expected to undermine the conditions in which human rights can be fulfilled.[33] If the temperature ceiling is to be 2°C above pre-industrial levels, then the remaining quota is 0.5 Tt C. If someone can believe that higher temperatures, and generally more severe climate change, are tolerable, her conception of the remaining quota will be correspondingly larger. But some ceiling—whether or not 2°C above pre-industrial levels—must be imposed unless climate change is to be allowed to grow indefinitely worse. Whatever the total emissions within the quota, questions of fairness arise about how those emissions are distributed. So the second major ethical issue, after harm, is fairness. The requirement to distribute fairly is no less fundamental than the requirement to do no harm (and to compensate for harm done), but its specification in this case is considerably more complicated.

As we have already seen, the central question is intergenerational at its core. Fairness—or 'intergenerational equity', as the lawyers and economists like to call it—is then not an additional peripheral aspect of the question that we may optionally take up or not, as we choose. A single budget for carbon emissions, whatever its total size, is shared by us and every foreseeable generation to come. Consequently there simply is no such thing as doing what is fair 'except for the intergenerational part'; in the concrete instance we face, what is fair is a pervasively intergenerational issue.[34]

---

[33] Tolerable for whom? The climate change that has already occurred has undermined the conditions necessary for the survival of many other species—most notoriously the polar bear, but various other animal and plant species as well. See *BBC News*, 'Climate Focus "Ignores Wildlife,"' 18 February 2008. Other species are of great value—I am concentrating on humans in the text merely for simplicity. Few biological scientists would accept the idea that any level of additional human emissions is tolerable.

[34] I think most issues of justice are fundamentally intergenerational and international at bottom, but I do not seem to have convinced many people yet. For the central argument, see H. Shue, 'The Burdens of Justice', *Journal of Philosophy*, 80, 10 (October 1983), 600–8. For one

Since all humans from now on share the same emissions budget and the budget is zero-sum, if one wants to be fair, one needs to leave for others their fair share and use only one's own fair share. But how can we think about this apparently novel issue of what are going to become fair shares of carbon emissions in the international institutional context we are about to create? One way to reduce carbon emissions is to tax them, but most of the plans currently discussed involve the creation and trading of permits to emit carbon—carbon trading, 'cap-and-trade', which will be a transformative international and intergenerational institution. Under trading the only way to create an incentive to reduce carbon emissions is to require the possession of a permit to emit carbon and to charge some people for carbon permits (and progressively to reduce the number of permits available—the 'cap'—driving up the price of permits).[35] At present the world economy is dependent upon fossil fuels and therefore dependent upon emitting carbon, but many people in the world are too poor to pay for not only what would otherwise be the price of fossil fuel but also the premium that will be added to the price by the cost of permits to release the amount of CO_2 generated by burning that much fossil fuel.

The basic requirement of human rights for any institution of carbon trading, then, is that it not make it impossible for people to survive by pricing those in the market for fossil fuels out of that market as long as so many people are still dependent on fossil fuel for lack of any affordable and sustainable alternative energy.[36] We are of course trying to reach a point at which none of us is dependent on fossil fuel, but we cannnot make the transition by simply

---

theorist who is convinced about the international aspect at least, see K.-C. Tan, *Justice Without Borders: Cosmopolitanism, Nationalism and Patriotism* (Cambridge University Press, 2004), 175.

[35] The political right wing in the USA is complaining mindlessly that 'cap-and-trade' is 'just another tax'. The whole point, of course, is to make fossil fuel consumption more expensive so that it will be reduced and—within forty years, if possible—eliminated. Naturally the fossil fuel sector and the politicians subservient to it oppose such measures which literally threaten their customary livelihood.

[36] The energy supplies of significant numbers of poor people would not be affected because they do not participate in markets for fossil fuel but instead, for example, gather sticks and dung to dry and burn in highly inefficient stoves. These practices, however, are unsustainable in their own way in that they also contribute substantially to climate change when, in particular, millions of primitive cookstoves generate massive clouds of black carbon ('soot') that settle in polar regions and reduce the albedo of the snow. 'Black carbon might account for as much as half of Arctic warming'—see E. Rosenthal, 'Third-World Stove Soot Is Target in Climate Fight', *New York Times*, 15 April 2009; and International Bank for Reconstruction and Development, *Development and Climate Change*, Box 7.10, 312. Higher market prices for coal, oil, and gas might tend to trap people in these other environmentally highly destructive technologies unless complementary action is taken to create opportunities for them to use energy in less damaging ways, such as inexpensive but far more efficient cookstoves in the short term.

I will in the remainder, then, describe those who must purchase fossil fuel in the market as the 'market-dependent poor'. Strictly speaking, they are the fossil fuel market-dependent poor, but to avoid worse clumsiness I will generally use the shorter label. I am grateful to Denis G. Arnold for raising this important distinction.

pretending we are already there and ignoring the fact that most people are now dependent on fossil fuels.[37] So the fundamental issue of fairness that arises under carbon trading is: Who must pay for permits and who receives free permits? The distributive principle for free carbon emissions (emissions without purchase of a permit) needs to be a distributive principle appropriate to an intergenerationally zero-sum resource that is a necessity of life for as long as the current predominantly fossil fuel energy regime retains its grip on us. Suppose for concreteness, as before, that the trading regime has a budget of 0.5 Tt C emissions remaining before the dominance of fossil fuels must be ended (if warming beyond 2°C above pre-industrial levels is to be avoided). So the principle must be appropriate to the distribution of those 0.5 Tt C. The logic is the same whatever the numbers are.

One point that is perfectly obvious—unless one believes that it is acceptable to create social institutions that cause widespread deaths—is that any acceptable distribution of the intergenerationally zero-sum quota of 0.5 Tt C must be compatible with every individual's benefiting from the minimal amount of carbon emissions made necessary for a decent life by the then existing fossil fuel energy regime until the regime's environmentally destructive dominance can be broken.

The institutional arrangements currently under discussion include no plan for a (hopelessly impractical) distribution of emissions permits directly to individuals around the world. Most proposed trading schemes involve the distribution (free or at auction) of permits to firms and/or nations; the additional cost created by the requirement to have permits will then of course be passed on to individuals by firms (just as it would if firms faced carbon taxes). But the logic of the situation is clearest if we simplify by describing the situation as if there were going to be permits for individuals, some sold and some provided free of charge. We would then need a priority list specifying who gets (free) emissions first from the remaining, but rapidly diminishing, budget of 0.5 Tt C. Obviously, unless some people are to be condemned to death for lack of benefiting from a minimal amount of carbon emissions, those who can least afford to pay for emissions ought to be at the top of the list of those who do not have to pay for emissions. For example, some people can grow adequate food only by using petroleum-based fertilizer or fossil fuel-powered irrigation pumps. Unaided, people who now can barely afford the

---

[37] How the globe came to be dependent on an energy economy centred on fossil fuels is an important historical question. For a fascinating account of how it happened, see K. Pomeranz, *The Great Divergence: China, Europe, and the Making of the Modern World Economy* (Princeton University Press, 2000). For an introduction to the military implications of the dependency of the 'superpower' on fossil fuel, see M. T. Klare, *Blood and Oil: The Dangers and Consequences of America's Growing Dependency on Imported Petroleum* (New York: Henry Holt, 2004); and for the role of the American car in particular, see I. Rutledge, *Addicted to Oil: America's Relentless Drive for Energy Security* (New York: I. B. Tauris, 2005).

fertilizer or the fuel for the pump might not be able to afford any increase in prices driven by permit charges.

The next step in the argument will not be obvious, but it seems to me to be the only prudent approach, given how dangerous extreme climate change will be and how vital it therefore is to enforce a relatively low cap on total cumulative emissions (such as 1 Tt C) by the time fossil fuel use is eliminated completely (in order to avoid a temperature rise exceeding 2°C above pre-industrial levels). We do not know for how long the remaining budget consisting of the second 0.5 Tt C of possibly 'tolerable' emissions—0.5 Tt C have already been emitted as of now[38]—will have to supply the for-the-meantime-unavoidable carbon-emission needs of many of the poor. As things are going now, the budget consisting of the second half of the total of 1 Tt C will likely be exhausted in less than forty years—well before 2050.[39] The longer that many of the poor people on the planet must rely for survival on carbon emissions within a dominant fossil fuel energy regime, the longer they will need to draw from whatever remains of this budget at any given time. If we are serious about not making the lives of the market-dependent poor impossible, and we accept the science, we must, in effect, reserve enough of the remaining budget of 'tolerable' emissions for the fossil fuel market-dependent poor to use to maintain themselves at a decent level of existence for the duration of the period during which they must depend on the fossil fuel regime. Obviously, the longer they are dependent on fossil fuels, the longer they will need to draw upon the budget and the more of it that will be needed strictly for them. On the one hand, the remaining budget of carbon emissions could be enlarged only by allowing warming beyond 2°C above pre-industrial levels, which is yet more dangerous. On the other hand, the time period of the dependence of the poor on carbon emissions can be shortened by making affordable alternative energy without carbon emissions available to them sooner, which is one of the actions most urgent to be taken, for this and other reasons.

It is vital not to confuse the number of Tt C withdrawn from the remaining budget of 'tolerable' emissions by the poor without charge (without requiring purchase of a permit), *a*, with the total number of metric tons of carbon still able to be withdrawn from the budget altogether, *N*. Amount *a* will be only a fraction of total sum *N*, quite possibly a very small fraction, depending on all the variables affecting the purchase of permits, including how many are

---

[38] The best estimate is that, as of 20 February 2010, we have emitted more than 535,157,625,000 metric tons of carbon into the atmosphere. See the results of the University of Oxford's calculations at <http://trillionthton.org/>.

[39] See Allen et al., 'The Exit Strategy', 2. As I write this, the computer-based meter at <http://trillionthton.org/> shows the trillionth metric ton probably being emitted very early in 2045—thirty-five years from now.

created, to whom they are distributed, what percentage are auctioned, etc., etc. The consumption of N, the pool of 0.5 Tt C in 'tolerable' future emissions still remaining today, will consist of (a), the consumption by any market-dependent poor who are not required to have permits, and (b), the consumption by those who purchase, steal, or otherwise acquire permits. Any emissions regime less flagrant than business as usual in the past has been could extend the life of the total remaining budget of 0.5 Tt C beyond 2050 somewhat—the farther the arrangements depart from business as usual, the farther beyond 2050 the viability of this total remaining budget of emissions compatible with a 50 per cent chance of the temperature rise not exceeding 2°C above pre-industrial levels could be extended.

Clearly it is an empirical question how long it will take to exhaust the emission budget consisting of the possibly 'tolerable' 0.5 Tt C of emissions into the atmosphere,[40] given that both (a) free and (b) priced emissions will be coming out of the same budget. My suggestion—this is the non-obvious next step in the reasoning—is that *all* the free emissions should at least tentatively and temporarily be reserved entirely for the market-dependent poor.[41] Everyone else pays for permits. If and when further investigation provides solid grounds for a confident judgement about the total number of people across foreseeable generations who will both be unable to afford to purchase emission permits but will also need to benefit from carbon emissions because they do not yet have affordable alternative sources of energy, we might choose to provide free emissions to some larger group, if poor people are still subject to dependence on fossil fuel. How long the budget of possibly 'tolerable' additional carbon emissions (0.5 Tt C) will last depends heavily upon (*b*) the emissions by those who purchase permits, but also upon the numbers of those people across the next generations who cannot afford permits for carbon emissions but nevertheless need to rely on carbon emissions and therefore must be allowed (a) emissions free of charge, unless they are to be condemned to death or desperation. The latter number in turn depends on how long it takes for the dependence of the poor on fossil fuels to be ended by the affordability of alternatives to fossil fuels. If and when the numbers become

---

[40] Further consideration may show that fewer emissions can be treated as not constituting dangerous interference with the climate system—see note 21.

[41] A definite specification of who precisely count as the poor is clearly needed, but this is a familiar problem to which I can contribute nothing special here. It may be, depending on how the free permits would in effect reach the poor (continuing our simplification of permits for individuals), that the free permits ought also to be inalienable or non-tradable in order to make it pointless for 'unreliable trustees', such as rapacious governments, to steal them—for an interesting discussion, see Simon Caney, 'Equality in the Greenhouse'. If the mechanics of cap-and-trade seem to be becoming unworkably complex, I would be perfectly happy to see carbon taxes with compensating tax credits for the poor, except perhaps for the giant utilities and the other largest emitters, who could be allowed to trade permits among themselves. The EU has such a mixed system. I am grateful to Ed Page for discussion of this point.

clearer, it might, depending upon what the actual numbers are, no longer be necessary to reserve all free emissions for the poor. Until evidence to the contrary appears, however, the priority list for free emissions from the budget should contain, I would suggest, absolutely no one other than the market-dependent poor. Otherwise, we risk pricing living people into desperation while trying to eliminate carbon emissions for the sake of future people by way of a trading mechanism to which some living people simply cannot afford to adjust.

What percentage of the permits should be distributed free to the poor? Whatever percentage they need to fulfil their basic rights. Otherwise, our institutional arrangements to protect the poor of the future would be arrangements for starving the poor of the present, which would be a crime against humanity. This will of course reduce the supply of permits to be purchased and drive up their price. The only way to reduce the price pressure is to bring affordable alternative energy on line more quickly.

Those familiar with the climate change debates will realize that my argument here has reached at least as strong a conclusion as the conclusion ordinarily reached by those who appeal to historical responsibility, but I have not relied upon their controversial premise that there is a universal right to equal per capita emissions. The premise playing the analogous role in this reformulation is merely that we could not in decency condemn those who need carbon emissions but cannot afford to purchase permits, by refusing to guarantee them within our new carbon trading institution the emissions they cannot do without, thereby violating their rights to life, health, and/or subsistence. I do not claim to have disproven anywhere the premise about equal per capita rights to emissions underlying the standard versions of the argument about historical responsibility; I have simply not needed that strong a premise because of what the nature of the challenge is empirically turning out to be, namely, the distribution of an intergenerational zero-sum carbon budget.[42] Basically, I have not needed the stronger premise about equal rights to emissions because our situation is so direly constrained by the necessity of remaining within a zero-sum emissions budget. Extreme situations are in some respects clearer—their very starkness simplifies our choices. All that we need to be committed to is the protection of the rights of the current, and soon to be born, most vulnerable against the workings of the very permit system we may feel compelled to create for the sake of the most vulnerable of the more distant future. If we do not act vigorously now, we will be unfair to people who will live in our shadow.

[42] Compare H. Shue, 'Historical Responsibility', Technical Briefing for Ad Hoc Working Group on Long-Term Cooperative Action Under the Convention (AWG-LCA), SBSTA, UNFCC, Bonn, 4 June 2009. <http://unfccc.int/files/meetings/ad_hoc_working_groups/lca/application/pdf/1_shue_rev.pdf>.

The significance of the recommendation that all free emissions be reserved for the market-dependent poor can be brought out by contrast with another proposal. In his excellent chapter on the competitiveness effects of various climate policies, Richard D. Morgenstern notes that one proposal is for free allowance allocations under the cap-and-trade system for 'carbon-intensive' industries.[43] In my terminology this would amount to granting some free emissions to these industries. An implicit appeal to fairness seems to me to be being made by these industries, who are, in effect, arguing as follows: 'We understand that emissions must be reduced overall, and we are willing to do our fair share. But as luck would have it, our industry depends on much higher emissions than most. So a fair share for us would be, not an equal share, but a larger share more in line with our need for more "carbon-intensive" emissions.' But what, after all, is a 'carbon-intensive' industry? It is an industry that emits an especially large quantity of carbon for the value of what it produces! If the solution to the competitiveness challenges for industries with high emissions of carbon is to concede them special privileges to emit, we simply continue the rush toward emission of the trillionth metric ton and use up portions of the quota of remaining emissions that could be protected for the market-dependent poor. It is the 'carbon-intensive' industries whose emissions it is most important to challenge, not to indulge.

A SPECTACULAR LEGACY: TRANSCENDING THE STANDARD CRUEL DILEMMA

Clearly, then, the third reason for urgent vigorous action is that for now, but not indefinitely, we face an opportunity to arrange for the protection of two sets of human rights that will become more and more difficult to protect simultaneously. On the one hand, we can protect against undermining by severe climate change the ability of people of the more distant future to enjoy their rights to life, subsistence, and health by avoiding the emission of the trillionth metric ton of carbon. On the other hand, we can protect against undermining, by means of the very cap-and-trade institution being created for the first purpose, the ability of the market-dependent poor of the present and the near future to enjoy their rights by guaranteeing them carbon emission permits without charge. As time goes by, we are liable to be told, as we often are, that we must choose between the 'present poor' and the 'future poor'. As the remaining pool of carbon emissions possibly 'tolerable' by the planetary climate system shrinks, we are likely to be told that everyone must in order to

---

[43] Denis G. Arnold (ed.), *The Ethics of Global Climate Change*), ch.10.

drive down carbon emissions pay more to emit carbon, which could price the then current poor out of the energy market even for what have sometimes been called 'subsistence emissions', carbon emissions essential to survival and subsistence.[44] This would sacrifice the present poor to the future poor. Or, we will be told, we must relax the ceiling on total cumulative carbon emissions and let them run on beyond 1 Tt C, which will likely produce more severe climate change and greater obstacles to the fulfilment of the rights of the future poor, sacrificing them to the present poor (and whoever else is emitting carbon!).

The most significant point is that we do not need to face any such dilemma between present rights and future rights if—and, as far as I can see, only if—we take robust action immediately that cuts carbon emissions sharply (so the future poor are not threatened by a deteriorating environment) and does it while protecting the urgent interests of the current poor, which are the substance of their same rights. The longer we continue to fiddle with our current casualness, the closer we will approach a dilemma in which a sudden crackdown on carbon emissions, designed to forestall the trillionth metric ton, which would threaten the subsistence emissions of the then current poor, will seem to be the only alternative to an abandonment of the ceiling of 1 Tt C, which would threaten the future poor (and possibly everyone else as well, not to mention innumerable other species). But there is no need to put ourselves— or, rather, the current and future poor—into this box by continuing to delay facing reality.[45]

Instead, action is urgent on two converging fronts. First, carbon emissions need to be cut back sharply and aggressively. The atmospheric concentration of carbon will not stop growing until emissions are zero, as the language quoted twice above from the latest IPCC report indicates. Probably the maximum carbon concentration will determine the maximum climate change. Second, alternative energy technologies need to be developed as quickly as humanly possible, aiming at an early day when prices of the alternative technologies are competitive with the prices of fossil fuel and become afford- able for the poorest. Fossil fuels are notoriously cheap, of course, which is the main reason we need the cap-and-trade (or carbon tax) institutions to drive up their price by political choice. We must aim for the point of crossover at which declines in the prices of alternative technologies and rises in the prices of fossil

---

[44] The distinction between subsistence emissions and luxury emissions was introduced in the first influential discussion of the ethics of climate change: A. Agarwal and S. Narain, *Global Warming in an Unequal World* (New Delhi: Centre for Science and Environment, 1991), 5. I carried their suggestion forward in 'Subsistence emissions and luxury emissions', this volume. Unlike my usage, the two categories are treated as mutually exhaustive of all emissions in Vanderheiden, *Atmospheric Justice*, 67–73 and 242–3.

[45] For discussion of related issues, see S. Humphreys (ed.), *Human Rights and Climate Change* (Cambridge University Press, 2010), *passim*.

fuels mean that fossil fuels lose their competitive price advantage. The farther we move on either front—making fossil fuels more expensive and making alternative energy technologies less expensive—the less far we need to move on the other front. Once the crossover occurs, even the purely selfish who care nothing for the environment and nothing for the rights of others will simply find it efficient to use alternative fuels. At that point humanity might be out of the woods, provided that we have meanwhile not emitted the trillionth metric ton, or whatever the rapidly advancing science tells us is the outer boundary of environmentally 'tolerable' carbon emissions. If we act vigorously and creatively now, we can invent institutions that will provide a priceless legacy of rights protection for multiple generations. Blinkered commission of the 'unforgivable sin' of myopic self-indulgence or farsighted creation of invaluable institutions of rights protection—which choice will your generation make? To its undying shame, mine appears to have chosen.

It was an exaggeration to say that we must 'eliminate completely the use of fossil fuel if the temperature is not to exceed 2° C' [307, 313, and note 35 on 311]. We must eliminate completely *net additions* to the atmospheric concentration of CO_2 before the concentration exceeds the level compatible with a 'tolerable' degree of climate change. The atmospheric concentration would not rise if carbon emissions on the surface were once again, as they were in 1850 and for millennia earlier, small enough to be absorbed by vegetation and the oceans. Carbon absorption by vegetation is basically fine, but carbon absorption by the oceans is causing acidification. Ocean acidification is technically a problem separate from climate change, but ocean acidification is severe, a worsening cause of species extinctions, dangerous to the human food supply, and very likely irreversible. So for all practical purposes we must cease to rely on fossil fuels as a primary source of energy because their use must be radically reduced, but not literally eliminated. Coal cannot be a significant source of electricity, unless the coal firms perfect carbon capture and storage. Meanwhile, the coal ought to be left in the ground. And petroleum cannot be a main energy source of transportation.

17

Climate hope: implementing the exit strategy*

> The goal of any action must, in the end, be to reduce the worldwide supply of fossil fuels.[1]

> The past is never dead. It's not even past.[2]

INTRODUCTION

Eric A. Posner and David Weisbach propose the following thesis: 'The problem of widespread poverty is urgent. So is the problem of reducing emissions . . . But no principle of justice requires that these problems be addressed simultaneously or multilaterally.'[3] I aim to show why it is not possible to accept their thesis of the separability of the solutions to poverty and climate. Rather than repeat here arguments that I have made over the last two decades, I shall very briefly sketch a single constructive approach that might make a substantial contribution on both fronts, in the process suggesting further reasons why climate and poverty are inextricably entangled.[4]

---

* 'Climate Hope: Implementing the Exit Strategy', copyright: *Chicago Journal of International Law*, 13:2 (Winter 2013), 381–402.

I am grateful for the invitations to conferences at which this chapter evolved and for the perceptive comments from participants, especially Simon Caney.

[1] Eric A. Posner and David Weisbach, *Climate Change Justice* 32 (Princeton, 2010).

[2] Lawyer Gavin Stevens in William Faulkner, *Requiem for a Nun*, Act I, scene iii, 92 (Random House, 1951).

[3] Posner and Weisbach, *Climate Change Justice* at 197 (cited in note 1). They go on to assert: 'Events in Copenhagen teach that trying to do so risks all . . . goals.' For any lessons from Copenhagen, I would recommend Lavanya Rajamani, *The Making and Unmaking of the Copenhagen Accord*, 59 Intl & Comp L Q 824 (2010).

[4] I argued in 1992 that it might be irrational for a state with inadequate wealth for dealing with adaptation to climate change on its own to sacrifice the speed of its economic development in order to cooperate with an international scheme for abating climate change, as long as

We will cause severe damage if we do not rapidly get a grip on our greenhouse gas (GHG) emissions, but we will also cause severe damage if we do get a grip exclusively by the means that is currently most discussed, raising the price of fossil fuels. Fortunately, we have available a realistic complementary means, the combination of which with the standard policy options of trading in emissions permits and carbon taxes would avoid both kinds of damage. This is what I shall recommend.

My primary focus here is on the nature of the problem of the mitigation of climate change because I think that there is a strong general tendency to misconstrue it. The wrong way to think of the political and moral problem is to consider it exclusively a problem of international distributive justice, when it is primarily a matter of avoiding doing damage by either of two routes, exacerbating climate change by continuing to emit large quantities of GHGs or placing insuperable obstacles in the way of Third World development by the method we create in order to reduce the emissions of GHGs.[5] The most discussed methods, 'cap-and-trade' and carbon taxes, both raise energy prices. If all we do is raise energy prices, we will make it more difficult for the global poor to escape the 'energy poverty' that is a critical obstacle to their development.[6] Thus, we have an example of how taking current empirical findings seriously can transform one's understanding of both what the relevant normative issues are and how to respond to them adequately.

---

wealthier states refused to commit themselves to a complementary international scheme to enable all states to adapt to climate change. In short, cooperation in mitigation may be irrational in the absence of cooperation in adaptation. See 'The unavoidability of justice', this volume. Posner and Weisbach would, I suspect, consider this an improper mixing of 'distributive justice' with 'climate change'. In general, they tend to criticize a generic position rather than grappling with the specific arguments in the now fairly extensive literature on climate and justice. See, for example, the relevant early essays collected in Stephen M. Gardiner et al., eds, *Climate Ethics: Essential Readings* (Oxford, 2010) and others cited in the seminal review of the literature by Stephen M. Gardiner included in that collection.

[5] Posner and Weisbach extensively discuss whether 'a climate treaty has to serve distributive justice'. Posner and Weisbach, *Climate Change Justice* at 86 (cited in note 1). Strictly speaking, I am not arguing here that what is done about climate change has to serve distributive justice, but that it has to avoid wreaking havoc upon the current or future poor. Not depriving the poor can be viewed as either not harming them or not treating them unfairly.

[6] Most proposals about what to do concerning climate change are highly complex and so accordingly are the corresponding justifications for doing what they propose. This chapter is an attempt to zero in on the considerations that, I believe, ought to be primary for the USA, boiling them down to their essentials. I concentrate on the USA because, at the national level, it is the state that is the farthest from beginning to fulfil its minimum responsibilities. My hope is that in a relatively simple account of the basics, it is less likely that the forest will be lost in the trees. This is, then, an exercise in simplicity. Some wise reflections on the value, and limits, of simplicity, as well as a valuable framework for surveying the relative merits of various proposals, are available from the Climate Action Network–International, *Fair Effort Sharing* Discussion Paper (Climate Action Network, 2011), online at <http://www.climatenetwork.org/publication/can-discussion-paper-fair-effort-sharing-jul-2011> (visited 14 October 2012).

It is well established by decades of analysis that the single greatest system-atic threat to human society in its familiar forms is human-induced climate change,[7] which is modifying our environment in ways that will, in the very next few decades, make many of our familiar practices unsustainable (and render many familiar species of plants and animals unsustainable, that is, headed for extinction). Meanwhile, global poverty continues to grow more tragically massive, and starvation and malnutrition annually take millions of lives.[8] For fairly obvious moral reasons, we ought to deal with climate change in a way that does not make it impossible to deal with global poverty and deal with global poverty in a way that does not make it impossible to deal with climate change.[9] Taking the separate 'whys' to be evident, I focus here on the practical question of the 'how', especially the necessity of tackling both these interacting threats together.[10] This is the inextricability that Posner and Weisbach assume can be escaped whenever they advocate tackling climate change and poverty separately.[11]

---

[7] Four accessible and responsible accounts, written from different perspectives, are Bill McKibben, *Eaarth: Making a Life on a Tough New Planet* (Times Books, 2010); Mark Herts-gaard, *Hot: Living Through the Next Fifty Years on Earth* (Houghton Mifflin, 2011); Clive Hamilton, *Requiem for a Species: Why We Resist the Truth about Climate Change* (Earthscan, 2010); Mark Lynas, *Six Degrees: Our Future on a Hotter Planet* (Fourth Estate, 2007). An engaging introduction to the science is Tyler Volk, *CO_2 Rising: The World's Greatest Environ-mental Challenge* (MIT, 2008). A disturbing sketch of a few of the richly funded, perverse, and ongoing political efforts by the fossil fuel industries to obscure the truth by making scientific questions that are effectively closed seem still to be wide open is Naomi Oreskes and Erik M. Conway, *Merchants of Doubt: How a Handful of Scientists Obscured the Truth on Issues from Tobacco Smoke to Global Warming*, pp. 169–215 (Bloomsbury, 2010). For another account of the fossil fuel Leviathan's thrashings, see Michael E. Mann, *The Hockey Stick and the Climate Wars: Dispatches from the Front Lines* (Columbia, 2012).

[8] See Thomas Pogge, *Politics as Usual: What Lies Behind the Pro-Poor Rhetoric*, p. 205 n 10 (Polity, 2010). Pogge explains why measures that suggest an optimistic interpretation of global-ization are misleading at pp. 93–109.

[9] While I am not nearly so convinced as Posner and Weisbach that International Paretianism (IP) is necessary, see Posner and Weisbach, *Climate Change Justice* at pp. 6–7, 87, 93–6, 132, 143, 178–81 (cited in note 1), I believe that avoiding exacerbating poverty is the only way to fulfil IP while reducing emissions as fast as required to avoid danger. I suspect that they are attempting to require of each part (a single treaty) a criterion—IP—that would be applicable only to a whole (international cooperation generally), a criticism they frequently make of partisans of distribu-tive justice. See Posner and Weisbach, *Climate Change Justice* at pp. 73–98.

[10] I have given my understanding of the reasons why we have a responsibility to deal with climate change in a number of places, most recently in 'Deadly delays, saving opportunities: creating a more dangerous world?', this volume; 'Human rights, climate change, and the trillionth ton', this volume. My primary reasons for thinking we have responsibilities toward the global poorest are in Henry Shue, *Basic Rights: Subsistence, Affluence, and US Foreign Policy* (Princeton 2nd edn 1996); Henry Shue, *Mediating Duties*, 98 Ethics 687 (1988). For more recent data and arguments, see Pogge, *Politics as Usual* at pp. 10–56 (cited in note 8). Also see the discussion of the 'method of isolation' and the 'method of integration' in Simon Caney, *Just Emissions*, 40:4 Phil & Pub Aff 1, 5 (2012).

[11] Posner and Weisbach, *Climate Change Justice*, pp. 73–98 (cited in note 1) (arguing that 'these claims improperly tie valid concerns about redistribution to the problem of reducing the effects of climate change').

'Creating the third human revolution: changing the context and opening new options' suggests the need for a third historic revolution in human affairs to deal with the climate effects of the Industrial Revolution, and 'The carbon budget: cumulative and limited' explains the meaning and significance of the concept of a budget of total cumulative carbon emissions since before the Industrial Revolution. 'Searching for an exit' examines alternative exit strategies from carbon-based energy in light of the cumulative nature of the carbon budget, while 'The moral significance of the carbon budget: using up what others need' shows that the fact that the cumulative carbon budget is zero-sum has striking moral significance. 'The other half of the story: not exacerbating poverty' sketches the ineliminable other half of the story that Posner and Weisbach try to set aside, and 'Conclusion: the past is not even past' argues that past emissions are not past history but continuing grounds for present and future responsibility.

CREATING THE THIRD HUMAN REVOLUTION: CHANGING THE CONTEXT AND OPENING NEW OPTIONS

Why is making progress on both climate change and poverty simultaneously, although morally inescapable, so difficult? Sustainable forms of the economic development necessary to eliminate the worst poverty will clearly depend on additional affordable energy as a necessary condition.[12] Energy is critical to economic development, and poverty cannot be reduced without reducing 'energy poverty'. Overcoming energy poverty means making adequate energy for economic development affordable. And what is, by far, the most affordable source of energy? Fossil fuel energy.[13]

But it is precisely fossil fuel energy—energy derived from burning coal, gas, and oil—that is the primary driver of human-induced climate change, so the burning of fossil fuels must be rapidly and severely curtailed if climate change is not to become uncontrollable. However much we might wish to simplify for the sake of practical manageability, we must consider climate change and global poverty together. How can we increase the affordability of energy in the

[12] See, for example, Peter Meisen and Irem Akin, *The Case for Meeting the Millennium Development Goals through Access to Clean Energy* *12–15 (Global Energy Network Institute, 2008), online at <http://www.geni.org/globalenergy/research/meeting-mdgs-through-access-to-electricity/MDG_Final_1208.pdf> (visited 14 October 2012).

[13] Catherine Mitchell et al., *Policy, Financing and Implementation*, in Ottmar Edenhofer et al., eds, *Renewable Energy Sources and Climate Change Mitigation: Special Report of the Intergovernmental Panel on Climate Change* 865 (Cambridge, 2011), online at <http://srren.ipcc-wg3.de/report/IPCC_SRREN_Full_Report.pdf> (visited 10 November 2012).

poorest countries without greatly increasing fossil fuel emissions—indeed, while aggressively reducing fossil fuel emissions? One general negative conclusion is that global poverty must not be reduced by means that increase, or even contribute to maintaining, current levels of the burning of fossil fuels, which are spewing out the carbon dioxide that is the main driver of rapid climate change. But the means of reducing poverty must nevertheless include increases in the energy available to the poorest members of humanity. Half a billion citizens of India, for instance, have no access to electricity; they cannot develop without electricity and other forms of energy.[14] The poor need vastly increased amounts of energy for the sake of adequate development, but, for the sake of slowing climate change, that affordable energy must not be fossil fuel energy.

Fairly obviously, then, we need as quickly as possible to create both sources of energy alternative to fossil fuels and feasible institutional arrangements to deliver them affordably to the poorest. We must construct an exit strategy away from fossil fuels into alternative energy. Otherwise, either climate change will continue to become worse or we will continue to abandon the poorest. In the words of the International Council on Human Rights Policy, 'making clean energy universally available is vital to protect human rights as climate change encroaches'.[15]

Humanity has so far undergone two great revolutions. First came the Agricultural Revolution, ten thousand years ago, which allowed stable human settlements and eventually grand cities, supplied with food by the productivity of agriculture. Second was the Industrial Revolution, two hundred years ago, which brought some of us our contemporary levels of wealth and consumption. But the Industrial Revolution is turning out to have been a

---

[14] Lavanya Rajamani and Shibani Ghosh, *India*, in Richard Lord et al., eds, *Climate Change Liability: Transnational Law and Practice* 139, 140 (Cambridge, 2012). For further complications, see Vikas Bajaj, 'No Power, No Boom', *New York Times*, B1 (20 April 2012). See also Sruthi Gottipati, 'A Silver Lining in India's Coal Crisis', *New York Times* Global Edition: India (20 April 2012), online at <http://india.blogs.nytimes.com/2012/04/20/a-silver-lining-in-indias-coal-crisis/> (visited 14 October 2012).

[15] *Climate Technology Policy and Human Rights: Protecting Rights in a Climate-Constrained World, Summary and Recommendations* *3 (International Council on Human Rights Policy, 2011), online at <http://www.spidh.org/uploads/media/64697446-Climate-Technology-Policy-and-Human-Rights-Protecting-Rights-in-a-Climate-Constrained-World-Executive-Summary_1___01.pdf> (visited 14 October 2012). The report goes on to say: '[S]uch a policy goal is affordable, manageable and urgent . . . It will ensure that development priorities do not have to be sacrificed to climate change exigencies.' See also *Beyond Technology Transfer: Protecting Human Rights in a Climate-Constrained World*, pp. 106–26 (International Council on Human Rights Policy, 2011), online at <http://www.ichrp.org/files/reports/65/138_ichrp_climate_tech_transfer_report.pdf> (visited 13 October 2012). In the words of Bert Bolin, founding chairman of the Intergovernmental Panel on Climate Change (IPCC): 'A sustainable supply of renewable energy is the prime long-term goal.' Bert Bolin, *A History of the Science and Politics of Climate Change: The Role of the Intergovernmental Panel on Climate Change*, p. 250 (Cambridge, 2007). The long term is looking shorter and shorter.

Faustian bargain because the fossil fuel energy that created modern wealth is undermining the very environment that supports our economy and especially our agriculture. *The second revolution, we now realize, threatens to undercut the first.* The third great revolution must, therefore, be the creation of *both* an escape route from fossil fuel energy *and* a path to the most rapid possible transition to alternative sources of energy in order to preserve the ecological preconditions for sustainable development. How should we make this great transition from the carbon-based fossil fuels with which we are now cutting our own throats to safer sources of energy?

We could ask developed countries to reduce annual GHG emissions enough to make space for increased annual emissions by the developing countries. This would cap the annual global total of emissions and change the distribution away from the developed world toward the developing world. This is essentially what I, and a number of other normative theorists, have advocated in the past.[16] But this is misguided, because it fails to appreciate the most recent understanding of the empirical reality of climate change, most notably, the importance of, not annual emissions, but total cumulative emissions since 1750.

THE CARBON BUDGET: CUMULATIVE AND LIMITED

The atmospheric scientists (not the normative theorists!) have recently made an important conceptual breakthrough. Because emissions of CO_2 are the single most important contributor to rapid climate change, it has been generally obvious for some time that if climate change is to be limited, emissions of CO_2 must be limited. Calculations have now been made that show that for any given increase in global average temperature beyond pre-industrial levels, there is a cumulative amount of emissions of CO_2 that will more than likely produce it, starting the calculation of the cumulative total from around the year 1750, which is before so much CO_2 began to be extracted from fossil fuels and injected into the oceans and the atmosphere by the burning of these fuels as the Industrial Revolution took off.[17] Thus, increase the total cumulative

---

[16] See generally 'Subsistence emissions and luxury emissions', this volume.

[17] Two companion studies converged upon the conceptual point. Myles Allen et al., *Warming Caused by Cumulative Carbon Emissions Towards the Trillionth Tonne*, 458 Nature 1163 (2009); Malte Meinshausen et al., *Greenhouse-Gas Emission Targets for Limiting Global Warming to 2°C*, 458 Nature 1158 (2009). See also Myles Allen et al., *The Exit Strategy*, 3 Nature Rep Climate Change 56 (May 2009) (laying out the conceptual point in an accessible manner); *Solving the Climate Dilemma: The Budget Approach*, (German Advisory Council on Global Change, 2009) (proposing German policies based on the studies), online at <http://www.wbgu.de/fileadmin/templates/dateien/veroeffentlichungen/sondergutachten/sn2009/wbgu_sn2009_en.pdf> (visited

emissions of CO_2 built up in the atmosphere since 1750, and the temperature will have a calculable probability of rising by a specific additional amount.[18]

Take the analysis of Myles Allen and his colleagues: total emissions of more than one trillion tonnes of CO_2 between 1750 and 2500 will most likely result in an emissions-induced rise in average global temperature more than 2°C above the pre-industrial average.[19] In mid June 2012 we had already emitted more than 558 billion tonnes (558 Gigatonnes (Gt)), and we would at current rates emit the trillionth tonne in August 2043—about thirty years from now![20] Those who are twenty-five years old now will be fifty-five when emissions go through the planetary ceiling if political action continues to be as weak as it has been so far during their lives; and their children will be roughly their own current age, with most of their lives in front of them. One analysis predicts that for a better than even chance of not exceeding a temperature rise of 2°C, cumulative emissions would naturally need to be still lower (for a three-in-four chance of remaining below a 2°C rise in temperature, emissions must stop at 750 Gts).[21]

---

14 October 2012); Niel H.A. Bowerman et al., *Cumulative Carbon Emissions, Emissions Floors and Short-Term Rates of Warming: Implications for Policy*, 369 Phil Transactions Royal Socy A 45 (2011) (providing a policy-relevant follow-up study); H. Damon Matthews, Susan Solomon, and Raymond Pierrehumbert, *Cumulative Carbon as a Policy Framework for Achieving Climate Stabilization*, 370 Phil Transactions Royal Socy A 4365 (2012) (analysing the policy implications of carbon dioxide's lengthy lifespan relative to other greenhouse gases). For a more sophisticated discussion of the science than mine, see the article in *Chicago Journal of International Law*, 13:2 (Winter 2013) by R.T. Pierrehumbert, 'Cumulative Carbon and Just Allocation of the Global Carbon Commons'.

[18] The primary reason that it is the cumulative amount that matters is the extraordinarily long atmospheric residence time of CO_2 molecules, which is several centuries for most and a millennium for about 25 per cent. See David Archer and Victor Brovkin, *The Millennial Atmospheric Lifetime of Anthropogenic CO_2*, 90 Climatic Change 283 (2008). From a human perspective of generations or even a century, it is essentially the case that what CO_2 goes up into the atmosphere does not come down. Posner and Weisbach correctly say in the text 'that most greenhouse gases, once emitted into the atmosphere, remain there for a very long time—as far as climate policy is concerned, we can think of them as permanent'. See Posner and Weisbach, *Climate Change Justice* at 14 (cited in note 1). Unfortunately, the comment in the accompanying footnote that '[t]he IPCC states that more than half of the CO_2 is removed from the atmosphere within a century', see p. 201 n 10, is not correct—no such statement occurs on the page cited or anywhere else I am aware of. On the contrary, the latest IPCC analysis says that 'a[n] [atmospheric] lifetime for CO_2 cannot be defined . . . The behaviour of CO_2 is completely different from the trace gases with well-defined lifetimes.' Gerald A. Meehl et al., *Global Climate Projections*, in Susan Solomon et al., eds, *Climate Change 2007: The Physical Science Basis, Contribution of Working Group I to the Fourth Assessment Report of the Intergovernmental Panel on Climate Change* 748, 824 (Cambridge, 2007).

[19] Allen et al., *Warming Caused by Cumulative Carbon Emissions* at 1163 (cited in note 17).

[20] See *trillionthtonne.org* (Oxford e-Research Centre 2012), online at <http://trillionthtonne.org/> (visited 14 October 2012).

[21] Id. By no means does all CO_2 emitted go into the atmosphere, of course—much enters the land and its vegetation, and much enters the ocean, causing acidification, which is another terrible problem that will also affect human food sources but which is not strictly speaking part of climate change (but is yet another compelling reason to exit quickly from fossil fuels as an energy

For two independent reasons these numbers, frightening as they are, are deeply conservative. First, they ignore the effects of all GHGs other than CO_2—when the rest of the gases are considered, the temperature rise will probably be greater still and more rapid.[22] Second, the target of merely keeping the temperature rise from exceeding 2°C is very lax, chosen politically to try to entice cooperation from recalcitrant governments such as Washington. Plenty of bad results will come from a rise in average global temperature as great as 2°C.[23]

What matters here are not particular numbers, which are artificially precise computer projections based on many (quite plausible) assumptions and are constantly being refined, but the general idea. The conceptual breakthrough is the realization that the best way to conceive of the situation is as the planet's having a *cumulative carbon budget* for any particular probability of any specific amount of temperature rise.[24] More emissions yield a calculable probability of a higher temperature. Humans (largely) determine the cumulative emissions; the dynamics of the planet in response determine the temperature rise. We can choose the level of emissions we aim for, but we cannot choose the most likely temperature rise that will accompany whatever level of emissions we actually produce—Mother Nature chooses the resulting temperature. This obviously means that we ought to choose politically a level of temperature rise that is to some extent 'tolerable' and reason back to the total cumulative emissions to be permitted. If much of humanity cannot safely handle a temperature rise beyond (roughly) 2°C, then we are well advised not to emit as much as (roughly) a trillion cumulative tonnes of carbon. The

source). See Volk, *CO_2 Rising* at 167, 170 (cited in note 7). The one trillion tonnes is the total that humans may cumulatively emit that would not lead to more CO_2 entering the atmosphere than is compatible with a fifty-fifty chance of temperature rising 'only' 2°C.

[22] See Meinshausen et al., *Greenhouse-Gas Emission Targets* at 1160 (cited in note 17).

[23] I believe that 2°C beyond pre-industrial temperatures is far too weak a target, but I use it as an example because there is a broad rhetorical commitment to it. For the argument that it is too weak a target, see Hertsgaard, *Hot* at 251–7 (cited in note 7). See also Lynas, *Six Degrees* at 57–106 (cited in note 7). For the rhetorical commitment (for what it is worth), see *The Cancun Agreements: Outcome of the Work of the Ad Hoc Working Group on Long-Term Cooperative Action under the Convention*, Decision 1/CP.16, in *United Nations Framework Convention on Climate Change (UNFCCC), Report of the Conference of the Parties on Its Sixteenth Session, Addendum, Part Two: Action Taken by the Conference of the Parties*, UN Doc FCCC/CP/2010/7/ Add. 1, ¶ 4 (2011). For what the concrete Copenhagen commitments—which likewise stipulate a 2°C limit—are worth, see Joeri Rogeli et al., *Copenhagen Accord Pledges Are Paltry*, 464 Nature 1126 (2010); Michel den Elzen et al., *The Emissions Gap Report: Are the Copenhagen Accord Pledges Sufficient to Limit Global Warming to 2°C or 1.5°C? A Preliminary Assessment* (United Nations Environmental Programme, 2010), online at <http://www.unep.org/publications/ ebooks/emissionsgapreport/pdfs/GAP_REPORT_SUNDAY_SINGLES_LOWRES.pdf> (visited 14 October 2012).

[24] Allen et al., 3 Nature Rep Climate Change at 57 (cited in note 17) ('This is where acknowledging the principle of a cumulative budget could be helpful: the higher emissions are allowed to be in 2020, the lower they will need to be in 2050.').

numbers will be refined as the science progresses, but the fundamental point is the conceptual one: for any given probability of a given rise in temperature, there is a cumulative carbon budget. If humans go beyond that budget, the planet will probably go beyond that temperature.

SEARCHING FOR AN EXIT

Obviously carbon emissions cannot simply continue on a business-as-usual basis until August 2043 (or whenever) and then suddenly fall to net zero injected into the atmosphere.[25] Such a path is neither politically nor techno-logically plausible, even if it were economically sensible, which is also highly doubtful.[26] Let us put aside some extreme options in order to get our bearings and begin to look for plausible, acceptable pathways out of the danger that looms over our children, if not us. Some might suggest that because the accelerating climate change constitutes a planetary emergency, carbon emis-sions ought not only to be cut anywhere and everywhere they can be, but also that they ought not to be increased anywhere. This inflexible position, how-ever, would ignore uncontroversial facts and elementary moral commitments. The most salient fact is that although we ought now to be attempting to escape it as rapidly as we can, within the constraints flowing from our moral commitments, the utterly dominant energy regime on this planet is a fossil fuel regime.[27] If the vast majority of the poorest people on the planet have any

[25] Carbon emissions do not need to become absolutely zero—we do not need, for example, to stop breathing, even though we constantly emit CO_2. A certain amount of CO_2 can be dealt with on the surface by plants, the oceans, etc., as they were in the millennia prior to 1750. See Volk, *CO_2 Rising* at 22–5 (cited in note 7). We need as soon as possible to reduce emissions until net injections of CO_2 into the atmosphere become zero. Depending on what else is happening, this means basically ceasing soon to obtain any very significant portion of our energy from burning fossil fuels. We may continue to need oil, for example, for everything from 'computer chips, insecticides, anesthetics, and fertilizers, right through lipstick, perfume, and pantyhose, to aspirin and parachutes'. McKibben, *Eaarth* at 30 (cited in note 7), insofar as we need all these things. But we will have to stop burning oil (and coal) to generate electricity and to power transportation because it is the burning of fossil fuels to generate energy that puts emissions over the top. I did not make the distinction between zero emissions and net zero emissions into the atmosphere clear in *Human Rights, Climate Change, and the Trillionth Ton*, especially on page 307, where I incorrectly claimed that we must 'eliminate completely the use of fossil fuel'. Shue, 'Human rights, climate change, and the trillionth ton', this volume.

[26] See, for example, Nicholas Stern, *The Economics of Climate Change: The Stern Review* 282–96 (Cambridge, 2007).

[27] For example, of the ten largest listed companies by revenue in the world, six are oil companies, and two others are a general energy company and a car manufacturer. See 'State Capitalism: Special Report', *The Economist* 4 (21 January 2012). Only Walmart has more revenue than the next five, all oil companies, led by Royal Dutch Shell and Exxon Mobil. For a troubling analysis of the political power of Exxon Mobil, see generally Steve Coll, *Private Empire: ExxonMobil and American Power* (Penguin, 2012).

source of energy beyond burning what they themselves can gather (such as sticks and dung), they are dependent on fossil fuels. After all, fossil fuels are the cheapest sources of energy, so it is not surprising that they are what are relied on by the poorest (and, for now, by practically everyone else). Even with so much energy coming from the currently cheapest sources—fossil fuels— more than 1.5 billion people are still without any access at all to electricity.[28] To insist simply that carbon emissions not go up at all anywhere, and meanwhile do nothing else, would be to condemn most of those people to remain without electricity for some indefinite further period of time, and to condemn them and others to remain mired in the abject poverty they are in now. In the immediate future, the only choice for them is carbon-based electricity or none.

The position that no one may increase her carbon emissions beyond what they are now would be completely outrageous for a number of moral reasons, most of them obvious. Those whose subsistence rights are not fulfilled now because they lack sufficient electricity would be doomed to continue indefinitely with some of their most fundamental rights not honoured as a result of a policy (namely, no new carbon emissions) adopted by the rest of us to deal with climate change.[29] Whatever the status of the violation of their rights now,[30] the deprivations would in the future be the result of a conscious choice made by the rest of us to impose on them a policy incompatible with eliminating the deprivations and would thus constitute about as straightforward and massive a rights violation as is imaginable.[31]

It would be even more outrageous to enforce a no-new-carbon-emissions policy on the energy-deprived poor while doing nothing about the notoriously wasteful levels of emissions among affluent consumers. We could in the past have 'made room' under any aggregate ceiling on emissions for considerable

---

[28] United Nations, Department of Economic and Social Affairs, Division for Sustainable Development, *A Global Green New Deal for Climate, Energy, and Development*, Technical Note 8 (2009), online at <http://www.un.org/esa/dsd/resources/res_pdfs/publications/sdt_cc/cc_global_green_new_deal.pdf> (visited 31 October 2012).

[29] I assume here that, as acknowledged by major human rights treaties, a right to subsistence is a universal human right. See, for example, *Universal Declaration of Human Rights*, Art 25, General Assembly Res 217A (III), UN Doc A/810 (1948). Those who wish to push the philosophical question back farther might consult Shue, *Basic Rights* (cited in note 10); Elizabeth Ashford, *The Alleged Dichotomy between Positive and Negative Rights and Duties*, in Charles R. Beitz and Robert E. Goodin, eds, *Global Basic Rights* 92 (Oxford, 2009).

[30] See generally Thomas Pogge, ed., *Freedom from Poverty as a Human Right: Who Owes What to the Very Poor?* (Oxford, 2007).

[31] Someone could, I suppose, maintain that, like the 'collateral damage' from bombing legitimate targets, the sufferings of those left in energy poverty by our policy were not intentional but only foreseeable, though regrettable. However, even unintentional collateral damage in war must be proportional to some good to which it is also necessary, and in this case it is obscure both what that good might be and why maintaining energy poverty is necessary to its achievement. See note 32 and accompanying text.

additional carbon emissions generated by the poorest during the satisfaction of their subsistence needs by reducing totally unnecessary indulgences and pointless waste on the part of the richest by the same amount.[32]

Nevertheless, making room for subsistence emissions by reducing luxury emissions now would be no solution at all with the current high atmospheric concentration. We can easily see why if we frame our thinking in terms of the planetary cumulative carbon budget; and it is now clear that it was a mistake for me to suggest, in 1993, that the poor in the developing world 'be guaranteed a certain quantity of protected emissions, which they could produce as they choose'.[33] The crucial point to understand is that what the poor must be guaranteed is energy, not emissions.[34] It may be true, as I was assuming then, that in a world dominated by a fossil fuel energy regime the only means by which to achieve the goal of guaranteeing energy for now is to guarantee emissions. That is correct as far as it goes: as long as the only energy source affordable by the poorest generates emissions, their energy can be guaranteed only by protecting their emissions. But it does not go nearly far enough.

The same general thought as mine also lay behind the division of the world in the United Nations Framework Convention on Climate Change (UNFCCC)[35] into Annex I countries, which were supposed to begin reducing emissions immediately in 1992, and others, which were allowed to continue increasing emissions during a transition period intended to permit development.[36] If Annex I countries had begun to make significant emissions

---

[32] I tried to perpetuate the helpful concepts first introduced by Anil Agarwal and Sunita Narain in *Global Warming in an Unequal World: A Case of Environmental Colonialism* (Centre for Science and Environment, 1991) in 'Subsistence emissions and luxury emissions', this volume. My main contention was that 'it is not equitable to ask some people to surrender necessities so that other people can retain luxuries'. Shue, 'Subsistence emissions and luxury emissions', this volume. Alternatively, '[e]ven in an emergency one pawns the jewellery before selling the blankets'. Shue, 'The unavoidability of justice', this volume. Wherever exactly one ought to draw the lines among subsistence emissions, emissions from reasonable ordinary consumption, and luxury emissions—an important line-drawing problem—both extremes of this spectrum are abundantly clear.

[33] Shue, 'Subsistence emissions and luxury emissions', this volume. The suggestion was contingent on there being global trading in carbon emissions permits: 'If there is to be an international market in emissions allowances, the populations of poor regions could be allotted inalienable—unmarketable—allowances, for whatever use they themselves consider best.'

[34] Strictly speaking, just as emissions are one means to energy, energy is one means to fulfilment of subsistence rights (and other rights). So fulfilment of rights is what should properly be guaranteed. At this point it is difficult to imagine better provisions for subsistence that do not require more energy as an indispensable means, but this too is a contingent matter and could change—but only over a term much too long to be relevant now. The fundamental point here against my suggestion was made convincingly in Tim Hayward, *Human Rights Versus Emissions Rights: Climate Justice and the Equitable Distribution of Ecological Space*, 24 Ethics & Intl Aff 431, 440–3 (2007).

[35] UNFCCC (1992), 1771 UN Treaty Ser 107 (1994).

[36] The principle underlying the division is 'common but differentiated responsibilities' (CBDR). See UNFCCC (1992), 1771 UN Treaty Ser 107 at Art 4.1. For a thorough study of

reductions, as of course the USA—led by the reality-defying US Senate—has completely failed to do, the poorer countries would have been able to increase their emissions in the course of development without the global total of emissions increasing, because the reductions by the USA and other Annex I countries would have made space under the aggregate ceiling for the increases by others. Obviously that would have been preferable to the soaring global totals of emissions, and mounting cumulative emissions, that we have in fact produced in the two lost decades since the adoption of the UNFCCC in 1992. But it would not be nearly good enough now.

The reason why merely maintaining a constant global total of annual emissions, which we have not done, would have been a failure is evident from the fact that the global carbon budget is cumulative. What will determine the amount of the rise in temperature at the surface of the planet is total cumulative carbon emissions from 1750 until emissions reaching the atmosphere are reduced to net zero. Annual global emissions totals that held constant would be superior only to the actual rising annual global emissions totals that we now produce.[37] This is because each annual total exhausts more of what remains of any cumulative budget. The remainder of the carbon budget for (only) a fifty-fifty chance that temperature will not rise beyond 2°C is now being exhausted with remarkable speed: '[H]aving taken 250 years to burn the first half-trillion tonnes of carbon, we look set, on current trends, to burn the next half trillion in less than 40.'[38]

Again, the point is not that the budget is exactly one trillion tonnes, which is an approximate round number, or that there are exactly only thirty years left on the budget at current rates, which is a computer projection. The point is that even a constant annual rate of emissions would be gobbling up the remainder of absolutely any cumulative carbon budget, whatever its exact size. Providing only that what determines temperature rise (and the many other aspects of climate change) is the cumulative total of carbon emissions,

the status of the principle in international law, see Lavanya Rajamani, *Differential Treatment in International Environmental Law*, pp. 133–62 (Oxford, 2006). For doubts about CBDR, see Simon Caney, *Cosmopolitan Justice, Responsibility, and Global Climate Change*, in Gardiner et al, eds, *Climate Ethics* 122, pp. 138–9 (cited in note 4). For the current status of CBDR, see Lavanya Rajamani, *The Changing Fortunes of Differential Treatment in the Evolution of International Environmental Law*, 88 Intl Aff 605 (2012).

[37] The annual rises paused during the global economic recession of 2008–09. Global Carbon Project, *Rapid Growth in CO$_2$ Emissions after the 2008–2009 Global Financial Crisis*, 2 Nature Rep Climate Change 2 (2012), online at <http://www.nature.com/nclimate/journal/v2/n1/pdf/nclimate1332.pdf> (visited 13 October 2012).

[38] Allen et al., 3 Nature Rep Climate Change at 57 (cited in note 17). As mentioned earlier, the time remaining appears already to have shrunk to closer to thirty years because of the rapidly accelerating annual rates of emissions. See *trillionthtonne.org* (cited in note 20).

we have launched our descendants barrelling downhill faster and faster toward a stone wall. Whether the stone wall is a few feet farther away or a few feet closer—that is, whether one trillion tonnes and 2°C are precisely the right numbers—is, while not unimportant, distinctly secondary.

Therefore, total global annual carbon emissions must rapidly decline until net emissions into the atmosphere, that is, beyond the carbon that can readily be recycled on the surface, reach zero.[39] Nowadays, except during major economic recessions, carbon emissions rise every year—so fast that humanity is now set to emit in much less than forty years the amount of carbon we emitted in the previous 250 years! This cannot continue. If we do not turn to face reality, it will simply run over us from behind.

THE MORAL SIGNIFICANCE OF THE CARBON BUDGET: USING UP WHAT OTHERS NEED

The empirical concept of the carbon budget, that is, the understanding that the emissions still remaining inside the cumulative budget are all there will be for perhaps the next five hundred years for all nations and all generations, radically transforms the nature of the normative problem we face. Carbon emissions are zero-sum spatially and temporally, transnationally and transgenerationally, and therefore increasingly scarce. We—certainly I—have tended to view the problems about action to deal with climate change as exclusively distributive issues. That is, there is a problem to be dealt with and many different parties need to contribute positively to the solution of the problem. Such an approach raises several questions: How shall we distribute the responsibility? According to historical responsibility for producing the problem? According to extent of benefit from the processes that created the problem (roughly speaking, industrialization)? According to ability to pay?[40] But the nations who have already consumed the lion's share of the cumulative carbon budget are not being asked in the main to make a positive contribution—they are above all being asked to cease destructive, damaging action in the form of continuing to use up what others need: the dwindling capacity of the planet safely to dispose of carbon emissions. 'Excess encroaches': emitting more than one's share of emissions encroaches on the sink shares of others.[41]

---

[39] This again ignores the severe problem of acidification caused by the oceans' absorption of additional carbon.

[40] I have discussed these issues in 'Global environment and international inequality', this volume. Simon Caney has presented a critical alternative view in Simon Caney, *Human Rights, Responsibilities, and Climate Change*, in Beitz and Goodin, eds, *Global Basic Rights* 227 (cited in note 29).

[41] 'After you: may action by the rich be contingent upon action by the poor?', this volume.

Business as usual in the form of continuing emissions from the use of fossil fuel as an energy source is doubly damaging and destructive. Above all, it feeds the speeding up of climate change, which will produce more violent storms, more unmanageable rains, unpredictable crop conditions, sea-level rise, and so on, directly threatening the health and well-being of many random persons.[42] It also consumes any carbon budget: whatever the budget of cumulative carbon emissions compatible with any given temperature rise turns out to be, continuing emissions use it up and leave behind less sink space than was there before.

Each day there is less remaining sink space—that is, fewer places into which emissions can go without causing disruptions elsewhere in the climate system—than there was the day before. The sink is being used up, leaving the people of the future with fewer options. This makes planetary absorptive capacity for carbon (and other GHGs) a classic example of what Elinor Ostrom has called 'common-pool' resources, with each use of the capacity being 'subtractable' from the remaining total and 'rival' to all other possible uses, most notably, use by our descendants.[43] The budget of absorptive capacity is being consumed, depriving the people of the future of options.

The individual identities and the numbers of future persons are yet to be determined, but we know that we are leaving their situation worse than ours, making flourishing and perhaps even survival more difficult for them than it is for us. The imperative is not only to do our share to help, leaving us in need of a distributive principle for allocating our share of assistance; it is also an urgent imperative to stop inflicting the damage, stop making the environment of the planet less hospitable to human life and flourishing, and to the life and flourishing of other existing species of animals and plants, which are adapted to the only environment that they know, the one that we are undermining with our profligate emissions.

Finding sources of energy for our economy that do not undermine the environment is, therefore, not only a positive good that we might or might not be obligated to provide (and the share of our responsibility for which is contentious). Finding alternative sources of energy also is the only way to maintain our own lives while not damaging the lives of others to come by both exacerbating climate change and reducing available sink space. If we are to maintain our own lives without making the lives of whoever lives in the future much harder than ours are—if we are to live without undermining the conditions of life for whoever lives in the future—we must create alternative forms of energy, not merely vaguely hope that somehow our descendants will

---

[42] See Hertsgaard, *Hot* at 50–9 (cited in note 7).

[43] See Nives Dolšak and Elinor Ostrom, *The Challenges of the Commons*, in Nives Dolšak and Elinor Ostrom, eds, *The Commons in the New Millennium: Challenges and Adaptations* 3, 7–9 (MIT, 2003).

be able to do it for themselves in time. By then it will likely be too late to keep the cumulative total of carbon dioxide small enough to allow only a tolerable degree of climate change.

THE OTHER HALF OF THE STORY: NOT EXACERBATING POVERTY

Climate change is only half the story. The other half is that more than two billion human beings suffer from energy poverty right now. They need to be provided with access to energy, especially electricity. Ending, or even significantly alleviating energy poverty, while also ending net carbon emissions into the atmosphere may initially seem simply impossible—a doubly difficult combination of two individually challenging tasks. I believe, however, that the best solution to ending energy poverty may contain the seeds of the end of net carbon emissions. At present, the price of energy produced by burning any fossil fuel is much less than the price of energy produced from almost any renewable source. This is the fundamental economic problem. If the alternatives to fossil fuel produced cheaper energy than fossil fuel did, everyone would simply find it in her interest to abandon fossil fuel, and a huge step toward slowing climate change would be taken by individual consumers pursuing their own interest in the market. But as Nicholas Stern has observed, climate change is 'the greatest example of market failure we have ever seen'.[44] The Invisible Hand will not stop climate change.

It would be conceivable to try to follow a strategy that, in effect, reserved what remains of the cumulative carbon budget for the global poor, who cannot now afford any form of energy except fossil fuel energy, if indeed they can afford that, and insisted, by some means or other, that the better off pay for energy from renewables now in order not to add more emissions on top of the temporarily unavoidable carbon emissions by the poor.[45] Quite a few concerned individuals admirably already deliberately pay more for energy from renewable sources than they would have to pay for fossil fuel energy as part of their effort at dealing with the threat to the climate. But the current and foreseeable spontaneous supply of such a strong sense of individual moral responsibility, coupled with understanding of the problem, seems insufficient

[44] Stern, *The Economics of Climate Change* at xviii (cited in note 26).

[45] This option is wrestled with rather obscurely and inconclusively in Shue, 'Human rights, climate change, and the trillionth ton', this volume, 311–15. My discussion there painfully displays the inadequacy of any position that proposes an emissions trading scheme alone as a comprehensive solution, and demonstrates that separate direct provisions also need to be made to protect subsistence rights against being undercut, as I am arguing here but did not yet fully see there.

to the urgency and magnitude of the task; some powerful government bodies, most notably the US Congress, have completely failed to create economic incentives by passing climate legislation to put a price on carbon.[46] So whatever ought ideally to occur from a moral point of view, it is difficult to envision how individual decisions to forswear energy from fossil fuels and pay more for energy from renewable sources is going to come about. Government has failed to lead by creating sufficient economic incentives to motivate unconscientious and uncomprehending citizens, and highly conscientious and comprehending citizens are in short supply. These doubts about the sufficiency of likely individual voluntary sacrifice, however, are obviously somewhat speculative.

The most feasible kind of alternative seems to be a rapid enough increase in the supply of alternative energy to drive down its price a great deal very soon.[47] For example, a specific alternative proposal that seems workable to me is what is called the 'Global Green New Deal' (GGND).[48] Once again, the fundamental conception, not the specific details of the illustrative case, is what matters here.[49] The GGND relies on the well-documented tendency for prices to fall as, other things equal (of course), installed capacity grows.[50] The subsidies involved are designed to speed up the installation of capacity for producing electricity from renewable sources by paying a higher-than-market price for renewable electricity.[51] The current market price is extremely low because fossil fuel is so plentiful and cheap and does not cover the current costs of renewable energy. The mechanism that would be employed for government subsidies is the 'feed-in tariff', a guaranteed higher price for renewable electricity fed into the grid.[52] This mechanism is currently used with considerable success by Germany and others, including Denmark, and until the euro

---

[46] See Jonathan L. Ramseur and Larry Parker, *Carbon Tax and Greenhouse Gas Control: Options and Considerations for Congress* (Congressional Research Service, 10 March 2009), online at <http://www.fas.org/sgp/crs/misc/R40242.pdf> (visited 13 October 2012) (analysing the political feasibility of a carbon tax).

[47] Some movement in the right direction is occurring. See, for example, Christian Azar, Thomas Sterner, and Gernot Wagner, 'Rio Isn't All Lost', *Intl Herald Trib* 9 (20 June 2012).

[48] United Nations, *Global Green New Deal* (cited in note 28). For an updated argument, see Tariq Banuri and Niclas Hällstöm, *A Global Programme to Tackle Energy Access and Climate Change*, 61 Development Dialogue 264 (September 2012), online at <http://www.dhf.uu.se/publications/development-dialogue/climate-development-and-equity/> (visited 21 November 2012).

[49] See Mitchell et al., *Policy, Financing and Implementation* at 865 (cited in note 13).

[50] United Nations, *Global Green New Deal* at ii (cited in note 28) ('The key mechanism is a rapid increase in installed capacity. A "big push" in both public and private investment to scale up renewable energy will lead to rapid cost reduction, technology improvement, and learning by doing.').

[51] United Nations, *Global Green New Deal* at 14–15.

[52] United Nations, *Global Green New Deal* at 14–15.

crisis, Spain.[53] It has the great virtue of not requiring the government to choose which technology is likely to be successful—to 'pick winners'. The government simply guarantees to add to the grid whatever electricity is produced by whatever renewable means.[54] Private investors choose which technology to bet on. Entrepreneurs would also be encouraged to subsidize renewable electricity in other ways. The projections done for the UN Division of Sustainable Development in 2009 suggested that a total subsidy of $1.5 trillion—$100 billion per year for fifteen years—would produce sufficient installed capacity to bring the price of renewable electricity down to a level affordable by most of the world's poorest.[55] This counts any private contributions made. For such a revolutionary result, $100 billion a year shared across the globe is not an unreasonable amount of money. It is, for example, roughly the annual cost of either the US war in Iraq or the US war in Afghanistan.[56]

Most discussion has been devoted so far to how to make carbon emissions more expensive. Action is urgent on that front as well, but two initiatives need to converge. First, carbon emissions need to be cut back sharply and aggressively. An effective means by which to do that is probably by placing a price on carbon emissions, through either cap-and-trade or carbon taxes.[57] But pricing the carbon released by energy production does nothing to help the global poor—indeed, in itself it makes the problem of energy poverty worse by making what is now the most affordable energy less affordable. Cap-and-trade alone would simply make life worse for the poorest by driving up the price of fossil fuels. We cannot rely on cap-and-trade or carbon taxes alone because they do nothing positive about energy poverty. We must implement some plan, such as the GGND, that directly tackles energy poverty by driving down the price of renewables to a level that the poorest can afford.

Second, therefore, alternative energy technologies need to be developed as quickly as humanly possible, aiming at an early day when prices of the alternative technologies become, first, competitive with the prices of fossil fuel and then become, second, affordable for the poorest. Fossil fuels now are notoriously cheap, of course, which is the main reason we also need the cap-and-trade and/or carbon tax to drive up their price by political will. Here we

---

[53] See generally Environmental and Energy Study Institute, *Feed-in Tariffs* (Environmental and Energy Study Institute 2010), online at <http://www.eesi.org/files/feedintariff_033110.pdf> (visited 14 October 2012).

[54] Environmental and Energy Study Institute, *Feed-in Tariffs*.

[55] United Nations, *Global Green New Deal* at 3 (cited in note 28).

[56] See Neta C. Crawford et al., *Costs of War Project* (Watson Institute, 2011), online at <http://costsofwar.org> (visited 14 October 2012).

[57] It is of course absurd from an economic perspective that such an environmentally damaging practice as burning fossil fuels to supply general energy needs should be costless for those doing the damage—such costs should be internalized in fossil fuels' price.

need political action to supplement market mechanisms in order to protect the environment on which all economic activity depends. (This is ordinarily considered a fundamental purpose of government.) We must aim politically for the point of crossover at which declines in the prices of alternative technologies and increases in the prices of fossil fuels mean that the damaging fossil fuels lose their competitive price advantage. The farther we move on either front, the less distance we need to cover on the other front. Once the crossover occurs, even the purely selfish who care nothing for the environment and nothing for the rights of others will simply find it efficient to use alternative fuels.

It is critical to understand, however, that a competitive price and an affordable price are two different things. That the price of alternative energy could become competitive at some date with the price of fossil fuel energy does not mean that the price would be affordable for the poorest, especially if between now and then the price of fossil fuel energy has been driven up by cap-and-trade and/or carbon taxes. We have no guarantee in the unguided workings of the market that the crossover price, at which renewable fuels become competitive with fossil fuels, will not be higher than the affordable price for the poorest. This is an additional reason why market mechanisms must be supplemented by political action, such as the GGND, to create conditions in which the poorest have a chance to develop. (This too is ordinarily considered a fundamental purpose of government.) The greater the extent to which the crossover between the prices of fossil and non-fossil energy occurs from drops in price of the latter rather than rises in the former, the more likely energy will at that time be affordable for the poor. At the point when basic energy becomes affordable for everyone, humanity might be out of the woods concerning both desperate poverty and climate change, provided that we have meanwhile not emitted the trillionth tonne, or whatever the rapidly advancing science tells us is the outer boundary of environmentally tolerable carbon emissions.

If we act vigorously and creatively now, we can politically invent economic institutions that will provide a priceless legacy of rights protection for multiple generations. Funds spent on initiatives such as the GGND would not be some generous gift to other nations or to future generations. On the contrary, this kind of initiative is the only way to protect the global poor from being made even worse off than they are now by the efforts that we must make to slow climate change before it reaches even more dangerous levels than already are guaranteed. If we fail to soon provide renewable energy affordable for the poorest, we not only fail in our positive duties of justice toward the *current* poor but also fail to protect them against the economic consequences of our efforts to limit climate change. If we fail to limit climate change by limiting carbon emissions, we undermine the environmental conditions of life for the *future* poor. If we are not to damage the life prospects of either of these two vulnerable groups, we must promptly and aggressively tackle global poverty and climate change.

The most urgent imperative, then, is to cease doing damage, much of which may be irreparable and irreversible (in any normal timescale), through continuing high carbon emissions. Because what we are doing now is destructive, in that it is undermining the conditions of flourishing human life, the question is not, what would be our fair share of contribution to accomplishing a positive good, but how soon can we completely stop causing future damage? But we must also be careful that the institutional and policy choices that we make in order to cease our physical harm do not produce the social harm of making the prospects of development for the poorest even more bleak than they are now. Thus, we have two strong negative duties: a duty not to continue to undermine the climate, which would make life much more difficult for everyone in generations somewhat farther into the future, and a duty not to increase the obstacles to development for the current poor and their immediate successors. The two duties are additive, but the beauty of fulfilling them by developing affordable, renewable energy is that such measures would fulfil both at once.

Where one is doing damage, one's responsibility is normally simply to stop doing it. In the case of destructive behaviour, there are no difficult distributive issues that raise the question: What share of the damage that I am doing can I reasonably be expected to stop? The obvious answer to such a question is: All of it. Of course, there is the theoretically uninteresting but practically important issue of how rapid a transition counts as stopping. Economies must have energy. One cannot cease using energy sources that damage the environment with carbon emissions until sources that do not emit CO_2 can be made widely available.[58] One cannot in this case stop immediately. But the efforts to make alternatives available can either be desultory, lackadaisical, low-priority, and meagrely financed, as they are now at the national level in the USA, or they can be vigorous, urgent, high-priority, and well financed, as they are in many US states and cities and at the national level in a number of other countries.[59] Only robust and aggressive action can reasonably be counted as stopping with all deliberate speed the continuing process of driving the cumulative total of carbon emissions to dangerous levels.

---

[58] Except, of course, by the vital methods of eliminating waste, increasing energy efficiency, and ceasing needless activities.

[59] See the numerous examples in Hertsgaard, *Hot* at 79–81 (cited in note 7) (discussing the greening of Seattle and Chicago); see also Center for Climate and Energy Solutions (C2ES), *Multi-State Climate Initiatives*, online at <http://www.c2es.org/us-states-regions/re gional-climate-initiatives> (visited 13 October 2012). For other states' initiatives see Bill Hare et al., *Climate Action Tracker* (Climate Action Tracker Partners, 2011), online at <http://www. climateactiontracker.org/> (visited 14 October 2012).

CONCLUSION: THE PAST IS NOT EVEN PAST

The point of this analysis has been that the USA, like many other nations,[60] is making ongoing contributions to a progressively worsening danger. The worry now is, not damage done in the past, but damage to be done in the future unless currently unquestioned policies change. Emissions produced in the past are not primarily damage done in the past. The emissions were in the past, but the damage, while under way already, is mostly still to come because the emissions of carbon have collected in the atmosphere and will produce increasingly profound effects for centuries. For example, the CO_2 already in the atmosphere in June 2012 will continue to force sea-level rise for centuries![61] The ocean system has enormous inertia, changing extraordinarily slowly, but the atmospheric CO_2 is patiently working and will still linger centuries from now, quietly exerting its steady influence to raise the sea level.[62]

One may be inclined to think of the emissions accumulated in the atmosphere as (solely) a fact about the past—past history, water under the bridge. But the fact is that this water did not flow under the bridge—it is almost all still silently gathering, rising, deeper and deeper. And the rapidly building atmospheric accumulation is going to be a mushrooming problem in the future. When the USA tries to walk away from its own past it is also trying to walk away from everyone else's future.[63] In explaining why he so much liked the Durban Platform, the outcome of the latest meeting of the Conference of the Parties to the UNFCCC, a former advisor to the USA's chief climate negotiator crowed: 'There is no mention of historic responsibility or per capita emissions. There is no mention of economic development as the priority for developing

---

[60] The fact that other nations are doing damage as well does nothing to weaken the unconditional imperative for the USA to stop as quickly as possible. No one has the right, as US climate negotiators seem to believe, to bargain over who stops wreaking havoc first. For the argument that others' non-compliance makes action more, not less, imperative, see Anja Karnein, *Putting Fairness in Its Place: Why There Is a Duty to Take Up the Slack*, unpublished paper (on file with the author).

[61] Meehl et al., *Global Climate Projections* 828–31 (cited in note 18); Susan Solomon et al., *Irreversible Climate Change Due to Carbon Dioxide Emissions*, 106 Proceedings of the Natl Acad of Sci 1704, 1707–09 (2009); Michiel Schaeffer et al., *Long-Term Sea-Level Rise Implied by 1.5°C and 2°C Warming Levels*, Nature Climate Change, *1 (June 2012), online at <http://www.nature.com/nclimate/journal/vaop/ncurrent/full/nclimate1584.html> (visited 26 November 2012).

[62] Barring removal through geo-engineering techniques that may or may not work and may or may not do more good than harm. For much further complexity, see Stephen M. Gardiner, *A Perfect Moral Storm: The Ethical Tragedy of Climate Change* 339 (Oxford, 2011).

[63] Posner and Weisbach suggest: 'If a specified level of reductions will give significant benefits to India, but more modest benefits to the United States and Russia, the latter nations . . . might demand some kind of compensation.' Posner and Weisbach, *Climate Change Justice* at 142 (cited in note 1). The idea of compensation for what amounts to desisting from decades of knowingly causing monumental damage to the common environment of the whole planet is a *reductio ad absurdum* of taking a purely 'forward-looking view' as if a world with no history and no accountability begins anew with each dawn.

countries. There is no mention of a difference between developed and developing country action.'[64] This seems to mean, as the rich and powerful love to say, 'We are all in this together'—the rich and the poor are equally free to sleep under the bridges of Paris, and the USA will not exert itself a bit less on climate change than Burkina Faso and Ecuador.

The problem of climate change became clear in the 1980s and was acknowledged by the US Senate in 1992 when it ratified the UNFCCC, but to this day the USA has imposed no national limit whatever on its total carbon emissions! This shameful failure to have a national policy concerning the greatest challenge of the twenty-first century constitutes flagrant disregard for the interests of others (and of the US national interest). After two decades—an entire generation—of paralysing obstructionism at international negotiations and defiant inaction at home, a little American leadership at the national level in leaving behind a livable climate would not be out of place.[65] We are unmaking other people's futures. Our continuing high carbon emissions are now making the future hard for multiple generations to come in many nations. We need to act promptly to cease making life so difficult for those who will inherit the planet that our brief lives here are radically reshaping for the worse for the very long term. If we do, we can replace the currently darkening prospects with brightening hope.

---

[64] John M. Broder, 'Climate Change: A Battle on Many Fronts', *Intl Herald Trib* 12 (25 January 2012).
[65] See generally 'Face reality? After you! A call for leadership on climate change', this volume. On precisely why climate leadership is in fact imperative and urgent, see Aaron Maltais, *Failing International Climate Politics and the Fairness of Going First*, 1–16 (Political Studies, 2013), doi: 10.1111/1467-9248.12073 (early view, published online 13 September 2013).

Declaration on climate justice

All human beings are born free and equal in dignity and rights.[1]

Our vision

As a diverse group of concerned world citizens and advocates, we stand in solidarity for a global climate system that is safe for all of humanity. We demand a world where our children and future generations are assured of fair and just opportunities for social stability, employment, a healthy planet and prosperity.

We are united in the need for an urgent response to the climate crisis—a response informed by the current impacts of climate change and the science that points to the possibility of a global temperature increase of 4°C by the end of this century. The economic and social costs of climate impacts on people, their rights, their homes, their food security and the ecosystems on which they depend cannot be ignored any longer. Nor can we overlook the injustice faced by the poorest and most vulnerable who bear a disproportionate burden from the impacts of climate change.

This reality drives our vision of climate justice. It puts people at the centre and delivers results for the climate, for human rights, and for development. Our vision acknowledges the injustices caused by climate change and the responsibility of those who have caused it. It requires us to build a common future based on justice for those who are most vulnerable to the impacts of climate change and a just transition to a safe and secure society and planet for everyone.

Achieving climate justice

A greater imagination of the possible is vital to achieve a just and sustainable world. The priority pathways to achieve climate justice are:

Giving voice: The world cannot respond adequately to climate change unless people and communities are at the centre of decision-making at all levels—local, national and international. By sharing their knowledge, communities can take the lead in shaping effective solutions. We will only succeed if we give voice to those most affected, listen to their solutions, and empower them to act.

A new way to grow: There is a global limit to the carbon we can emit while maintaining a safe climate and it is essential that equitable ways to limit these emissions are achieved. Transforming our economic system to one based on low-carbon production and consumption can create inclusive sustainable development and reduce inequality. As a global community, we must innovate now to enable us to leave the majority of the remaining fossil fuel reserves in the ground—driving our transition to a climate resilient future.

---

[1] Article 1, The Universal Declaration of Human Rights.

To achieve a just transition, it is crucial that we invest in social protection, enhance worker's skills for redeployment in a low-carbon economy and promote access to sustainable development for all. Access to sustainable energy for the poorest is fundamental to making this transition fair and to achieving the right to development. Climate justice also means free worldwide access to breakthrough technologies for the transition to sustainability, for example, efficient organic solar panels and new chemical energy storage schemes.

Investing in the future: A new investment model is required to deal with the risks posed by climate change—now and in the future, so that intergenerational equity can be achieved. Policy certainty sends signals to invest in the right things. By avoiding investment in high-carbon assets that become obsolete, and prioritizing sustainable alternatives, we create a new investment model that builds capacity and resilience while lowering emissions.

Citizens are entitled to have a say in how their savings, such as pensions, are invested to achieve the climate future they want. It is critical that companies fulfil their social compact to invest in ways that benefit communities and the environment. Political leaders have to provide clear signals to business and investors that an equitable low-carbon economic future is the only sustainable option.

Commitment and accountability: Achieving climate justice requires that broader issues of inequality and weak governance are addressed both within countries and at a global level. Accountability is key. It is imperative that Governments commit to bold action informed by science, and deliver on commitments made in the climate change regime to reduce emissions and provide climate finance, in particular for the most vulnerable countries.

All countries are part of the solution but developed countries must take the lead, followed by those less developed, but with the capacity to act. Climate justice increases the likelihood of strong commitments being made as all countries need to be treated fairly to play their part in a global deal. For many communities, including indigenous peoples around the world, adaptation to climate change is an urgent priority that has to be addressed much more assertively than before.

Rule of law: Climate change will exacerbate the vulnerability of urban and rural communities already suffering from unequal protection from the law. In the absence of adequate climate action there will be increased litigation by communities, companies and countries. International and national legal processes and systems will need to evolve and be used more imaginatively to ensure accountability and justice. Strong legal frameworks can provide certainty to ensure transparency, longevity, credibility and effective enforcement of climate and related policies.

Transformative leadership

World leaders have an opportunity and responsibility to demonstrate that they understand the urgency of the problem and the need to find equitable solutions now.

At the international level and through the United Nations, it is crucial that leaders focus attention on climate change as an issue of justice, global development and human security. By treating people and countries fairly, climate justice can help to deliver a strong, legally binding climate agreement in 2015. It is the responsibility of leaders to ensure that the post-2015 development agenda and the UNFCCC climate negotiations support each other to deliver a fair and ambitious global framework by the end of 2015. Local and national leaders will implement these policies on the ground, creating an understanding of the shared challenge amongst the citizens of the world and facilitating a transformation to a sustainable global society.

As part of global collective action, greater emphasis should be given to the role of diverse coalitions that are already emerging at the community, local, city, corporate and country levels and the vital role they play in mobilizing action. These coalitions are already championing the solutions needed to solve the crisis and their effect can be maximized by supporting them to connect and scale up for greater impact.

Climate justice places people at its centre and focuses attention on rights, opportunities and fairness. For the sake of those affected by climate impacts now and in the future, we have no more time to waste. The 'fierce urgency of now' compels us to act.

Issued on 23 September 2013 at the United Nations by the High Level Advisory Committee to the Climate Justice Dialogue, an initiative of the Mary Robinson Foundation-Climate Justice and the World Resources Institute.

Index